AF709751

VOLUME EIGHTY SIX

Current Topics in MEMBRANES

Membrane Biomechanics

CURRENT TOPICS IN MEMBRANES, VOLUME 86

Series Editor

IRENA LEVITAN
Division of Pulmonary and Critical Care Medicine
University of Illinois at Chicago
Chicago, IL, United States

VOLUME EIGHTY SIX

CURRENT TOPICS IN MEMBRANES

Membrane Biomechanics

Edited by

IRENA LEVITAN

Department of Medicine, Division of Pulmonary, Critical Care and Sleep Medicine, University of Illinois at Chicago, Chicago, IL, United States

ANDREEA TRACHE

Associate Professor, Department of Medical Physiology, College of Medicine, Texas A&M University Health Science Center, Bryan, TX, United States; Department of Biomedical Engineering, College of Engineering, Texas A&M University, College Station, TX , United States

ELSEVIER

ACADEMIC PRESS

An imprint of Elsevier

Academic Press is an imprint of Elsevier
50 Hampshire Street, 5th Floor, Cambridge, MA 02139, United States
525 B Street, Suite 1650, San Diego, CA 92101, United States
The Boulevard, Langford Lane, Kidlington, Oxford OX5 1GB, United Kingdom
125 London Wall, London, EC2Y 5AS, United Kingdom

First edition 2020

ISBN: 978-0-12-821021-5
ISSN: 1063-5823

For information on all Academic Press publications
visit our website at https://www.elsevier.com/books-and-journals

Publisher: Zoe Kruze
Acquisitions Editor: Ashlie Jackman
Editorial Project Manager: Shellie Bryant
Production Project Manager: Abdulla Sait
Cover Designer: Alan Studholme

Typeset by SPi Global, India

Contents

Contributors

Sang Joon Ahn
Department of Medicine, Division of Pulmonary, Critical Care and Sleep Medicine, University of Illinois at Chicago, Chicago, IL, United States

Manuela A.A. Ayee
Department of Engineering, Dordt University, Sioux Center, IA, United States

Navid Bavi
Institute for Biophysical Dynamics, University of Chicago, Chicago, IL, United States

Omid Bavi
Department of Mechanical and Aerospace Engineering, Shiraz University of Technology, Shiraz, Iran

Alan Bowling
Department of Mechanical and Aerospace Engineering, University of Texas at Arlington, Arlington, TX, United States

Brendan C. Bunker
Department of Engineering, Dordt University, Sioux Center, IA, United States

Michael Cho
Department of Bioengineering, University of Texas at Arlington, Arlington, TX, United States

Charles D. Cox
Molecular Cardiology and Biophysics Division, Victor Chang Cardiac Research Institute; St. Vincent's Clinical School, University of New South Wales, Sydney, Australia

Jordan L. De Groot
Department of Engineering, Dordt University, Sioux Center, IA, United States

Marina Inés Giannotti
Biomedical Research Networking Center on Bioengineering, Biomaterials and Nanomedicine (CIBER-BBN), Madrid; Institut de Bioenginyeria de Catalunya (IBEC), The Barcelona Institute of Science and Technology (BIST); Departament de Ciència de Materials i Química Física, Universitat de Barcelona, Barcelona, Spain

Zhongkui Hong
Department of Biomedical Engineering, University of South Dakota, Vermillion, SD, United States

Elizabeth Le Master
Department of Medicine, Division of Pulmonary, Critical Care and Sleep Medicine, University of Illinois at Chicago, Chicago, IL, United States

James C. Lee
Department of Bioengineering, University of Illinois at Chicago, College of Medicine, Chicago, IL, United States

Irena Levitan
Department of Medicine, Division of Pulmonary, Critical Care and Sleep Medicine, University of Illinois at Chicago, Chicago, IL, United States

Caleb Liebman
Department of Bioengineering, University of Texas at Arlington, Arlington, TX, United States

Patricia Losada-Pérez
Experimental Soft Matter and Thermal Physics (EST) Group, Department of Physics, Université Libre de Bruxelles, Brussels, Belgium

Adam D. Martinac
Neuroscience Research Australia and Prince of Wales Clinical School, UNSW, Kensington, Australia

Boris Martinac
Molecular Cardiology and Biophysics Division, Victor Chang Cardiac Research Institute; St. Vincent's Clinical School, University of New South Wales, Sydney, Australia

Michael P. Massett
Department of Kinesiology and Sport Management, Texas Tech University, Lubbock, TX, United States

Andrew McColloch
Department of Bioengineering, University of Texas at Arlington, Arlington, TX, United States

Yoshitaka Nakayama
Molecular Cardiology and Biophysics Division, Victor Chang Cardiac Research Institute, Sydney, Australia

Yury A. Nikolaev
Cellular & Molecular Physiology, Yale University, New Haven, CT, United States

Manoochehr Rabiei
Department of Mechanical and Aerospace Engineering, University of Texas at Arlington, Arlington, TX, United States

Lorena Redondo-Morata
Center for Infection and Immunity of Lille, INSERM U1019, CNRS UMR 8204, Lille, France

Cynthia A. Reinhart-King
Department of Biomedical Engineering, Vanderbilt University, Nashville, TN, United States

Alex P. Rickel
Department of Biomedical Engineering, University of South Dakota, Vermillion, SD, United States

Paul Rohde
Molecular Cardiology and Biophysics Division, Victor Chang Cardiac Research Institute, Sydney, Australia

Valentin Romanov
Molecular Cardiology and Biophysics Division, Victor Chang Cardiac Research Institute, Sydney, Australia

Hanna J. Sanyour
Department of Biomedical Engineering, University of South Dakota, Vermillion, SD, United States

Samantha C. Schwager
Department of Biomedical Engineering, Vanderbilt University, Nashville, TN, United States

Jae-Won Shin
Department of Bioengineering; Department of Pharmacology, University of Illinois at Chicago, College of Medicine, Chicago, IL, United States

Giulia Silvani
Molecular Cardiology and Biophysics Division, Victor Chang Cardiac Research Institute, Sydney, Australia

Andreea Trache
Department of Medical Physiology, Texas A&M University Health Science Center, Bryan; Department of Biomedical Engineering, Texas A&M University, College Station, TX, United States

Christopher R. Woodman
Department of Health and Kinesiology, Texas A&M University, College Station, TX, United States

CHAPTER ONE

Lipid bilayers: Phase behavior and nanomechanics

Lorena Redondo-Morata[a], Patricia Losada-Pérez[b], and Marina Inés Giannotti[c,d,e,*]

[a]Center for Infection and Immunity of Lille, INSERM U1019, CNRS UMR 8204, Lille, France
[b]Experimental Soft Matter and Thermal Physics (EST) Group, Department of Physics, Université Libre de Bruxelles, Brussels, Belgium
[c]Biomedical Research Networking Center on Bioengineering, Biomaterials and Nanomedicine (CIBER-BBN), Madrid, Spain
[d]Institut de Bioenginyeria de Catalunya (IBEC), The Barcelona Institute of Science and Technology (BIST), Barcelona, Spain
[e]Departament de Ciència de Materials i Química Física, Universitat de Barcelona, Barcelona, Spain
*Corresponding author: e-mail address: migiannotti@ibecbarcelona.eu

Contents

Abstract

Lipid membranes are involved in many physiological processes like recognition, signaling, fusion or remodeling of the cell membrane or some of its internal compartments. Within the cell, they are the ultimate barrier, while maintaining the fluidity or flexibility required for a myriad of processes, including membrane protein assembly. The physical properties of *in vitro* model membranes as model cell membranes have been extensively studied with a variety of techniques, from classical thermodynamics to advanced modern microscopies. Here we review the nanomechanics of solid-supported lipid

Current Topics in Membranes, Volume 86
ISSN 1063-5823
https://doi.org/10.1016/bs.ctm.2020.08.005

membranes with a focus in their phase behavior. Relevant information obtained by quartz crystal microbalance with dissipation monitoring (QCM-D) and atomic force microscopy (AFM) as complementary techniques in the nano/mesoscale interface is presented. Membrane morphological and mechanical characterization will be discussed in the framework of its phase behavior, phase transitions and coexistence, in simple and complex models, and upon the presence of cholesterol.

1. Introduction

Cells can be thermodynamically defined as open systems in constant exchange of mass, energy and information with the environment. The membrane is the ultimate boundary for the cell, confining it from the medium and some of its internal compartments. The cell membrane is a fundamental structure of the cell, providing a support matrix for proteins and carbohydrates. As complex systems, cell membrane performance is the result of lipids and proteins working together, with main functions like being barriers, mediating the exchange of molecules and information, promoting signaling and adhesion, and being metabolically self-renewing structures.

Cell membranes are materials with unique physical properties allowing cells to rapidly change shape, squeeze, stretch, pinch off smaller units, fuse, and reseal. This is the result of being viscoelastic sheets with both fluid and elastic properties. Cell membranes are supramolecular structures, where the lipids and proteins interact through non-covalent bonding. Membranes are curved surfaces, vesicular in nature, whose curvature is deeply influenced by the lipid packing. The lipid bilayer, made of millions of building blocks held together by weak interactions, is in part responsible for the physical, dynamic and mechanical properties of the cell membrane. In the membrane, lipids are mainly organized in lamellar phases, where two leaflets of lipids are self-assembled exposing the polar headgroups to the aqueous interface and keeping the fatty-acyl chains aligned opposed and parallel to one another to form the hydrophobic core.

The enterprising scientist Alec Bangham was the discoverer in 1958 of the "multilamellar smectic mesophases" or, as himself referred to less seriously, the "bangasomes" (Bangham, Pethica, & Seaman, 1958)—which were eventually named "liposomes" to define microscopic lipid vesicles (Sessa & Weissmann, 1968). When the Babraham Institute (Cambridge, UK) acquired its first electron microscope in 1961, Bangham had the privilege to firstly observe the dispersions of phospholipids in water solutions of negative stains (Bangham & Horne, 1964). In the following years,

Bangham and his collaborators performed the key experiments to demonstrate that lipid bilayers maintain concentration gradients of ions such as potassium and sodium. Indeed, David D. Deamer wrote in his memoir "*It was the membrane equivalent of finding the double helix structure of DNA, another Cambridge discovery in the life sciences*" (Deamer, 2010). This was reflected in the well-known lecture by Bangham in the University of Bristol in 1975, entitled "Membranes came first," where he proposed that something similar to liposomes had been available to house the first forms of life. Indeed, all living organisms have a membrane.

Later in the early 1970s, Singer and Nicolson proposed the fluid-mosaic-model, depicturing the cell membranes as two-dimensional liquids where all lipid and protein molecules diffuse easily (Singer & Nicolson, 1972). Concerning the composition, besides all the proteins and carbohydrates, lipids are the main components of the cell membrane in terms of molar fraction. Lipids are a broad family which covers many different chemical structures: sphingolipids (ceramides, sphingomyelin, gangliosides, sphingosines), sterols (cholesterol and vitamins), and phospholipids, each of them contributing to different and crucial physicochemical properties.

Gradual progress was made in the knowledge of the complexity of the biological membranes. The most famous hypothesis in the field is the membrane raft proposed in 1997 by Simons and Ikonen (1997). The idea was based on the fact that portions of membrane were found to be detergent resistant, that may give rise to a virtual compartmentalization of the cell. This hypothesis generated a lot of literature and became a recursive concept in the field. Nowadays it is well known that there is a variety of nanostructures in the membrane of heterogeneous sizes and functions, and the methods that allow us to observe these nanodomains in vivo are only beginning to emerge (Goñi, 2019; Pinkwart et al., 2019). Nanodomains can be considered putative heterogeneous structures within the membrane which are due partly to phase separation.

The mechanical role of the lipid membrane in force-triggered and force-sensing mechanisms in cells is of significance and adds to the better-established role of the mechanosensitive proteins (Kechagia et al., 2019; Vogel, 2006). Membrane conformational changes such as bending, vesiculation or tubulation are involved in cellular processes including adhesion, signaling, endocytosis or membrane resealing (Hassinger et al., 2017; van Meer & de Kroon, 2011; van Meer et al., 2008). For instance, in endocytosis, the endocytic system needs to generate enough force to form an endocytic vesicle by bending the membrane. These mechanisms generally require the membrane separation from the cytoskeleton as well as strong bending, for which the membrane chemical composition and physicochemical properties, often highly localized and dynamic, are key players (Sheetz, 2001; van Meer et al., 2008). It is now clear that the lipid packing and composition are direct modulators of the membrane curvature and elasticity, at different scales, even locally at the nanometer scale (Vereb et al., 2003; Yeagle, 1989). Understanding the lipid bilayers structural and mechanical properties and the involvement and role of each individual component turns out to be essential in order to identify their contribution to the overall membrane traits. Structural and physical properties of lipid bilayers include shape and local distribution of components (phases and domains), and related mechanical stability in response to compression, bending or stretching, their ability to fuse, etc.

We review the nanomechanics of solid-supported lipid membranes and how these relate to lipid membrane phase behavior. To this end, we chose two complementary techniques with nano/mesoscale sensitivity to mechanical and viscoelastic properties, namely quartz crystal microbalance with dissipation monitoring (QCM-D) and atomic force microscopy (AFM). QCM-D is a label-free surface-sensitive technique, whose working principle is based on the inverse piezoelectric effect. Real-time simultaneous measurements of resonant frequency and energy dissipation make QCM-D very suitable for gravimetric and viscoelastic characterization of solid-supported nanoscale sized films. AFM is a technique that allows the observation of surfaces under controlled liquid environment, to resolve topographical features with nanoscale resolution and to measure and apply forces in the pN–nN range. Therefore, AFM is appropriate for imaging the topography of solid-supported lipid membranes and probing the physical and mechanical properties at the nanoscale by means of force spectroscopy, providing high spatial and force resolution. QCM-D is complementary to AFM, useful for probing changes in viscoelastic properties during solid-supported film formation. The complementarity of AFM and QCM-D manifests at different levels: (i) AFM measurements are performed at a local level, while QCM-D provides a global characterization of the solid-supported films; (ii) both techniques enable dynamic mechanical analysis by applying a small oscillatory stress, however, the AFM tip scans and deforms supported layers from the top, whereas QCM-D measures the sensor oscillation from the bottom. The combination of AFM and QCM-D for supported lipid membranes has been employed in previous works (see, for instance Richter et al., 2003; Van Lehn et al., 2014).

In Section 2 we briefly describe the membrane models and discuss the working principles behind both approaches, highlighting the quantitative and qualitative mechanical information that can be extracted using each technique. Specific examples on the usefulness to monitor phase transitions on one-component to more complex bilayers, and resolve domain coexistence are also provided in Sections 3 and 4, including the key role of cholesterol.

2. Experimental approaches to study membrane mechanics

2.1 Model systems: Supported vesicles layers (SVLs) and lipid bilayers (SLBs)

Due to the extreme complexity of biological membranes, cell membrane mimetics are excellent approaches to study membrane properties and biological processes at the cellular and subcellular level. A great deal of what we know about these processes comes from modeling lipid bilayers *in vitro*. Various membrane systems are established as biomimetic structures. They include vesicles: freely-suspended and supported liposomes, giant unilamellar vesicles (GUVs); micelles, bicelles; suspended and supported lipid films: Langmuir monolayers, phospholipid bilayers nanodiscs, black lipid membranes (BLM), supported lipid bilayers (SLB), tethered bilayer lipid membranes (tBLM), polymer-cushioned membranes, protein-tethered bilayer lipid membranes (ptBLM) and hybrid bilayers (Dimova, 2019; Doktorova et al., 2018; Rascol et al., 2016; Sebaaly et al., 2019; Siontorou et al., 2017). Among these, the giant unilamellar vesicles (GUVs) have been extensively used since they offer a perfect stage to study the mechanical, thermodynamic, electrical, and rheological properties of the overall GUV and lipid bilayer as a function of membrane composition, surrounding media and temperature (Dimova, 2019; Evans et al., 2003; Kahya et al., 2004). Being heterogeneous and dynamic at the nanoscale, nanotechnology-based techniques can further explore biomembranes locally with resolution at the nanometric level. For many of these techniques, like surface analytical techniques, solid-supported models are the most adequate, including supported lipid monolayers, supported vesicles layers (SVLs) or supported lipid bilayers (SLBs).

SVLs are the precursor solid-supported systems to planar SLBs formed by vesicle fusion and rupture. They result from spontaneous adsorption of small vesicles of diameter $d \leq 200\,\text{nm}$ onto solid surfaces. The geometry of SVLs embodies membrane curvature, tension and osmotic stress within the supported layer. This makes SVLs useful biomimetic platforms to probe membrane deformation using surface-sensitive techniques and a testing ground for adhesion, budding and lipid membrane exchange and fusion (Hurley et al., 2010; Lipowsky & Seifert, 1991; Steinkühler et al., 2019; Tabaei et al., 2016). SVLs are generally formed by small soft adsorbates which are mimics to small endosomes or exosomes. These are optically inaccessible systems and the experimental investigation of their deformation is not straightforward. Adsorbed vesicles onto inorganic surfaces also serve as model systems relevant to biocompatibility studies. Vesicle adsorption is very often accompanied by vesicle deformation upon contact with the solid support. The extent of vesicle deformation depends on several factors such as vesicle size, mechanical properties of the vesicles, adhesion strength of the surface, osmotic pressure difference over the vesicle, etc. The fate of vesicles upon adsorption on a solid support depends both on vesicle-vesicle and vesicle surface interactions. In the absence of osmotic stress and at a concentration where vesicle surface coverage enhances fusion probability, the rupture of vesicles will depend strongly on the adhesion strength. Specifically, on how this imbalances the energetic competition between adhesion energy and an elastic stretching of the membrane (Lipowsky & Seifert, 1991). If the adhesion strength W is large enough and exceeds the lysis tension of the membrane, the vesicles will rupture. SVLs are typically formed on surfaces whose adhesion strength is not sufficient to induce vesicle rupture, such as Au, TiO_2, Pt (Tero, 2012).

SLBs are very manageable platforms relatively simple to obtain and that retain two-dimensional order and lateral mobility. They are ideal to study lipid lateral interactions and membrane local mechanical properties, growth of lipid domains, as well as interactions between the lipid membrane and proteins, peptides and drugs, cell signaling, etc. They also offer an excellent environment for inserting membrane proteins. There are many different methods that can be used to obtain SLBs, like the spin-coating and hydration (Mennicke & Salditt, 2002), microcontact printing (Strulson & Maurer, 2011), solvent-exchange deposition (Hohner et al., 2010; Tabaei et al., 2016), and the most widely used methods like the Langmuir–Blodgett/Schäfer deposition to prepare mono and bilayers (Kurniawan et al., 2018), and the liposome fusion and rupture method for bilayers (Hardy et al., 2013; Mingeot-Leclercq et al., 2008). The liposome rupture method remains the most popular and simple one, based on depositing small unilamellar vesicles (SUVs) from a suspension onto a flat substrate where the adhesion strength is high, generally mica or silicon oxide, but can also be formed on bare or gold-coated glass, following the needs of the analysis technique (Choi et al., 2016; Gumi-Audenis et al., 2018; Mingeot-Leclercq et al., 2008; Richter & Brisson, 2005; Seeger et al., 2010; Winkler, Campelo, Giannotti, & García-Parajo, 2020). Once in contact with the substrate the SUVs start fusing between them, deforming, flattening, and finally rupturing to form a continuous film. It is important to have in mind that the final SLB structure is affected by variables like the lipid vesicles composition, concentration, and size, the physicochemical environment, i.e., pH, temperature, and ionic strength, as well as the surface roughness and charge density (Reimhult et al., 2003).

2.2 Quartz crystal microbalance with dissipation monitoring (QCM-D)

Quartz crystal microbalance with dissipation monitoring (QCM-D) is an acoustic-based surface-sensitive technique that enables label-free, real-time simultaneous measurements of wet-mass and viscoelastic properties of solid-supported nano/meso-scaled adlayers.

The QCM-D sensor consists of an AT-cut quartz crystal being sandwiched between two electrodes. When an AC voltage is applied across the electrodes with a frequency close to the resonant frequency of the quartz crystal, a mechanical deformation is induced, resulting in a standing shear wave. The surfaces of the electrodes coincide with the antinodes of the standing shear wave with wavelength $\lambda = 2d/n$, with d the quartz thickness and n the (odd) overtone number. The resonance frequency at each overtone is $f_n = nv/2d$, with v the speed of sound in quartz. The shear waves propagate as evanescent waves decaying across the boundary between the crystal and the fluid environment (air or liquid) with a penetration depth $\delta = \sqrt{\eta_L/\pi f_n \rho_L}$, which depends on the overtone frequency f_n and both the viscosity η_L and density ρ_L of the fluid in contact with the sensor surface (Reviakine et al., 2011). The penetration depth of a 5 MHz shear wave in water is $\delta \sim 250\,\text{nm}$, rendering QCM-D surface-specific. D is defined as the ratio between the dissipated energy during one vibration period and the total energy of the crystal at that instant $D = E_{lost}/2\pi E_{stored}$. In the so-called "ring-down method" measurements, energy dissipation D can be calculated when the AC voltage driving the quartz crystal oscillation is turned off. The working principle of QCM-D is reflected in Fig. 1, upon the presence of a soft, viscoelastic layer such as

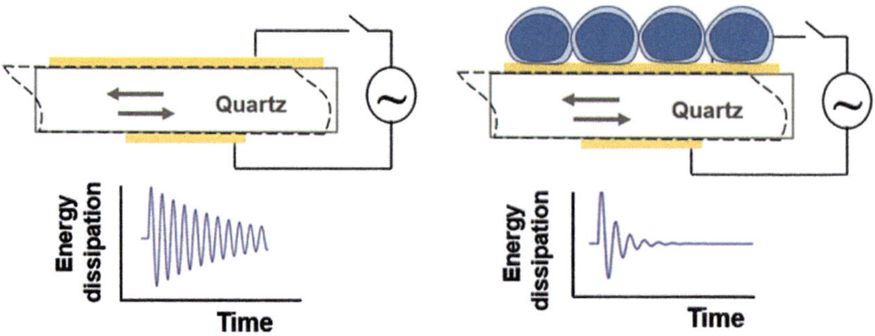

Fig. 1 Schematic view of a QCM-D sensor excited by an AC voltage. Left panel: Bare Au-coated sensor in liquid environment, the dampening of the oscillations takes place slowly. Right panel: The presence of an adsorbed lipid vesicle layer (SVL) induces a faster damping of the oscillations and thus a larger energy dissipation.

a layer of lipid SVLs. The amplitude of the crystal oscillation decays exponentially and much faster when in contact with a viscoelastic layer, characterized by a large energy dissipation response.

Commercial QCM-D instruments allow for monitoring changes in frequency and dissipation for several overtones ($n = 3, 5, 7, \ldots, 13$), Δf_n and ΔD_n, upon the presence of a nano/mesoscale layer. The spectroscopic data at multiple frequencies with varying detection range enables assessing the spatial homogeneity and rigidity of the adsorbed layer. For homogeneous thin and rigid films, $\Delta D_n \sim 0$, Δf_n overtones overlap and a simple relationship between mass adsorbed and change in frequency holds, the so-called Sauerbrey relation: $\Delta m = -C\Delta f/n$, with C being the Sauerbrey constant which for a 5MHz quartz crystal reads $C = d\rho_{quartz}/f_0 = 17.7ng/cm^2$ (Sauerbrey, 1959). For thicker and softer films $\Delta D_n > 0$, overtones do not overlap and the ensemble quartz-film-fluid can be represented by simple models combining elastic components and viscous dashpots. These models enable to extract effective viscoelastic parameters of the adsorbed film such as shear viscosity, thickness and shear storage modulus (Cho et al., 2007; Voinova et al., 1999).

The capability of monitoring the viscoelastic properties during film formation is particularly useful in the case of solid-SLB formation. QCM-D has been a valuable tool in deciphering the kinetic pathways of SLB formation from precursor vesicles and how those depend on relevant conditions such as solid surface adhesion strength, medium ionic strength, vesicle mechanics, etc. (Cho et al., 2010; Jing et al., 2013; Keller & Kasemo, 1998; Pramanik et al., 2016; Richter, 2006).

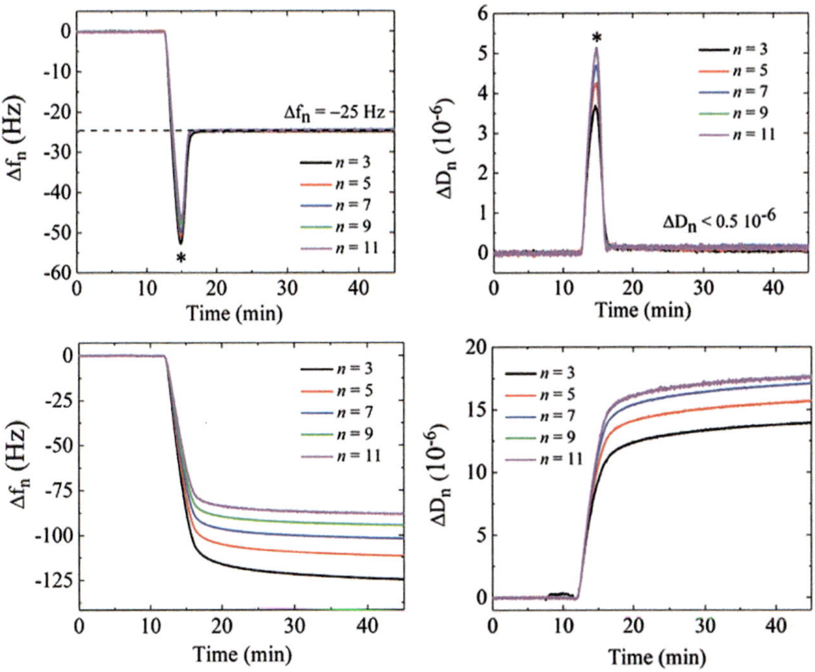

Fig. 2 Time dependence of Δf_n and ΔD_n during a QCM-D experiment of DOPC vesicle adsorption at 37 °C onto SiO_2 (upper panels) and Au (lower panels).

Fig. 2 shows a typical QCM-D experiment where real-time monitoring of the adsorption of zwitterionic DOPC vesicles onto SiO_2 and oxidized Au was carried out at 37 °C. The vesicles are dispersed in TRIS-buffer (10 mM TRIS, 100 mM NaCl, pH 8) at a concentration of 0.1 mg/mL and their hydrodynamic diameter $d = 79 \pm 20$ nm. As initially discussed by Keller and Kasemo (1998), the adsorption kinetics is surface-specific. The mechanistic picture of Δf and ΔD changes is governed by a delicate balance between the adhesive contribution to the free energy from lipid-surface interactions and the opposing effect of bending and stretching the membrane (Lipowsky & Seifert, 1991). DOPC vesicles adsorbing on SiO_2 follow a two-step adsorption process consisting of (i) adsorption of a critical number of vesicles and (ii) fusion, rupture and formation of an SLB. This is reflected in the Δf_n and ΔD_n signals depicted in the upper panels of Fig. 2. After a ~10 min baseline in buffer (Δf_n and $\Delta D_n = 0$), DOPC vesicles added at a very small flow rate adsorb onto the SiO_2-coated quartz sensor (Δf_n decreases and ΔD_n increases) until a maximum number (minimum in Δf_n and maximum in ΔD_n). Adsorbed vesicles then fuse and rupture as a consequence of the large adhesion strength of SiO_2, thus releasing the entrapped aqueous buffer. As a consequence, Δf_n increases (mass loss) to a constant plateau of $\Delta f_n \sim -25\,Hz$ and the ΔD_n signal decreases back to a very small value ($\Delta D_n < 0.5 \times 10^{-6}$). The final Δf_n and ΔD_n values and the fact that overtones signals overlap is consistent with the formation of a homogeneous rigid and thin DOPC SLB. The pathway of SLB formation is not unique and depends strongly on the ionic conditions (head group charges of the constituent lipids, buffer ionic strength presence of divalent cations, pH). Vesicles containing positively charged lipids like DOTAP SUVs rupture individually on SiO_2 as a consequence of stronger vesicle-support electrostatic interactions (Richter et al., 2003).

When DOPC vesicles adsorb onto an oxidized Au surface, a monotonic Δf_n decrease and ΔD_n increase can be observed reaching constant non-zero plateau values with non-overlapping overtones. Such time-dependent responses provide evidence that oxidized Au facilitates non-ruptured vesicle adsorption toward the formation of acoustically non-rigid vesicle layers with saturated coverage. As pointed out by Lind and Cárdenas (2016), it is worth mentioning that local vesicle rupture events and formation of small bilayer patches cannot be ruled out. As a matter of fact, QCM-D is very sensitive to hydrodynamic (wet) mass and the local, partial formation of SLBs might be masked by the adsorption of vesicles on top or in between the bilayer patches. In this case, complementary QCM-D and AFM measurements are particularly useful to obtain a complete picture of the heterogeneous layers SLBs with co-adsorbed vesicles formed. For example, the complementarity of AFM and QCM-D has proven successful in experimentally confirming defect (protrusion)-driven nanoparticle interactions with SLBs (Van Lehn et al., 2014).

Apart from kinetic information, QCM-D can be also useful to estimate the extent of vesicle deformation upon adsorption following a model-free approach introduced by Tellechea et al. (2009). This method consists in plotting $-\Delta D/\Delta f$ ratio vs $-\Delta f$ for all overtones during initial adsorption, which shows a linear decrease over a large range of frequency shifts. Extrapolation of this linear decrease to a frequency-independent intercept with the $-\Delta f$ axis (where overtones intersect) provides a value of the thickness of the adsorbed vesicle layer h or Sauerbrey thickness: $h = -\Delta f\, C/\rho$, where $C = 17.7\,ng/cm^2\,Hz$ and $\rho = 1\,g/cm^3$ is the density of the film. This approach assumes a complete surface coverage, where the presence of trapped buffer has been diminished to occupy only the void spaces between densely packed vesicles (the $-\Delta D/\Delta f$ ratio is close to zero and

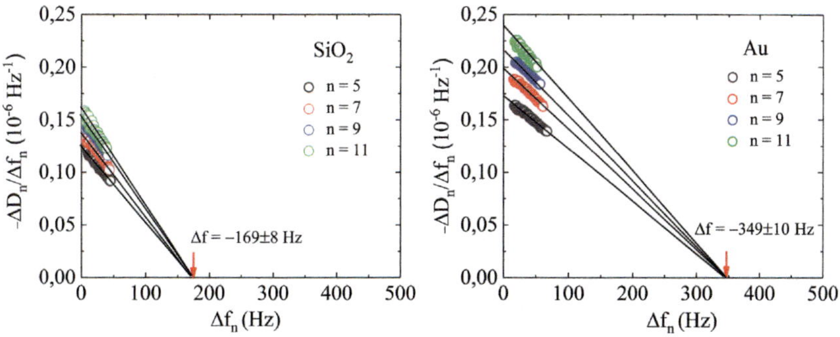

Fig. 3 $\Delta D_n/-\Delta f_n$ ratio as a function of frequency at different overtones for DOPC vesicles adsorbed onto SiO_2 and Au.

the $-\Delta f$ intercept values were the same on the extrapolation of a linear regression) (Olsson et al., 2013; Reviakine et al., 2012; Tellechea et al., 2009). Fig. 3 displays the extrapolated Sauerbrey thickness for DOPC SUVs adsorbed onto SiO_2 and Au surfaces. The respective effective frequencies read $h(SiO_2) = 30 \pm 2$ nm and $h(Au) = 62 \pm 5$ nm. If one compares these values with the original vesicle diameter in bulk measured by DLS ($d = 79 \pm 20$ nm), vesicles were deformed to a greater extent onto SiO_2 surfaces. A more exact approach based on a hydrodynamic model was recently introduced by Cho and coworkers (Gillissen et al., 2017) to estimate the deformation of small vesicles at low surface coverage. The model enables to estimate adsorbed vesicle shape and bending energy by using hydrodynamic spectroscopy (frequency shifts at several overtones). It treats vesicles as ellipsoids and relates the so-called QCM-D force to the scaled viscous penetration depth δ/a with δ the viscous penetration depth, and a the non-deformed vesicle radius, in order to probe the adsorbed particle aspect ratio. Further details on the model can be found in Gillissen et al. (2017).

2.3 Atomic force microscopy (AFM)-based methodology

Classical structural techniques such as X-ray, NMR or electron microscopy rely on ensemble averaging of molecular properties. Yet, by these means, the underlying molecular dynamics can be hidden, since the signal comes from the average of many unsynchronized molecules in the bulk. The picture changes when using single-molecule techniques. The measured signal of one—or a few—molecules reflects the stochasticity of thermally induced processes when randomly crossing free-energy barriers. Single-molecule optical microscopy techniques are particularly well suited for the study of the dynamics of molecules. However, these latter methods rely on the detection of fluorescent markers attached to proteins or lipids, which means that the resolution is limited to ~200 nm due to diffraction of light and to a few tens of nm thanks to super-resolution techniques. The atomic force microscopy (AFM) was since its invention in 1986 (Binnig et al., 1986) quickly positioned among the single-molecule, high-resolution structural analysis techniques. It provides information concerning structure, function-related conformational changes and supramolecular assemblies, crucial for a complete understanding of biological processes. Nowadays, AFM is in its thirties and has become an invaluable tool for studies at the micro- and nano-scale. As a stand-alone, high-resolution imaging technique and force transducer, it is today an established methodology among the biophysical community. The AFM uses a nanometer-sharp tip—resembling a record player needle—attached to a (micro)cantilever to sense the sample surface while scanning it using piezoelectric elements. An optical system is used to detect the cantilever deflection (or force), thanks to a laser diode beam reflected on the back of the cantilever and detected by a photodiode. Changes in deflection due to sample protrusions while the tip scans the sample are translated into electrical signals thanks to a feedback mechanism. This is schematically represented in Fig. 4.

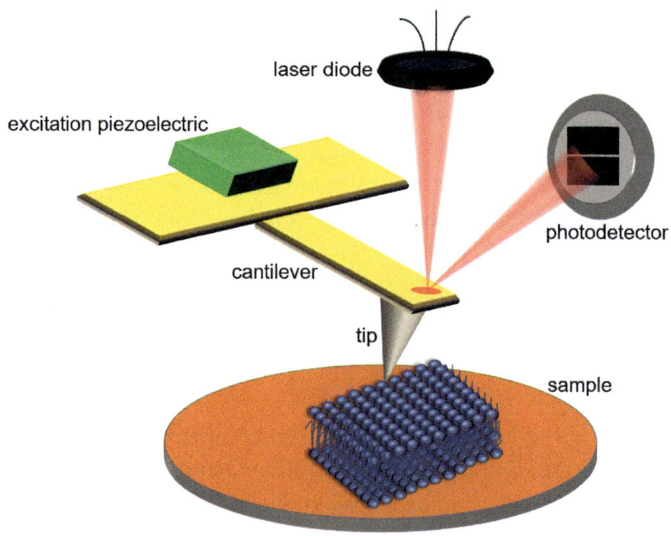

Fig. 4 Schematic representation of an AFM instrument. The deflection of the cantilever in the normal direction is monitored by a photodetector which collects the reflection of a laser beam focused at the back of the AFM cantilever. In the tapping mode, thanks to a piezoactuator, the AFM cantilever is excited near its resonance frequency.

One of the main strengths of AFM relies on the possibility of working in a controlled environment (medium composition and temperature) and with great spatial resolution. AFM-imaging mode provides submolecular resolution on a large variety of biological samples: from single molecules, that is, DNA or proteins, to supramolecular assemblies such as SLBs, or even entire cells (Dufrêne et al., 2017; Giles et al., 1993; Parot et al., 2007). Therefore, AFM has become a well-established technique for imaging the topography of lipid membranes that show homogeneous or phase-separated morphology, permitting then to increase the bilayer complexity, from bilayers of one component to multicomponent ones (El Kirat et al., 2010; Gumí-Audenis et al., 2016; Morandat et al., 2013; Redondo-Morata et al., 2014).

For the study of biological samples, it is common to use AFM-imaging dynamic modes that have an intermittent tip-sample contact. Among them, the most often used is oscillation or tapping mode, in which the AFM cantilever is excited near its resonance frequency. The resulting oscillating tip is intermittently contacting—tapping—the surface, giving rise to a damped oscillation amplitude. Additionally, the resonance frequency is also shifted due to the probe-surface interaction. The topography of the surface is reconstructed by monitoring the amplitude of the oscillating cantilever, which is kept constant by adjusting the Z-piezo position through a feedback loop. While scanning in the X–Y direction, the amplitude of the cantilever oscillation is kept constant thanks to a feedback control. However, biomolecules are dynamic in essence; hence, to understand how biomolecules work it was only natural to think about increasing the temporal resolution of conventional AFMs. The first time that the concept of High-Speed (HS) AFM was mentioned—to our knowledge—was in 1991 by Barrett and Quate (Barrett, 1991), who did a fairly attempt with the technology at that time. Other research groups, as Paul Hansma's, would challenge the practical speed limitations. It was only in 2010 when the laboratory leaded by Toshio Ando in Kanazawa University (Japan), filmed individual myosin molecules walking on an actin filament (Ando et al., 2008), operating their AFM instrument at a speed about 1000 times faster than the conventional instruments of that time. Besides the visual impact and scientific insight of those movies, these experiments illustrated that HS-AFM could obtain concomitantly structural and dynamic data, providing insights inaccessible by any other method (Ando, 2017; Chiaruttini et al., 2015; Kodera et al., 2010; Mierzwa et al., 2017). The basic principles, advantages and limitations of the most common AFM bioimaging mode are nicely detailed in a recent review from Dufrêne et al. (2017).

Following the capability of the AFM for mechanical manipulation, to sense and apply small forces accurately (pN–nN range), it rapidly developed into an excellent technique to study interactions—inter and intramolecular forces—at the molecular level, the so-called AFM-based force spectroscopy (AFM-FS). The manipulation of single molecules (single molecule force spectroscopy) has led to fascinating new insights in the mechanics of proteins, polysaccharides, synthetic polymers, and DNA at the molecular level (Clausen-Schaumann et al., 2000; Florin et al., 1994; Giannotti & Vancso, 2007; Hugel & Seitz, 2001; Lee et al., 1994; Marszalek et al., 1999, 2001; Noy et al., 1997; Radmacher, 1997; Rief et al., 1997).

In the case of lipid bilayers, AFM offers the unique opportunity to probe local physical and mechanical properties, determining the interaction forces with nanoscale lateral resolution, thereby providing new insights into membrane molecular mechanisms. For the nanomechanical characterization with AFM, an SLB is generally first imaged and then a series of force-distance curves is recorded over the same area. In each force-distance curve, the AFM tip away from the surface is approached and retracted at constant velocity while the cantilever deflection is recorded as a function of the Z-piezo position. Each portion of the force-distance curve provides information about the physical and mechanical properties of the bilayer, schematized in Fig. 5. Upon mechanical contact during approach, the cantilever deflection increases and the SLB is elastically compressed by the AFM probe until the tip suddenly breaks through the SLB, jumping into a contact with the substrate (Fig. 5B). The first part of the curve, corresponding to small elastic deformation upon compression, can be used to estimate elastic descriptors of the membrane, like the Young modulus (E), the area compression-expansion modulus (K_A) and bending modulus (k_C) (Dufrêne & Lee, 2000; Picas et al., 2012; Redondo-Morata et al., 2016). When performing very small elastic deformations (far from the bilayer rupture) it is also often used for this calculation the retraction part of the curve. To estimate the elasticity of the lipid film, the most common model used is the Hertz model, where the surface is approximate to an infinitely half-space of an isotropic elastic solid and the indenter as non-deformable. The parameter describing the sample is the Poisson's ratio (ν), which depends on the nature of the material. For soft biological samples, the Poisson's ratio is generally set to 0.5 (as for incompressible materials). The geometry of the indenter determines the contact area. The original Hertz model considers the shallow contact between two spherical bodies, but

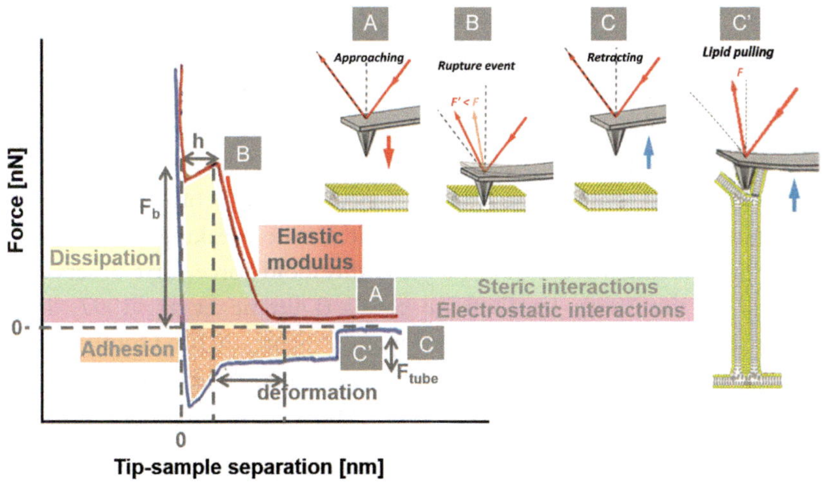

Fig. 5 Schematic representation of a representative force curve typically plot in vertical force vs tip-sample separation. The red curve represents the tip approaching (A). Right after the contact area, short-range tip-sample interactions from different nature appear, predominantly electrostatic (highlighted in pink) and steric (highlighted in green) interactions. When keeping the tip velocity constant, elastic deformation of the membrane is followed by a sudden jump (B) which is interpreted as the breakthrough force (F_b) or the maximum force the bilayer is able to withstand before breaking. F_b has been extensively used as a hallmark of the bilayer stability, some of these related works are reviewed in this chapter. The AFM tip reaching the hard substrate underneath is represented as a vertical in the force-separation curve. From the distance in the curve between the breakthrough force and the substrate, the minimal bilayer thickness (h) can be estimated (fully compressed). Afterwards, the tip retracts at constant velocity (C). The elastic modulus is usually estimated from the slope of the curve which displays membrane deformation. It is often observed a hysteresis between the approach and retract curves (yellow area) which is related to dissipation work and therefore to the viscoelasticity of the material. During retraction, negative force values are used to estimate the adhesion work between the sample and the AFM tip. Sometimes, a force plateau when retracting the AFM tip away from the SLB can be observed. It characterizes a tube pulling process (C′), defined by the tube growing force (F_{tube}) until eventually the AFM tip completely detaches.

several extensions were made for different indenter geometries (Lin, Dimitriadis, & Horkay, 2007). Again, when performing experiments with small deformations far from the onset of the lipid bilayer rupture, the hysteresis found between the approach and the retract curves is considered as dissipation work and can be used to obtain intrinsic viscoelastic parameters of the viscoelastic material, by the adaptation of common viscoelastic models as power-law rheology or the Kelvin-Voigt model (Garcia, 2020).

Moreover, different types of short-range interactions can be measured—DLVO, hydration or steric forces—for further information review here (Redondo-Morata, Giannotti, & Sanz, 2012).

The penetration of the AFM tip through the bilayer appears as a discontinuity in the approaching force-separation curve (Fig. 5B). The vertical force at which this discontinuity happens corresponds to the maximum force the bilayer is able to stand before breaking and is defined as the breakthrough force (F_b) or yield threshold force, first observed in the end of the 1990s by Dufrêne et al. (Dufrene et al., 1997, 1998) (Dufrêne & Lee, 2000; Garcia-Manyes & Sanz, 2010; Gumí-Audenis et al., 2016; Redondo-Morata, Giannotti, & Sanz, 2012; Schneider et al., 2000). F_b is generally of several nanonewtons (nN) and it is considered as a direct measurement of the lateral interactions within the lipid bilayer, at a specified loading rate. From the step in the separation before and after the breakthrough, the SLB thickness at different compression forces (and at zero force) can be extracted.

The mechanical rupture of lipid bilayers is of thermal fluctuation nature, whose destructive action is facilitated and directed by the application of the external force. The penetration of the AFM tip through the SLB has been modeled and conceived as a two-state process with an associated energy barrier (Butt & Franz, 2002; Franz et al., 2002; Loi et al., 2002). Specifically, two models describe the activation process. The first is the continuum nucleation model, which considers the bilayer as a molecular thin homogeneous film—a two-dimensional fluid layer—between the solid substrate and the solid surface of the AFM tip. The second model ponders the molecular nature of the lipid bilayer and proposes that each molecule in the SLB has specific binding sites corresponding to energetically favorable positions. These sites are energetically equivalent, and as the SLB is pressed by the AFM tip, the energy of the molecules significantly increases, leading them to jump apart and create a hole under the tip. After a critical number of molecules have jumped out of the contact area, the tip indents the SLB due to the high pressure of the remaining molecules, breaking the bilayer. Determining the energy barriers, represented by the Arrhenius law, governing the lipid membranes rupture process contributes to the understanding of the extent of the lateral interactions in the bilayer, and this is generally achieved by following the variation of the F_b with the loading rate (r) by varying the tip-sample approaching velocity (v) in what is called the dynamic force spectroscopy (DFS). Out of equilibrium, the F_b increases linearly with the logarithm of r (and of v as well) (Butt & Franz, 2002;

Loi et al., 2002). The loading rate dependence with the thermomechanically activated nature of the bilayer rupture kinetics leads to the evaluation of the activation energy of the bilayer failure in the absence of an external force (ΔE_0). The activation energy ΔE for the formation of a hole in the bilayer that is large enough to initiate rupture and lead the tip breakthrough can then be calculated following the expression proposed by Butt et al.:

$$\Delta E(F_b) = -k_B \cdot T \cdot ln\left[\left(\frac{0.693 \cdot k_S}{A}\right) \cdot \frac{dv}{dF_b}\right] \tag{1}$$

where k_B is the Boltzmann factor, T is the absolute temperature, k_s is the cantilever spring constant, A is the frequency at which the AFM tip attempts to penetrate the bilayer, generally approximated to the resonance frequency of the cantilever, v is the tip velocity, and F_b the breakthrough force.

When retracting the AFM tip away from the SLB, it sometimes remains connected to the bilayer through a lipid tube (Fig. 5C′). While the tip moves further away, the membrane tube grows longer until it breaks at a certain distance, and the cantilever returns to the equilibrium position. This tube pulling process occurs at constant force, the tube growing force (F_{tube}) (Armond et al., 2011; Gumí-Audenis, Costa, Ferrer-Tasies, et al., 2018; Maeda et al., 2002; Richter & Brisson, 2003), and it is observed as a force plateau in the retract part of the force-separation curve (Fig. 5), at several tens of pN from which F_{tube} and the tube growing distance (d) can be recorded. Pulling lipid tubes with an AFM tip out of SLBs is a simplified but analogous situation to some biological processes involving mechanical and conformational adjustments—bending, vesiculation, tubulation-like formation of membrane tethers, endocytic vesicles, etc. (Schmidtke & Diamond, 2000; Shao et al., 1998; Sheetz, 2001; van Meer et al., 2008). Separation of a membrane segment from the cytoskeleton as well as strong membrane bending are both involved, for which the membrane chemical composition and physicochemical properties are key players. Pulling tethers from cells with an AFM probe is a widely explored approach to evaluate the membrane mechanics on entire cells (Brochard-Wyart et al., 2006; Marcus & Hochmuth, 2002; Nawaz et al., 2015; Sun et al., 2005). Recently, it has also been established as a method to locally probe SLB mechanics and the contribution of the underlying substrate to the measured properties on SLBs (Gumí-Audenis, Costa, Ferrer-Tasies, et al., 2018).

When pulling tethers from cells, the F_{tube} depends on the membrane properties like bending stiffness (k_c) and in-plane membrane tension (σ), and also on the adhesion between the membrane and the cytoskeleton (Sheetz, 2001). In regions where the membrane has separated from the cytoskeleton—bleb or free membrane—as there is no direct interaction with the cytoskeleton, F_{tube} strictly relies on k_c and σ (Dai & Sheetz, 1999). However, the cytoskeleton adhesion and σ are difficult terms to separate, and therefore the concept of apparent membrane tension, σ_{app}, has been proposed to include the adhesion energy parameter γ (Sheetz, 2001), $\sigma_{\text{app}} = \sigma + \gamma$. When the lipid tube grows under thermodynamic equilibrium, at the limit of zero velocity (static thermodynamic analysis), the following mathematical expression relates the membrane parameters mentioned above with the F_{tube} (Armond et al., 2011; Canham, 1970; Daillant et al., 2005; Evans & Yeung, 1994; Gumí-Audenis, Costa, Ferrer-Tasies, et al., 2018; Hochmuth et al., 1996; Marcus & Hochmuth, 2002; Roux, 2013): $\sigma_{app} = \frac{F_{tube}{}^2}{8k_c\pi^2}$.

The plus of this approach, AFM-based pulling lipid tubes from SLBs, is that it combines the advantage of the AFM local probing with nanoscale lateral and force resolution, with the simplicity of the SLB model preparation. Besides, the local nanomechanical properties of SLBs can be explored through the combination of the tubing force spectroscopy approach and the F_b analysis exposed before, within the same experimental procedure (Gumí-Audenis, Costa, Ferrer-Tasies, et al., 2018).

Although it will not be extensively discussed here, we find important to mention that the constant force approach can also be used to evaluate the membrane rupture kinetics in SLBs and supported lipid multibilayers (SLMs). The AFM tip is approached at constant velocity, and then compresses the membrane at a constant force (force clamp, FC). Through measuring the time it takes to penetrate the bilayer—time to breakthrough (t_b)—distribution at different compression forces, the AFM-FC mode allows to estimate the rate of bilayer failure and calculate the location of the energy barrier maximum—distance to the transition state—along the reaction coordinate (Δx) and the activation energy (E_0) that characterize the energy barrier (Redondo-Morata, Giannotti, & Sanz, 2012a; Relat-Goberna et al., 2017). Both E_0 as well as Δx were found to be higher for SLBs than for supported lipid multibilayers, endorsing previous suspicion that the mechanical properties of lipid bilayers obtained from nano-indentation are strongly influenced by the underlying hard substrate.

3. Phase behavior and nanomechanics. From one-component membranes to higher complexity

3.1 One-component membranes

3.1.1 The gel and the fluid phase

To start with the simplest system and increase the complexity in this revision—which does not intend to be exhaustive—consider a one-component phospholipid bilayer, which is thermalized at a temperature far from its T_m, the temperature of the main phase transition, at which a disordering of the hydrophobic tails happens followed by a sudden increase of the lateral mobility. We are then able to distinct two different main phases, gel-like or solid ordered phase (s_o) below T_m, and fluid-like phase or liquid disordered (l_d) above T_m.

A mechanical descriptor of lipid bilayers that shows significant differences between both phases is the bending modulus k_c. At $T < T_m$, k_c is an order of magnitude larger than at $T > T_m$, rendering lipid vesicles quite stiff, while softer and more deformable above T_m. In the case of SVLs, where probing deformation is not straightforward, QCM-D has proved to be a useful tool to estimate the extent of vesicle deformation upon adsorption onto surfaces. Reviakine et al. (2012) reported deformation measurements of small DMPC vesicles on TiO_2 above and below T_m, and experimentally found out that adsorbed vesicle deformation correlates with the bilayer bending modulus. Liposomes in the fluid phase bear a smaller bending modulus and deform to a greater extent.

When imaged with the AFM, the SLBs show typically 4–6 nm of height. The measured height is also influenced by both the elasticity of each bilayer and the particular AFM methodology, i.e., acoustic-modulation "tapping" mode, contact mode, etc., where the representation of the interface of the lipid bilayer changes. In terms of its mechanical properties, whether we are measuring the elastic deformation or the yield threshold force, a one-component SLB yields a monomodal distribution. While the distribution of deformations or breakthrough forces of the bilayer, being a stochastic process, have been often approached to Gaussian distributions, other distributions based on different theoretical frameworks can also be found in the literature, like the models proposed by Franz et al. (Butt & Franz, 2002) for the two-state activation process of the bilayer rupture, mentioned in Section 3.2.

In the first place, it could be said that, in a general way, the bilayer thickness depends on the nature of the lipids—solid-like bilayers are usually higher than liquid-like counterparts of similar number of carbons in the hydrocarbon chains. In an s_o phase, the hydrocarbon chains are stretched, so the apparent height is greater, and this is observed when imaging SLBs by AFM. Consequently, stretched acyl chains lead to stronger van der Waals interactions between neighboring molecules, which translates into a shorter intermolecular distance between phospholipids, and a greater mechanical resilience, both elastically and upon AFM tip penetration.

For example, let's briefly consider the elastic deformation of DOPC and DPPC, as models of l_d and s_o SLBs, respectively. The slope of the curve in the contact zone (elastic regime), as we have commented in Section 3.2, is related to the elastic modulus of the material. At room temperature, the slope for the DOPC SLB, which is in the fluid phase, is lower than that for the solid DPPC SLB. This indicates that DPPC bears a larger Young modulus than DOPC and thus the membranes in the s_o phase are therefore stiffer than in the l_d phase. Both slopes are smaller than the one that can be measured on the bare substrate (often assumed as an infinitely stiff material, such as mica).

Besides the evidenced difference in elasticity for the fluid and gel-like membranes, l_d SLBs show in general lower F_b values than s_o SLBs. A good example are homologous series phospholipids; while at room temperature DLPC ($T_m \sim -2\,°C$) SLBs break at force values between 2 and 13 nN, according to the ionic environment, DPPC ($T_m \sim 41.5\,°C$) SLBs are pierced by the AFM probe, in similar conditions, in the force range that goes from 7 to 30 nN (Redondo-Morata, Oncins, & Sanz, 2012). It is also noticeable that the F_b increase with the logarithm of the loading rate is lower for the fluid than for the gel membranes (Redondo-Morata, Oncins, & Sanz, 2012). This can be expected, if we consider the Bell-Evans or assimilable model, the most likely breaking force depends linearly on the logarithm of the loading rate r, $F(r) = \left(\frac{k_B T}{x}\right) \ln \frac{rx}{K_0 k_{BT}}$, where k_B is the Boltzmann constant, x is the distance to the transition state and K_0 is the off-rate constant at 0 force (Evans & Ritchie, 1997). In this case, x is directly related to the distance between neighboring phospholipid molecules. This intermolecular distance, therefore x, is smaller in s_o than in l_d.

Early work showed how AFM-FS is able to explore quantitatively the molecular determinants that provide mechanical stability to the supported membrane (Garcia-Manyes, Oncins, & Sanz, 2005a; Garcia-Manyes et al., 2010). By systematically and independently changing the chemical composition of the lipid headgroup, the length of the hydrophobic tail, and the number of chain unsaturations, it was possible to describe the mechanical stability of each bilayer as a direct signature of membrane lateral organization (Garcia-Manyes et al., 2010). The distribution of F_b values as a function of the length of the hydrophobic chain showed that on average, increasing each hydrocarbon tail by two extra $-CH_2-$ groups results in ~7 nN increase of the F_b, equivalent to an enthalpy increase by roughly 2 kJ/mol (Berg et al., 2002). In the same study, the mechanical properties of the linear DPPC were compared with its analogous branched phospholipid, DPhPC, to investigate the effect of acyl chain branching. The distribution of F_b showed a decrease from ~20 to ~12 nN, revealing an important contribution of the lateral $-CH_3$ groups on the molecular packing of the bilayer. This observation can be rationalized in terms of the larger area per lipid for the branched DPhPC (76.8 $Å^2$) compared to that for the linear DPPC (62.0 $Å^2$), as revealed by molecular dynamics simulations conducted in the fluid phase (Shinoda et al., 2007). The nanomechanical effect of the –cis double bonds present in the acyl chains was also investigated. The F_b values distributions for each particular bilayer showed an inversely proportional trend between mechanical stability and number of unsaturations. It was estimated that upon introduction of one additional unsaturation in each acyl chain, F_b values decrease by ~5 nN. Interestingly, introducing an asymmetric unsaturation resulted in an even further mechanical destabilization. The resulting molecular tilt reduces the effective packing between molecules, giving rise to a lower T_m (Berg et al., 2002). The findings regarding the mechanical implications of the headgroup chemistry were, however, far less straightforward and highly unanticipated. Comparing phospholipids with the same hydrocarbon tails but different headgroups, a huge range of forces required to indent the bilayer in the same experimental conditions, spanning from ~3 nN (DPPA) all the way up to ~66 nN (DPPG).

Furthermore, the lateral packing differences observed between fluid and gel-like bilayers is also reflected in the tube growing force, pulled as the AFM probe moves away from the SLBs, as this approach has demonstrated to be sensible to the membrane chemical composition in one-component SLBs (Gumí-Audenis, Costa, Ferrer-Tasies, et al., 2018; Maeda et al., 2002). Higher F_{tube} values for DPPC with respect to DMPC have already been observed by Maeda et al. (2002) and, as reported in Gumí-Audenis, Costa, Ferrer-Tasies, et al. (2018), F_{tube} is lower for l_d mica-SLBs

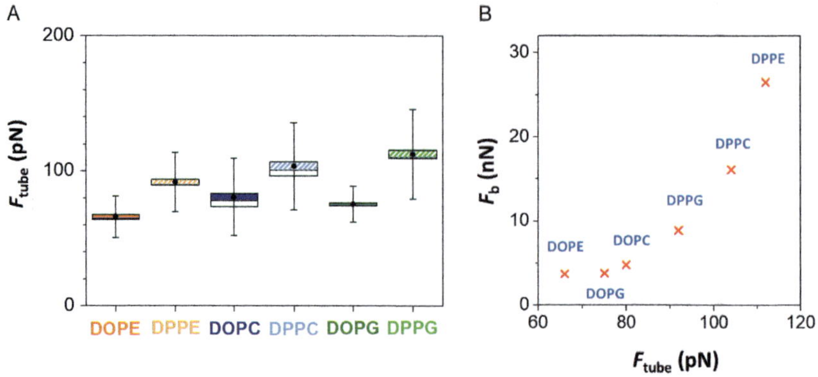

Fig. 6 AFM-pulling lipid tubes sensible to the membrane composition in one-component SLBs. (A) F_{tube} values for single component SLBs of different composition (DOPE, DPPE, DOPC, DPPC, DOPG and DPPG). Boxchart for F_{tube} values show the mean (●), SE (box) and SD (bars). (B) Correlation between F_b and F_{tube} values for single component SLBs. The individual phospholipids contain a constant tail length of 18C with 1 unsaturation (DOPE, DOPG and DOPC) or 16C and fully saturated (DPPE, DPPC and DPPG), in fluid and gel state at room temperature (RT), respectively. All the measurements were performed in 150mM NaCl, 20mM $MgCl_2$, 20mM HEPES (pH 7.4) buffer solution and at RT. *Adapted from Gumí-Audenis, B., Costa, L., Ferrer-Tasies, L., Ratera, I., Ventosa, N., Sanz, F., & Giannotti, M. I. (2018). Pulling lipid tubes from supported bilayers unveils the underlying substrate contribution to the membrane mechanics.* Nanoscale, *10, 14763–14770. https:/doi.org/10.1039/C8NR03249A, with permission from the Royal Society of Chemistry.*

(in the range of 65–80pN) than for s_o bilayers of the same headgroup (in the range of 90–110pN) (Fig. 6A). Additionally, while in the l_d state the F_{tube} slightly augments from PE to PC (66±15pN for DOPE, 80±25pN for DOPC and 75±13pN for DOPG), it clearly increases when changing the headgroup for s_o membranes from PE to PC and then PG (92±20pN for DPPE, 104±30pN for DPPC and 112±30pN for DPPG). Hence, F_{tube} is generally higher for s_o than for l_d and goes along with the lateral packing associated to different phospholipid headgroups, in a way that is comparable to the trend observed on the F_b approach, as can be seen in Fig. 6B.

Taken together, force spectroscopy experiments on SLBs are in qualitative agreement with the literature that correlates the physicochemical properties of membranes with the chemical composition of their phospholipid hydrophobic chains (Rawicz et al., 2000). Noteworthy, thermodynamic quantities such as the melting temperature or the order parameter can be translated into the realm of mechanical stability. The order parameter is usually calculated from NMR-based measurements and provides information about the overall membrane order and more particularly the conformations that the atoms within the lipid tails adopt (Piggot et al., 2017).

3.1.2 The thermal transition

3.1.2.1 Main transition

A useful application of QCM-D is the detection of thermotropic phase transitions within SLBs and SVLs, with the asset that it is label-free, it does not require long temperature equilibration time nor a large amount of sample. When temperature is varied in the absence of any transition, Δf and ΔD typically show a regular, linear temperature dependence, as a reflection of regular, temperature-dependent changes of the viscosity and density of the surrounding liquid medium. When approaching the main phase transition temperature, both frequency and dissipation responses will differ from a regular temperature dependence, displaying anomalies that reveal that structural changes are taking place (Losada-Pérez et al., 2014; Neupane, De Smet, Renner, & Losada-Pérez, 2018). Fig. 7 displays changes in the frequency and its derivative of an SVL of DMPC of vesicles of around 130 nm, with main transition temperature $T_m \sim 24\,°C$ in a rather narrow temperature range. The shape of frequency shift curves and its respective temperature derivative are reminiscent of the enthalpy jump and its heat capacity derivative typically obtained from calorimetric measurements (Heimburg, 1998; Losada-Pérez et al., 2014). The dissipation typically shows jumps and minima in the same temperature range. Therefore, extrema in the first-order derivative of frequency and dissipation curves stand as a

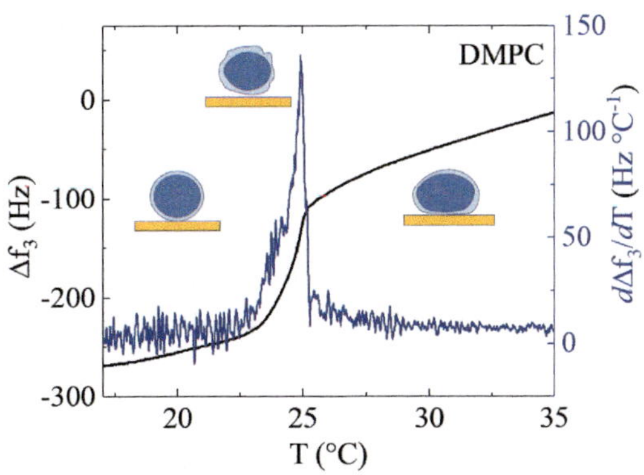

Fig. 7 Temperature dependence of the frequency shift $d\Delta f_3/dT$ (black solid line) and its first-order derivative (blue solid line) of a DMPC SVL upon heating at 0.2 °C/min.

token of the layer thermotropic phase transitions. Calorimetry is a state-of-the art technique for the detection of phase transitions based on changes in thermal properties when lipids undergo the main (first-order) transition (heat absorption or release), while QCM-D is based on changes in viscoelastic properties of the adsorbed vesicles as a result of changes in bilayer thickness, rigidity and vesicle shape. Although also detectable in SLBs, the size of the jumps in Δf and ΔD is several orders of magnitude smaller than the size of the jumps observed for SVLs (Hasan & Mechler, 2015; Wargenau & Tufenkji, 2014).

Reviakine and coworkers provided a comprehensive interpretation of the mechanisms by which QCM-D detects phase transitions in lipid membranes and the related relative contribution of thickness, stiffness and hydrodynamic channels to the Δf and ΔD fingerprints (Peschel et al., 2016). Upon heating, lipid bilayers in both SVLs and SLBs change from a thicker and stiffer s_o phase to a thinner and less stiff l_d one. Both stiffness and thickness changes in the lipid layer properties across the transition encompass changes in the Δf signal (decrease upon heating). In the case of the ΔD, the stiffness and the thickness contributions oppose each other since stiffer layers dissipate less. In the SLBs, this leads to an overtone dependence of the transition peak (Peschel et al., 2016). In thicker and softer SVLs, changes in the shape of the adsorbed vesicles might induce further hydrodynamic channels that contribute to ΔD. As a consequence, the size of the maximum in SVLs is several orders of magnitude larger than the one observed in SLBs.

The thermotropic phase behavior of pure SLBs has been widely related with the nanomechanical properties of the bilayer. Combining the previously exposed approaches, the AFM-FS sensitivity to the lipid phase of SLBs is evidenced, positioning it as an appropriate tool to determine the phase thermal transitions. Temperature controlled AFM has been established as a suitable tool to analyze both the topographical and mechanical evolution of biologically relevant issues that are temperature dependent at the nanometer scale.

Seminal work of Leonenko et al. demonstrated the existence of several phase transitions in DPPC and DOPC mica-supported bilayers (Leonenko et al., 2004). Data of both topographical imaging and force measurements provide invaluable insight into membrane behavior during phase transitions. For DPPC SLBs, F_b values decrease as temperature increases, and beyond the main transition, DPPC shows similar force plots as the fluid-phase DOPC bilayers. This dependence of the F_b on the phase state of the SLB below and above T_m has been further reported for different lipids, like DMPC ($T_m \sim 24\,°C$) that changes F_b from 13 to 14 nN at 20 °C below T_m to about 6 nN at or 40 °C DPPC ($T_m \sim 42\,°C$) that shows a switch of F_b from around 20–25 nN at 27 °C to around 3–5 nN at 55 °C (Garcia-Manyes, Guell, et al., 2005; Redondo-Morata, Giannotti, & Sanz, 2012b). When the DPPC SLB is maintained a temperature slightly above T_m, coexistence of the gel and fluid phase occurs and a bimodal F_b distribution is obtained (Fig. 8), as reported in

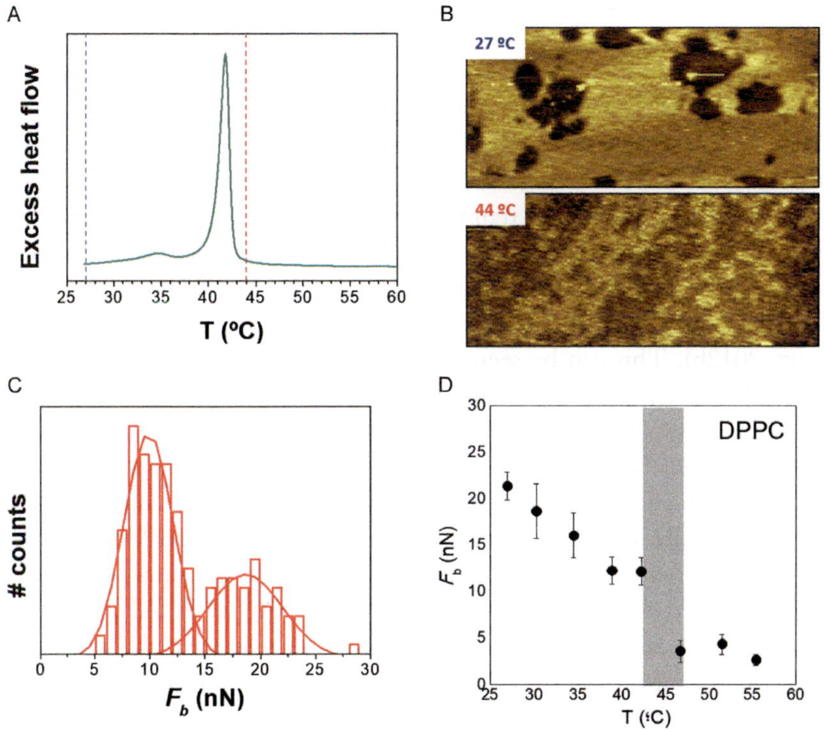

Fig. 8 Phase transition of DPPC bilayers: (A) DSC thermogram of DPPC vesicle suspension; (B) 6 μm × 2.5 μm AFM images of DPPC silicon-SLB at 27 °C and 44 °C; (C) F_b histogram at 44 °C and Gaussian fits to the bimodal distribution. All experiments performed in 150 mM NaCl, 20 mM $MgCl_2$, 20 mM HEPES (pH 7.4) buffer solution. (D) Mean F_b value of DPPC mica-SLB as a function of temperature, in 10 mM HEPES, 20 mM $MgCl_2$ and 150 mM NaCl (pH 7.4). The shadowed vertical line marks the observed transition temperature range. *Panel (C) adapted from Gumí-Audenis, B., Costa, L., Carlá, F., Comin, F., Sanz, F., & Giannotti, M. I. (2016). Structure and nanomechanics of model membranes by atomic force microscopy and spectroscopy: Insights into the role of cholesterol and sphingolipids.* Membranes, *6(4), 58. https://doi.org/10.3390/membranes6040058, Panel (D) adapted from Redondo-Morata, L, Giannotti, M. I., & Sanz, F. (2012b). Influence of cholesterol on the phase transition of lipid bilayers: A temperature-controlled force spectroscopy study.* Langmuir, *28(35), 12851–12860. https://doi.org/10.1021/la302620t.*

Gumí-Audenis et al. (2016). This is because the thermal transitions in SLBs generally happen at temperatures slightly higher and in a broader range than in vesicle suspensions as a consequence of the underlying substrate affecting lateral order and interleaflet coupling (Garcia-Manyes, Guell, et al., 2005; Leonenko et al., 2004; Redondo-Morata, Giannotti, & Sanz, 2012b; Seeger et al., 2010). In Garcia-Manyes et al. it was shown that the main phase transition occurs in a range of temperatures of ca. 10–13 °C (Garcia-Manyes, Oncins, & Sanz, 2005b).

There is an effect of temperature on the mechanical stability for the SLBs because the increase of temperature directly affects the molecules fluctuations provoking an increment in the area per molecule—lower packing density—especially noticeable in the gel phase, with the consequent decrease of the F_b values. At temperatures around the T_m of the lipid, and for a temperature range of a few °C, a pronounced drop in F_b takes please, as the SLB gradually transitions to the fluid phase, thereby softening. Several degrees over the T_m, the F_b values are on the order of the ones typically observed for l_d SLBs at room temperature (Alessandrini & Facci, 2012; Garcia-Manyes, Oncins, & Sanz, 2005b; Redondo-Morata, Giannotti, & Sanz, 2012b). This can be seen in the example shown in Fig. 8D, where the DPPC mica-SLB shows an F_b value around of 21 nN at 27 °C and slight decreases with temperature as the 42–45 °C range is reached. At higher temperatures, the F_b remains approximately constant at ca. 3.5 nN. The F_b-temperature trend displays a break around the T_m value of DPPC.

Additionally, Mouritsen and coworkers proposed that AFM topography at a varying temperature could decouple the phase transition of the two different leaflets of the mica-supported bilayer (Keller et al., 2005). The low-temperature transition was proposed to be related to the melting of the leaflet far from the substrate surface, and the second transition to the phase transition of the leaflet in contact with the mica. Such decoupling would respond to the stronger interaction between the lipid headgroups of the proximal monolayer and the mica support.

3.1.2.2 Pretransition

Some lipids present a pretransition before the main first-order transition which gives rise to an intermediate order in the lipid bilayers referred to as ripple phase. Ripple phases are curved structures with undulating surface topography. AFM imaging revealed the structure of the ripple phase in hydrated conditions at the nanometer scale (Kaasgaard, Leidy, Crowe, Mouritsen, & Jørgensen, 2003), and such particular structure has been recently explored nanomechanically (Majewska, Mrdenovic, Pieta, Nowakowski, & Pieta, 2020). It can be found when solid phase lipids experience an increase in temperature and go into the fluid phase, and alternatively it occurs when the fluid phase is cooled down. The observation of this phase transition process is characterized by a melting point temperature of the ripple phase to fluid phase, nucleation point and growth directionality for the formation of the ripple phase. HS-AFM coupled to a temperature-controlled system can report directly ripple to fluid phase transitions (reversibly) in real time and at high resolution (Takahashi, Miyagi, Redondo-Morata, & Scheuring, 2016). Transition processes from ripple phase to fluid phase and from ripple phase to metastable ripple phase to fluid phase, could be reversibly, phenomenologically and quantitatively studied. In Takahashi et al., results showed phase transition hysteresis in fast cooling and heating processes, while both melting and condensation occurred at 24.15 °C in quasi-steady state situation. A second metastable ripple phase with larger periodicity was formed at the ripple phase to fluid phase transition when the buffer contained Ca^{2+}. Thermodynamically, a lipid system containing small domains does not necessarily correspond to a phase coexistence regime, but they may constitute a one phase regime with micro- or nano-scale heterogeneities[a] (Heimburg, 2000). Altogether, these findings may have relevance in more complex cellular systems where lipid nanodomains form and dissociate reversibly.

3.2 Phase coexistence

Phase segregation and coexistence in lipid bilayers generally happens in the simplest models when mixing a fluid and a gel-like phospholipids, and in the more complex mixtures, like those containing phospholipids, sphingomyelins, ceramides, and cholesterol. At room temperature the bilayers composed of a mixture of a gel and a fluid phospholipid generally segregate into different domains, showing a continuous fluid phase enriched with l_d lipid and thicker dispersed domains associated to a phase enriched in the s_o lipid. When the T_m of the gel-phase lipid is overcome, the binary phospholipid system becomes homogeneous and fluid. Among the many that can be found in the literature, we here mention some examples of these binary systems.

The capability of QCM-D to detect lateral phase coexistence and map phase diagrams for binary lipid mixtures of saturated phospholipids will be

[a] We will further visit this topic in Section 5.

provided for binary mixtures of homologous phosphatidylcholines DMPC and DSPC, whose chemical structures differ in four methylene groups. This system has been thoroughly investigated experimentally and theoretically in free standing lipid systems (Ehrig et al., 2011; Fidorra et al., 2009; Mabrey & Sturtevant, 1976). DMPC:DSPC mixing behavior is driven by the large hydrophobic mismatch between the two components and departs from ideality as a consequence of the large difference in T_m of the pure constituents.

Fig. 9 (left panel) shows representative QCM-D signature of phase transitions we now revisit taking place upon heating pure SVLs of pure DMPC and three DMPC:DSPC mixtures (Losada-Pérez, Mertens, et al., 2015). The diameter of the precursor vesicles forming the SVLs was around 130 nm. For pure DMPC, the main transition takes place in a rather narrow temperature range around $T_m \sim 24\,°C$. The first-order derivatives of Δf curves capture the typical behavior of two-component lipid mixtures displaying s_o and l_d phase coexistence. As very low DSPC concentration (DSPC mole fraction = 0.1), the transition is shifted toward higher temperatures and takes place in a broader temperature range indicating a loss of intermolecular cooperation between DMPC molecules to undergo the main transition due to the presence of DSPC. For the mixtures close to the equimolar concentration two small peaks can be observed indicating the coexistence of two phases, s_o and l_d. The peaks are not simple additions of the peaks of the individual constituents, the size of each peak scales

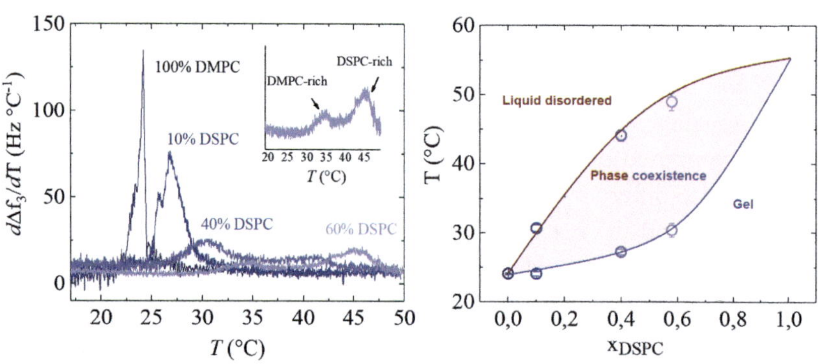

Fig. 9 Left panel: temperature profiles of the first-order temperature derivative of the frequency shifts for pure DMPC and DMPC:DPSC mixtures 10%, 40% and 60% mole fraction (in percentage) of DSPC. Right panel: phase diagram obtained from extracting onset and completion temperatures from the peaks (Losada-Pérez, Mertens, et al., 2015).

with the concentration of each lipid in each coexisting phase. For instance, the inset of Fig. 9 displays the peaks of the mixture DMPC:DSPC (40:60). The low-temperature peak corresponds to the DMPC-rich phase, while the high-temperature peak to the DSPC-rich phase. As expected, the size of the latter is higher, since the concentration of DSPC in the mixture is larger.

The analysis of the peaks using the tangent method (Ehrig et al., 2011) to determine the onset and completion temperatures enables to map the phase diagram of this mixture. The solid lines are added as a guide to the eye, delimiting the phase coexistence region, dark red (liquidus line), light blue (solidus line) and follow experimental heat capacity values by Maubrey et al. (Mabrey & Sturtevant, 1976) obtained for multilamellar vesicles of the same concentration.

As discussed in Sections 2.3 and 3.1, the nanomechanical properties determined by AFM-FS on SLBs are distinctive of the chemical structure and molecular organization of a bilayer at specific environmental conditions. This allows for a direct correlation of the elastic modulus, F_b, or F_{tube} values with the composition in phase-segregated membranes. In most cases, the AFM topographical characterization easily identifies the segregation of an SLB in different domains, due to their difference in thickness and compressibility. However, it is possible to reveal the presence of different domains in an SLB on the basis of their different mechanical properties even when the domains have a height difference which is too small to be detected by conventional imaging.

Bernchou et al. studied, using combined fluorescence microscopy and AFM, the nucleation and growth of domains in equimolar binary SLBs composed by DOPC and DPPC (Bernchou et al., 2009), a mixture that shows coexistence of l_d and s_o domains at room temperature (Leonenko et al., 2004). These authors could observe that during cooling the shape of the domains evolved from compact to a branched morphology, which they discussed as the domain growth being controlled by the diffusion of DPPC from the liquid phase toward the solid domain interface. In the late stages of the growth, the size and overall shape of the domains depend on the position of the nucleation points relative to the surrounding nucleation point positions.

A mechanical contribution concerning the elastic deformation of the membrane before the breakthrough point should be considered to fully comprehend the force-distance curve. A method to study the elastic properties consists of quantitative nanomechanical mapping (Picas et al., 2012).

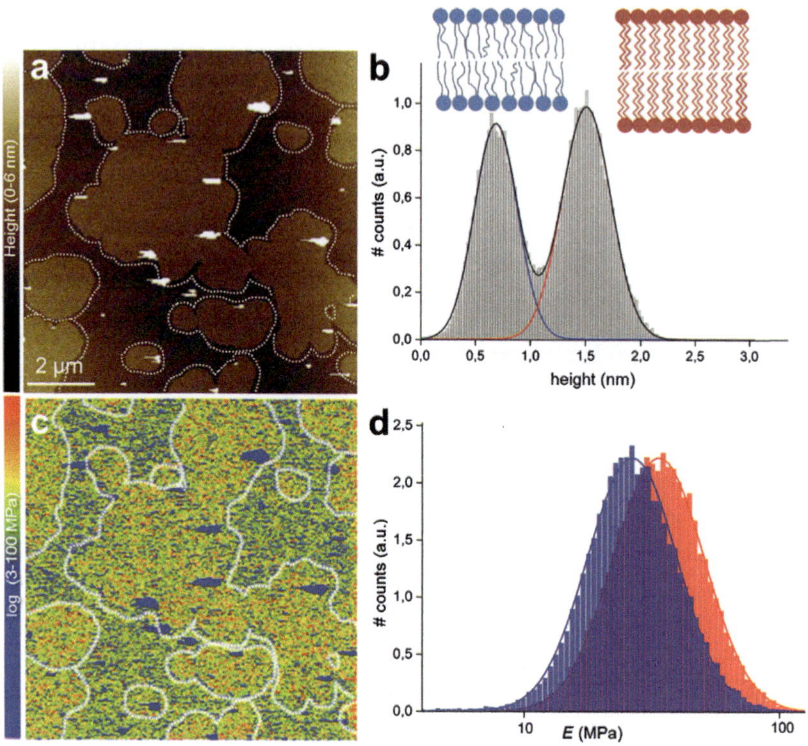

Fig. 10 PF-QNM AFM topography and elasticity mapping of DOPC:DPPC (1:1) SLBs. (A) Topography image (nm) and (B) height histogram analysis of the topography (A). The dashed lines in (A) represent the domain edges defined by edge detection. The solid lines in (B) are Gaussian fits to the height distribution, indicating an average height difference between the DOPC l_d (blue line) and DPPC s_o (red line) domains of 0.82 nm. (C) The corresponding stiffness map (log scale, MPa) and (D) Young's modulus values histograms (log scale, MPa) of the regions outlined in the topography (A) and stiffness (C) maps, corresponding to the DOPC (blue) and DPPC (red) domains. The solid lines are Gaussian fits to the distributions peaking at $26\pm^{58}_{12}$ MPa (DOPC) and $33\pm^{75}_{15}$ MPa (DPPC). *From reference Redondo-Morata, L, Sanford, R. L., Andersen, O. S., & Scheuring, S. (2016). Effect of statins on the nanomechanical properties of supported lipid bilayers.* Biophysical Journal, *111(2), 363–372. https://doi.org/10.1016/j.bpj.2016.06.016 with permission.*

In (Redondo-Morata et al., 2016), the equimolar binary lipid mixture DPPC:DOPC was studied, where l_d and s_o coexisting domains in SLBs are large and directly visually discernable by AFM imaging (Fig. 10A) with clearly distinguishable heights (~0.8 nm height difference, Fig. 10B). Elastic modulus nanomechanical mapping (Fig. 10C) shows good agreement with the E values in the literature, characterized mechanically by AFM

(Picas et al., 2012; Rawicz et al., 2000) and micropipette aspiration (Dieluweit et al., 2010). The mean E values are 26^{+58}_{-12}MPa for the l_d domain and 33^{+75}_{-15} MPa for the s_o domain, respectively (Fig. 10D).

Using the Young modulus for each domain of the DOPC:DPPC mixture together with the experimental bilayer thickness (h), the thin shell theory can be used to calculate the area compression-expansion modulus (K_A) and bending modulus (k_C) as (Rawicz et al., 2000):

$$K_A = \frac{Eh}{(1-\nu^2)} \text{ and } k_c = \frac{Eh^3}{24\,(1-\nu^{2\cdot})}$$

When lipid domains separate in lipid mixtures, or when lipids surround membrane embedded proteins there is, in general, a hydrophobic mismatch at the domain interfaces (Wallace et al., 2006) or protein/bilayer interfaces (Mouritsen & Bloom, 1984), with an ensuing compression or stretching of the hydrocarbon chains in the vicinity of the interface. This hydrocarbon tail stretching or compression may influence the distribution of domain sizes and the kinetics of domain separation, as well as membrane protein function. The compression/stretching of the acyl chains of the s_o and l_d domains can be determined by minimizing the elastic energy of the system, $G_s(x)$, expressed assuming volume conservation (Wallace et al., 2006) as

$$G_s = \frac{K_o A_o}{2h_o^2}(x-h_o)^2 + \frac{K_D A_D}{2h_D^2}(x-h_D)^2 \tag{2}$$

where K_o and K_d are the stretch moduli for s_o and l_d, respectively. A_o and A_d are the lipid area per molecule for s_o and l_d, respectively. A values are well-documented in literature (notably through X-ray diffraction and Langmuir methods), being $0.72\,\text{nm}^2$ for l_d DOPC and $0.49\,\text{nm}^2$ for s_o DPPC at room temperature (Nagle & Tristram-Nagle, 2000). h_o and h_d are the measured bilayer thickness of s_o and l_d, respectively. The estimated G_s values for the DOPC l_d and the DPPC s_o domains are 0.27 k_BT. Hence, the experimental determination of E allows to estimate the stretch modulus and the bending stiffness of the lipid membrane, as well as the energetic cost of the lipid domains to accommodate to different thickness.

For the SLB system composed of a POPE:POPG mixture (Picas et al., 2009), AFM images revealed the existence of two separated phases, the higher showing a region with protruding subdomains. AFM-FS was then used to clarify the nature of each phase. The values of F_b, adhesion force and height extracted from the force curves were assigned to the corresponding s_o (the mechanically stable domains) and l_d phase (the mechanically labile phase). In Crespo-Villanueva et al., while the SLB composed of equimolar DSPC:DOPC was shown to phase segregate, clearly demonstrated in the AFM topographical images and the correlative F_b maps and bimodal distributions around ~40 and ~5 nN, the equimolar mixture of the two l_d lipids DOPC:DOPS led to a homogeneous topography and narrow F_b distribution around 3.5 nN (Crespo-Villanueva et al., 2018). In Domenech et al., they studied ternary mixtures of POPC:POPE:CL, a minimum system that mimics the mitochondrial inner membrane (Domenech et al., 2007). They could observe laterally segregated domains whose height decreased with increasing temperatures.

Pulling lipid tubes with the AFM probe also allows to characterize phase-segregated bilayers and laterally resolve the different domains in multicomponent SLBs (Gumí-Audenis, Costa, Ferrer-Tasies, et al., 2018). This is exemplified in Fig. 11, a bilayer composed of the l_d lipid DOPE and the s_o lipid DPPG in ratio DOPE:DPPG 25:75, that at room temperature forms SLB patches with coexisting domains: gel-phase rich in DPPG, dispersed in fluid-phase rich in DOPE. When indented by the AFM tip at a constant velocity of $1\,\mu m\,s^{-1}$, a bimodal F_b distribution is obtained, with differentiated mean F_b values associated with the fluid (~5 nN) and gel (~12 nN) domains. Analyzing the retract force-separation curves that showed the tube pulling event allowed to represent the F_{tube} distribution and map, where the lower force values (red pixels) are distributed in the peripheral part of the SLB patch and higher forces (yellow pixels) in the center, and the F_{tube} distribution is resolved when sorting according to the corresponding F_b, giving a mean value of 75 ± 20 pN for the DOPE-rich phase, and 112 ± 25 pN for the DPPG-rich phase. The fact that F_b and F_{tube} for the domains slightly differ from the values for pure DOPE and DPPG SLBs allows to infer partial miscibility among them.

Higher complexity models, like the ternary lipid mixtures containing an unsaturated phosphatidylcholine (PC), a saturated PC or sphingomyelin (SM), and cholesterol (Chol) have been probably the most studied due to the physiological abundance of these lipids. Depending on the ratio of the components, these mixtures have shown to exhibit two coexisting fluid phases: an l_o phase that is enriched in Chol and SM or saturated PC, and an l_d phase that is predominantly composed of unsaturated PC (Johnston, 2007), and the possibility of a coexisting s_o phase. The group of Schwille has extensively analyzed the domain organization in DOPC/SM/Chol model membranes by means of AFM in combination with fluorescence

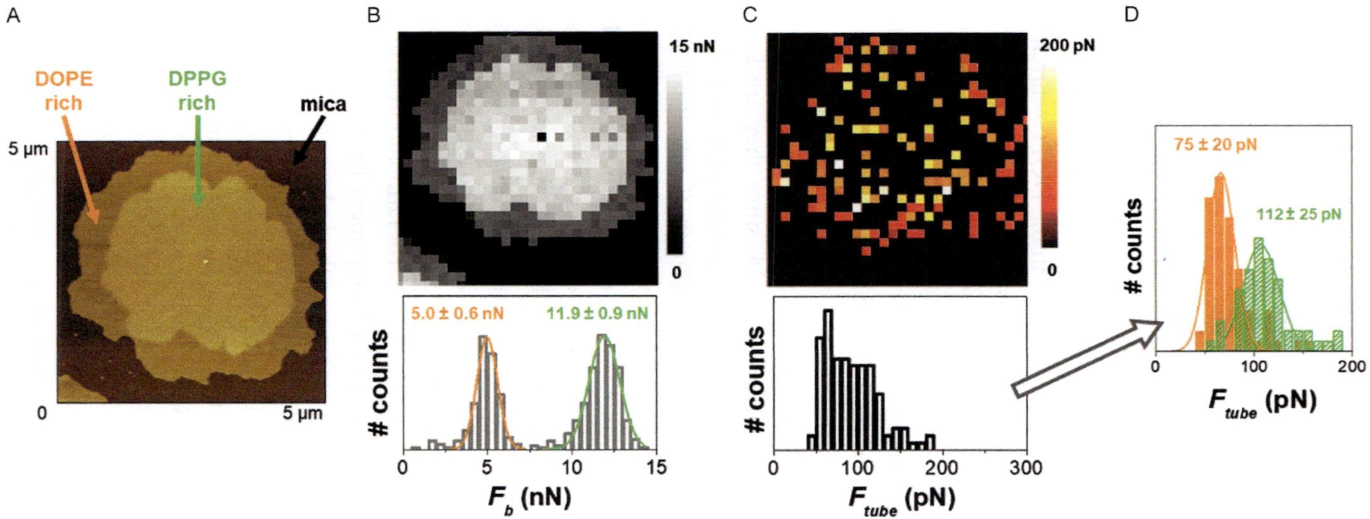

Fig. 11 DOPE:DPPG (25:75). (A) AC-mode AFM topographical image; (B) F_b map and distribution; (C) F_{tube} map and distribution; (D) F_{tube} distributions resolved by filtering according to the F_b distribution. *Reproduced from Gumí-Audenis, Berta, Costa, L., Ferrer-Tasies, L., Ratera, I., Ventosa, N., Sanz, F., & Giannotti, M. I. (2018). Pulling lipid tubes from supported bilayers unveils the underlying substrate contribution to the membrane mechanics.* Nanoscale, *10 14763–14770. https://doi.org/10.1039/C8NR03249A, with permission from the Royal Society of Chemistry.*

correlation spectroscopy (FSC) (Chiantia, Kahya, et al., 2006; Chiantia, Ries, et al., 2006; Chiantia et al., 2007, 2008) and by performing AFM-FS experiments on the different phases they have shown that the average force needed to pierce the l_o phase of the DOPC/SM/Chol bilayer is about 4 nN higher than the one needed to pierce the l_d phase. In a parallel study, Zou and coworkers explored the influence of different cholesterol contents on the morphology and nanomechanical stability of phase-segregated DOPC/SM/Chol lipid bilayers by means of AFM imaging and force mapping (Sullan et al., 2010). They observed consistently higher F_b values in the SM/Chol-enriched l_o domains than in the DOPC-enriched l_d phase, and that the nanomechanical stability of both coexisting phases counter-intuitively decreased with increasing Chol content. By following the dependence of the F_b value with the loading rate in both phases, they calculated the activation energies (ΔE_0) of bilayer rupture at zero applied force. ΔE_0 values obtained were in the range 75–125 kJ/mol for all tested Chol contents and did not show a dependence on the different lipid phases.

Besides sphingolipids, ceramides (Cer) have also gained attention in physicochemical studies by means of AFM, since they are known to be predominant in physiologically relevant nanodomains—which have been controversially called for many years lipid rafts (Goñi, 2019). Longo and coworkers observed that DLPC/GalCer/Chol ternary mixtures only display l_d/s_o coexistence and no l_o phase even at high Chol content (Blanchette et al., 2006). By measuring the area/perimeter ratios of the segregated domains, the authors concluded that Chol was able to decrease the line tension between the two phases. In another study (Lin, Blanchette, & Longo, 2007), they could also observe a phase coexistence attributed to l_d/l_o phases with POPC and DOPC at 10 mol% Chol, an indication that fatty-acyl chain unsaturation, cholesterol ratio and lipid hydrophobic mismatch in the organization of multicomponent lipid systems.

Ceramide incorporation on coexisting fluid phase and ordered domains in phase-separated binary and ternary lipid mixtures was shown to affect the lipid spatial organization in domains, with the appearance of a Cer-enriched gel-like phase and the displacement of cholesterol from rafts (Chiantia, Kahya, et al., 2006; Chiantia et al., 2007; Ira et al., 2009). In order to directly probe and quantify the (nano) mechanical stability and rigidity of the Cer-enriched platforms, Zou et al. performed AFM topographical images and force mapping that allowed the determination of the F_b value and elastic modulus of the different phases (Sullan et al., 2009a, 2009b; Zou & Johnston, 2010). They confirmed the expulsion of cholesterol from sphingolipid/Chol-enriched domains as a result of Cer incorporation, and observed an increased mechanical stability attributed to the influence of Cer in the lipid organization and packing behavior, also reported later on GalCer rich domains on model membranes with PC and Chol (Gumí-Audenis et al., 2015).

The group of Goñi has systematically explored the effect of the acyl compositions of sphingomyelins and the corresponding ceramide counterparts in ternary mixtures with PC and Chol (García-Arribas et al., 2016; González-Ramírez et al., 2019; Jiménez-Rojo et al., 2014). Combining confocal microscopy, DSC, AFM imaging and FS, they compared the effects of C24:1 Cer (nervonoyl ceramide, nCer) with those of C16:0 Cer (palmitoyl ceramide, pCer) in bilayers composed basically of DOPC, SM (either C24:1, nSM or C16:0, pSM) and Chol (García-Arribas et al., 2017). AFM-FS showed that nCer has a lower stiffening effect than pCer, while the presence of nSM reduces the stiffness. When the proportion of phospholipid increases beyond 60 mol%, a lateral phase separation occurs at the micrometer scale (González-Ramírez et al., 2019). The authors interpreted this data as a pCer:Chol interaction, that would predominate at the lower phospholipid concentrations. The putative pCer:Chol complexes (or nanodomains) would mix well with the phospholipid. At higher SM concentrations pSM: pCer and pSM:Chol interactions would become more important, giving rise to the coexisting s_o and l_o phases, respectively. Heterogeneity, or lateral phase separation, occurs more easily with pSM than with DPPC, indicating a higher affinity of SM over DPPC for Chol or Cer. Altogether, the findings of Goñi and coworkers show the sharp increase in complexity when membranes exhibit different sphingolipids of varying N-acyl chains, which should be a common issue in an actual cell membrane environment.

3.3 The role of cholesterol

Cholesterol is a fundamental constituent of the eukaryotic cell plasma membranes. Eukaryotic cell membranes are not entirely in the conventional l_d phase. Instead, they are, at least partially, in the Chol-rich liquid-ordered (l_o) phase or a phase with similar properties. l_o phase is characterized by a high degree of acyl chain order and is favored by high-T_m lipids with saturated acyl chains such as sphingolipids, when they are mixed with Chol, as exposed in the previous section. Chol plays the essential function of regulating the physical properties of the cell membrane as it controls the lipid organization and phase behavior, regulates the membrane fluidity and its mechanical strength, influencing the passive permeability of water and other small molecules (Pan et al., 2008; Pinkwart et al., 2019). From the molecular perspective, Chol is considered to produce a condensing effect and induce ordering to fluid phase lipids forming the membrane and to have the opposite effect on phospholipids present in the gel phase (Hung et al., 2007; Rog et al., 2009). However, several experiments pointed out the complex effect of Chol on lipid bilayers, which depends on the molecular structure of the neighboring lipids: the degree of chain unsaturation, the length of the hydrophobic tail and the chemistry of the headgroup (Garcia-Manyes et al., 2010; Mills et al., 2009; Pan et al., 2008). In view of acknowledging the particular role of cholesterol, let us briefly outline results on the phase behavior of SVLs and SLBs consisting of binary mixtures of saturated phospholipids and cholesterol using mechanical descriptors such as shear viscosity of SVLs and nanomechanics of SLBs.

We first describe the phase behavior of SVLs containing binary mixtures of DMPC:Chol using QCM-D. Upon the addition of Chol to phospholipid bilayers, the sharp main phase transition in pure DMPC gradually diminishes and eventually disappears over a given Chol concentration. In the case of DMPC, SVLs showed no transition from 30% cholesterol. Fig. 12 shows normalized shear viscosity $\eta_{norm} = [\eta(t)/\eta(t=0)$, with t the time] for pure DMPC and DMPC:Chol (20 mol% Chol) calculated using a Voigt-based viscoelastic model (Voinova et al., 1999). At intermediate Chol concentrations, asymmetric peaks can be observed. These peaks encompass two main contributions, a sharp, narrow peak and a broader one that overlaps with the

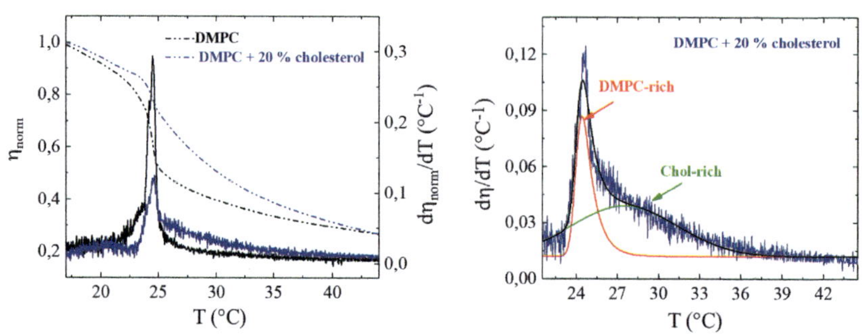

Fig. 12 Left panel temperature dependence of the normalized effective shear viscosity (dash dotted lines) and its first-order derivative (solid line) of DMPC and DMPC +20% cholesterol SVLs on a supported vesicle layer on a gold-coated QCM-D quartz sensor upon heating. Right panel closer view of the normalized shear viscosity first-order derivative for DMPC +20% cholesterol SVLs. Solid lines denote multiple peak Gaussian fitting.

former (see right panel in Fig. 12 for a more detailed view, including Gaussian fitting). Two domains can be identified, namely, the sharp contribution can be ascribed to the pure DMPC-rich domains undergoing the main transition and the latter to the Chol-rich ones. The broader peak appears as a result of the reduced mobility of the hydrophobic chains of DMPC in Chol-rich domains. The presence and the size of these domains depend strongly on the concentration of cholesterol and the temperature. On increasing the amount of cholesterol, the temperature range of the overall transition increases as a result of the extension of the second broad contribution, while the size of the maximum corresponding to the melting of the single lipid decreases. No phase transition is detected at Chol concentration over 30 mol% indicating the formation of a homogeneous liquid-ordered l_o phase at a macroscopic level. Such a large cholesterol concentration inhibits the formation of extended domains. A similar pattern of behavior was observed for DPPC:Chol SVLs, however, for this later system, the completion temperature of the overall transition at intermediate cholesterol concentrations was not attained, due to the (upper) limited temperature range of commercial QCM-D systems (Losada-Pérez, Khorshid, et al., 2015). These results are in agreement with recent calorimetric studies of free-standing membranes of the same systems (Mannock et al., 2010).

The role of Chol in the structure and nanomechanics of SLBs has been extensively studied by AFM (Adhyapak et al., 2018; Al-Rekabi & Contera, 2018; Garcia-Manyes et al., 2010; Redondo-Morata, Giannotti, & Sanz, 2012b; Sullan et al., 2010), as well as its interplay with sphingolipids or other lipids (Ahyayauch et al., 2002; García-Arribas et al., 2016; Guyomarc'h et al., 2014) and further reviewed in Gumí-Audenis et al. (2016).

By means of temperature-controlled AFM imaging and AFM-FS to assess the influence of Chol on the membrane ordering and stability (Redondo-Morata, Giannotti, & Sanz, 2012b), this work analyzed a DPPC:Chol in representative range of compositions up to 50 mol% Chol, studying the phase evolution upon heating (from room temperature to temperatures high above the T_m of DPPC) and the corresponding nanomechanical stability. This provided a correlation between the mechanical behavior and composition with the lateral order of each phase present in the bilayers. It proved that low Chol contents lead to a phase-segregated system, while high contents of Chol can give a homogeneous topography, although a bimodal F_b distribution was observed, possibly due to the presence of small heterogeneities that could not be resolved in the images. In all cases, it was demonstrated that Chol enhances the mechanical stability of the binary membrane, and an extraordinarily stable system was observed for equimolar fractions (50 mol% Chol). In addition, even when no thermal transition was detected by the traditional bulk analysis techniques for vesicles with high Chol content (40 and 50 mol%), it was shown that temperature-controlled AFM-FS is capable of identifying a thermal transition for the SLBs. As detailed in Fig. 13, for high-Chol contents—40 and 50 mol%—the DPPC:Chol system shows a double F_b distribution at room temperature, probably due to the presence of (space or time) non-resolved heterogeneities for these compositions. Although no thermal transition was observed in DSC thermograms of free-standing systems, when AFM-FS experiments were performed at different temperatures for these systems, a gradual decrease of the mean F_b value was observed until temperature reached ca. 40 °C. The continuous decrease in the F_b value with increasing temperature may be ascribed to an increase in fluidity of the system, and it is evident that, although not detected in DSC of DPPC:Chol vesicles, a transition occurs for DPPC:Chol SLBs with 40 and 50 mol% Chol around 42–45 °C. Below this temperature, the system in the l_o phase presents extraordinary lateral order and may have properties that resemble a gel-like bilayer, and as temperature increases, it gradually becomes a system with higher lateral mobility (but still with higher order than the DPPC l_d phase). This behavior indicates that for DPPC:Chol SLBs, a thermal transition occurs even for high-Chol compositions, in a temperature range between *ca.* 42–47 °C independently of the membrane composition, and that it may probably be of second-order for the high-Chol DPPC:Chol bilayers (40–50 mol%), which generally appear as a very broad transition or it is even not detected in DSC experiments of free-standing vesicles or QCM-D of SVLs.

Strikingly, information about the lateral in-plane structure of such systems can be instead obtained by grazing incidence XR diffraction (GIXD). In (Gumí-Audenis, Costa, Rendondo-Morata et al. (2018)), a silicon-SLB system was designed in order to deal with the requirement of the wetting preservation to guarantee the membrane stability while minimizing the background from the liquid environment. Using this methodology, it was confirmed that DPPC:Chol with low contents of Chol up to 30 mol% lead to separation into two coexisting phases at room temperature and concentrations higher than 30 mol% Chol appear to be a unique l_o phase. This condition was consistent with the earlier work (Redondo-Morata, Giannotti, & Sanz, 2012b) demonstrating that such increase of nanomechanical resistance is associated to a strong lateral interaction mediated by Chol molecules placed between the DPPC ones, due to a highly stable structure with most probably an equimolar DPPC:Chol ratio.

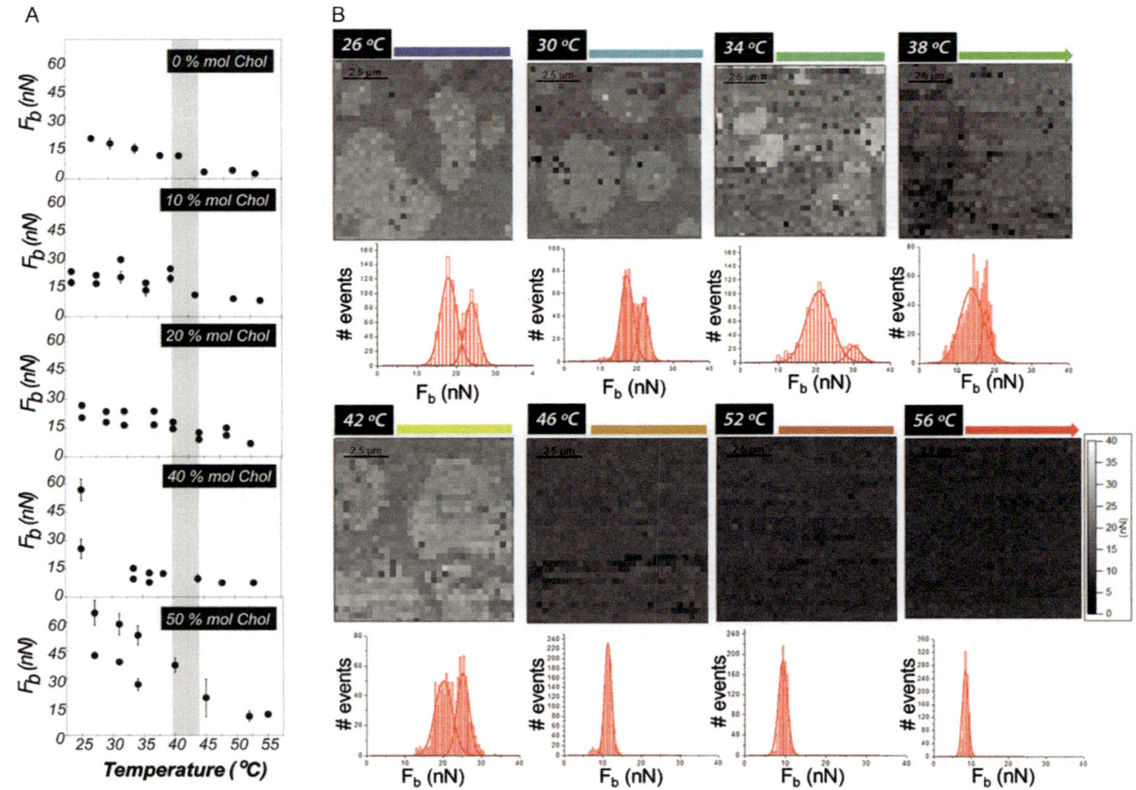

Fig. 13 (A) Mean F_b value of DPPC:Chol SLBs in 10 mM HEPES, 20 mM $MgCl_2$ and 150 mM NaCl, pH 7.4, with various Chol contents, as a function of temperature. The shadowed vertical line marks the temperature range where the main transition in pure DPPC occurs. For DPPC:Chol SLBs with 40 and 50 mol % Chol, although not detected in DSC of DPPC:Chol vesicles, a transition occurs around 42–45 °C. (B) F_b maps and distributions for DPPC:Chol SPB with 10 mol% Chol, in 10 mM HEPES, 20 mM $MgCl_2$ and 150 mM NaCl, pH 7.4, with increasing temperature. *Reprinted from Redondo-Morata, L, Giannotti, M. I., & Sanz, F. (2012b). Influence of cholesterol on the phase transition of lipid bilayers: A temperature-controlled force spectroscopy study.* Langmuir, *28(35), 12851–12860. https://doi.org/10.1021/la302620t.*

To date, efforts continue to understand the complex role of cholesterol in biomembranes. Still, these discoveries cannot always be directly translated to the behavior of living cell membranes. By means of advanced fluorescence imaging and spectroscopy, Sezgin and colleagues (Pinkwart et al., 2019) have shown that cholesterol diffuses faster than phospholipids in live membranes, but not in model membranes. Interestingly, a detailed statistical diffusion analysis suggested two-component diffusion for cholesterol in the plasma membrane of live cells.

Recent work based on a systematic AFM-FS study on DOPC:Chol SLBs at varying ratios suggested, however, that cholesterol may not just be responsible of increasing the mechanical stability but also introduce irregularities across the leaflets (Adhyapak et al., 2018). At Chol ratios above 20%, they could observe two sudden jumps in the force-distance curve, the second jump always displaying shorten separation. They interpreted that cholesterol induced asymmetry across the (in the inner and outer leaflets) bilayer, related to the phenomena of interleaflet coupling and depending on Chol concentration.

Recently, Contera and coworkers took a different approach to study the effect of Chol in the mechanical properties of lipid bilayers (Al-Rekabi & Contera, 2018). Applying different frequencies simultaneously in an AFM topography mode, namely amplitude modulation–frequency modulation (AM–FM), they investigated the effect of Chol on DPPC bilayers in concentrations from 0% to 60%. The lipid bilayers probed with this methodology displayed both elastic and viscous behavior. Cholesterol showed to modulates both, increasing elastic and viscous components as a function of Chol concentrations. The authors interpreted that their results may have a physiological relevance, where such viscoelastic properties may be used by cells to modulate the propagation (elastic) or attenuation (viscous) of mechanical signals across the cell membrane.

4. Connection between nanoscale measurements and thermodynamic descriptors of membranes

When studying phase coexistence in lipid membranes a distinction between thermodynamically stable phases and transient local domains in lipid bilayer at equilibrium should be made. Let us briefly discuss this in line with the above-exposed results on phospholipid-cholesterol binary mixtures. PC-Chol binary mixtures have been studied for many years using a plethora of experimental techniques. Yet, despite their apparent simplicity being two-component systems, their phase behavior remains controversial and, in many cases, departs from the canonical phase diagram proposed by Vist and Davis (1990). In fact, results depend strongly on the spatio-temporal resolution of the different experimental techniques. For instance, unlike predicted by most phase diagrams, l_d–l_o phase coexistence was not observed for micron-sized GUVs of DPPC-cholesterol mixtures using fluorescence microscopy (Veatch & Keller, 2003), while some other techniques captured this phenomenon for free standing MLVs vesicles (DSC (Mannock et al., 2010), volumetric (Miyoshi et al., 2014), NMR (Davis et al., 2009), SLBs (AFM) (Redondo-Morata, Giannotti, & Sanz, 2012b), and SVLs (QCM-D) (Losada-Pérez, Khorshid, et al., 2015)). These findings suggest that, instead of macroscopic phase separation, these systems are presumably filled with mesoscopic and nanoscopic domains, whose size and lifetime depends on temperature and mixture composition (Winkler et al., 2017).

Classically, when dealing with phase separation the Gibbs' phase rule is customarily invoked, assuming macroscopic phase coexistence and neglecting the role of phase boundaries, which comparably, hardly contribute to the free energy. If domains are small, the interfaces between domains constitute an important part of the system and its contribution to the free energy is not negligible. In this respect, using the term phase in the classical thermodynamic sense might not be appropriate (Almeida et al., 2005; Heimburg, 1998). Following Hill (2013) a system can no longer be called a phase when it ceases to behave like a macroscopic system and starts behaving like a small system.

In this respect, both AFM and QCM-D have demonstrated to be valuable tools to detect domain (nano/mesoscopic sized domains) coexistence within supported membranes of binary lipid mixtures.

5. Conclusions and future perspectives

As previously discussed, AFM-FS can interestingly resolve coexistence of domains at concentrations where height differences at domain boundaries are not detectable. Breakthrough force thus stands as a useful mechanical descriptor to resolve domain coexistence. Even though domains cannot be considered as phases in the purely thermodynamic standpoint, this does not mean that these systems cannot be treated in a rigorous thermodynamic way. As a matter of fact, the breakthrough force is able to detect a thermal transition at high cholesterol concentrations in binary PC:Chol SLBs, whose origin deserves to be revisited and further explored in the framework of either fluctuations arising in the vicinity of a critical point, the possibility of glassy-cholesterol-rich nanodomains or eventual cholesterol crystallization. In this sense, the development of HS-AFM will likely provide new insights due to its unprecedented spatio-temporal resolution, thanks to a direct correlation between structure and diffusion (Munguira et al., 2016). Similarly, HS force spectroscopy has the capacity to cover a wider force spectrum, which can help to validate theoretical frameworks of the field, as well as provide a direct correlation with molecular dynamics simulations (Rico, Gonzalez, Casuso, Puig-Vidal, & Scheuring, 2013).

There is indeed plenty of room in fundamental lipid biophysics and phase behavior in particular to use both purely mechanical and viscoelastic properties as descriptors to provide a more complete picture of the origins of domain formation and coexistence. The ultimate lipid phase coexistence to be fully understood are transient nanodomains, often (confusively) referred to as lipid rafts. There is to date no agreement about the size and lifetime of rafts, and the term is used to referred to a collection of different structures (Goñi, 2019). Based on the current knowledge, microdomains in equilibrium are no longer considered suitable models for the biological structure that rafts represent. Fortunately, advanced super-resolution microscopies and fast techniques are becoming available and will shed some light in one of the most discussed topics in biomembrane studies.

In light of this, multiscale spatio-temporal measurements of mechanical properties of supported membranes can help to experimentally address different scenarios where membrane micro- and nano-domain formation finds theoretical support. Starting from equilibrium ternary lipid mixtures in the vicinity of a critical point, it might be worth exploring how the presence of critical fluctuations reflects in the membrane mechanical properties, and if these properties can be used as physical descriptors to test critical-point universality laws. Secondly, tuning membrane (local and global) environment by actively controlling curvature, leaflet asymmetry, solid support topography and addition of external components would help complete the manifold where domain formation occurs. Finally, the growing spatio-temporal capabilities of AFM are promising tools to explore membrane organization in the realm of non-equilibrium phenomena.

Acknowledgments

We acknowledge financial support from the Generalitat de Catalunya (2017-SGR-1442) and the Instituto de Salud Carlos III, through "Acciones CIBER" and CIBER-BBN FlexQS-skin and FelxCAB projects to M.I.G. This work was also supported by a grant overseen by the French National Research Agency (ANR) as part of the "Investments d'Avenir" Programme (I-SITE ULNE/ANR-16-IDEX-0004 ULNE) to L.R.-M. Financial support from the project "SADI" by the programme Action Recherche Concertée, Université Libre de Bruxelles (ULB) to P.L-P. is also acknowledged.

Glossary of lipid acronyms

C16:0 palmitoyl Cer (aka palmitoyl Cer, pCer)	N-palmitoyl-D-erythro-sphingosine
C16:0 palmitoyl SM (aka palmitoyl SM)	N-palmitoyl-D-erythro-sphingosylphosphorylcholine
C24:1 nervonoyl Cer (aka nervonoyl Cer, nCer)	N-nervonoyl-D-erythro-sphingosine
C24:1 nervonoyl SM (aka nervonoyl SM)	N-nervonoyl-D-erythro-sphingosylphosphorylcholine-d9
Cer	ceramide
Chol	cholesterol
CL	cardiolipin
DMPC	1,2-dimyristoyl-sn-glycero-3-phosphocholine
DOPC	1,2-dioleoyl-sn-glycero-3-phosphocholine
DOPE	1,2-dioleoyl-sn-glycero-3-phosphoethanolamine
DOPG	1,2-dioleoyl-sn-glycero-3-[phospho-rac-(3-lysyl(1-glycerol))]
DOTAP	1,2-dioleoyl-3-trimethylammonium-propane
DPPA	1,2-dipalmitoyl-sn-glycero-3-phosphate
DPPC	1,2-dipalmitoyl-sn-glycero-3-phosphocholine
DPPE	1,2-dipalmitoyl-sn-glycero-3-phosphoethanolamine
DPPG	1,2-dipalmitoyl-sn-glycero-3-phospho-(1′-rac-glycerol)
DSPC	1,2-distearoyl-sn-glycero-3-phosphocholine
GalCer	galactosylceramides
PC	phosphocholine
PE	phosphoethanolamine
POPC	1-palmitoyl-2-oleoyl-glycero-3-phosphocholine
POPE	1-palmitoyl-2-oleoyl-sn-glycero-3-phosphoethanolamine
POPG	1-palmitoyl-2-oleoyl-sn-glycero-3-phospho-(1′-rac-glycerol)
SM	sphingomyelin

References

Adhyapak, P. R., Panchal, S. V., & Murthy, A. V. R. (2018). Cholesterol induced asymmetry in DOPC bilayers probed by AFM force spectroscopy. *Biochimica et Biophysica Acta (BBA)—Biomembranes, 1860*(5), 953–959. https://doi.org/10.1016/j.bbamem.2018.01.021.

Ahyayauch, H., Requero, M. A., Alonso, A., Bennouna, M., & Goni, F. M. (2002). Surfactant effects of chlorpromazine and imipramine on lipid bilayers containing sphingomyelin and cholesterol. *Journal of Colloid and Interface Science, 256*(2), 284–289. https://doi.org/10.1006/jcis.2002.8690.

Alessandrini, A., & Facci, P. (2012). Nanoscale mechanical properties of lipid bilayers and their relevance in biomembrane organization and function. *Micron*, *43*(12), 1212–1223. https://doi.org/10.1016/j.micron.2012.03.013.

Almeida, P. F. F., Pokorny, A., & Hinderliter, A. (2005). Thermodynamics of membrane domains. *Biochimica et Biophysica Acta (BBA)—Biomembranes*, *1720*(1), 1–13. https://doi.org/10.1016/j.bbamem.2005.12.004.

Al-Rekabi, Z., & Contera, S. (2018). Multifrequency AFM reveals lipid membrane mechanical properties and the effect of cholesterol in modulating viscoelasticity. *Proceedings of the National Academy of Sciences of the United States of America*, *115*(11), 2658–2663. https://doi.org/10.1073/pnas.1719065115.

Ando, T. (2017). High-speed atomic force microscopy and its future prospects. *Biophysical Reviews*, *10*(2), 285–292. https://doi.org/10.1007/s12551-017-0356-5.

Ando, T., Uchihashi, T., Kodera, N., Yamamoto, D., Miyagi, A., Taniguchi, M., et al. (2008). High-speed AFM and nano-visualization of biomolecular processes. *Pflügers Archiv—European Journal of Physiology*, *456*(1), 211–225. https://doi.org/10.1007/s00424-007-0406-0.

Armond, J. W., Macpherson, J. V., & Turner, M. S. (2011). Pulling nanotubes from supported bilayers. *Langmuir*, *27*(13), 8269–8274.

Bangham, A. D., & Horne, R. W. (1964). Negative staining of phospholipids and their structural modification by surface-active agents as observed in the electron microscope. *Journal of Molecular Biology*, *8*(5), 660–IN10. https://doi.org/10.1016/S0022-2836(64)80115-7.

Bangham, A. D., Pethica, B. A., & Seaman, G. V. (1958). The charged groups at the interface of some blood cells. *The Biochemical Journal*, *69*(1), 12–19. https://doi.org/10.1042/bj0690012.

Barrett, R. C. (1991). High-speed, large-scale imaging with the atomic force microscope. *Journal of Vacuum Science & Technology B: Microelectronics and Nanometer Structures*, *9*(2), 302. https://doi.org/10.1116/1.585610.

Berg, J. M., Tymoczko, J. L., Stryer, L., Berg, J. M., Tymoczko, J. L., & Stryer, L. (2002). *Biochemistry* (5th ed.). W H Freeman.

Bernchou, U., Ipsen, J. H., & Simonsen, A. C. (2009). Growth of solid domains in model membranes: Quantitative image analysis reveals a strong correlation between domain shape and spatial position. *The Journal of Physical Chemistry. B*, *113*(20), 7170–7177. https://doi.org/10.1021/jp809989t.

Binnig, G., Quate, C. F., & Gerber, C. (1986). Atomic force microscope. *Physical Review Letters*, *56*(9), 930–933. https://doi.org/10.1103/PhysRevLett.56.930.

Blanchette, C. D., Lin, W.-C., Ratto, T. V., & Longo, M. L. (2006). Galactosylceramide domain microstructure: Impact of cholesterol and nucleation/growth conditions. *Biophysical Journal*, *90*(12), 4466–4478. https://doi.org/10.1529/biophysj.105.072744.

Brochard-Wyart, F., Borghi, N., Cuvelier, D., & Nassoy, P. (2006). Hydrodynamic narrowing of tubes extruded from cells. *Proceedings of the National Academy of Sciences of the United States of America*, *103*(20), 7660.

Butt, H. J., & Franz, V. (2002). Rupture of molecular thin films observed in atomic force microscopy. I. Theory. *Physical Review E*, *66*(3), 031601. https://doi.org/10.1103/PhysRevE.66.031601.

Canham, P. B. (1970). The minimum energy of bending as a possible explanation of the biconcave shape of the human red blood cell. *Journal of Theoretical Biology*, *26*(1), 61–81.

Chiantia, S., Kahya, N., Ries, J., & Schwille, P. (2006). Effects of ceramide on liquid-ordered domains investigated by simultaneous AFM and FCS. *Biophysical Journal*, *90*(12), 4500–4508. https://doi.org/10.1529/biophysj.106.081026.

Chiantia, S., Kahya, N., & Schwille, P. (2007). Raft domain reorganization driven by short- and long-chain ceramide: A combined AFM and FCS study. *Langmuir*, *23*(14), 7659–7665. https://doi.org/10.1021/la7010919.

Chiantia, S., Ries, J., Chwastek, G., Carrer, D., Li, Z., Bittman, R., et al. (2008). Role of ceramide in membrane protein organization investigated by combined AFM and FCS. *Biochimica et Biophysica Acta—Biomembranes*, *1778*(5), 1356–1364. https://doi.org/10.1016/j.bbamem.2008.02.008.

Chiantia, S., Ries, J., Kahya, N., & Schwille, P. (2006). Combined AFM and two-focus SFCS study of raft-exhibiting model membranes. *ChemPhysChem*, 7(11), 2409–2418. https://doi.org/10.1002/cphc.200600464.

Chiaruttini, N., Redondo-Morata, L., Colom, A., Humbert, F., Lenz, M., Scheuring, S., et al. (2015). Relaxation of loaded ESCRT-III spiral springs drives membrane deformation. *Cell*, *163*(4), 866–879. https://doi.org/10.1016/j.cell.2015.10.017.

Cho, N.-J., Frank, C. W., Kasemo, B., & Höök, F. (2010). Quartz crystal microbalance with dissipation monitoring of supported lipid bilayers on various substrates. *Nature Protocols*, *5*(6), 1096–1106. https://doi.org/10.1038/nprot.2010.65.

Cho, N.-J., Kanazawa, K. K., Glenn, J. S., & Frank, C. W. (2007). Employing two different quartz crystal microbalance models to study changes in viscoelastic behavior upon transformation of lipid vesicles to a bilayer on a gold surface. *Analytical Chemistry*, *79*(18), 7027–7035. https://doi.org/10.1021/ac0709504.

Choi, S.-E., Greben, K., Wördenweber, R., & Offenhäusser, A. (2016). Positively charged supported lipid bilayer formation on gold surfaces for neuronal cell culture. *Biointerphases*, *11*(2), 21003. https://doi.org/10.1116/1.4945306.

Clausen-Schaumann, H., Rief, M., Tolksdorf, C., & Gaub, H. E. (2000). Mechanical stability of single DNA molecules. *Biophysical Journal*, *78*(4), 1997–2007.

Crespo-Villanueva, A., Gumí-Audenis, B., Sanz, F., Artzner, F., Mériadec, C., Rousseau, F., et al. (2018). Casein interaction with lipid membranes: Are the phase state or charge density of the phospholipids affecting protein adsorption? *Biochimica et Biophysica Acta—Biomembranes*, *1860*(12), 2588–2598. https://doi.org/10.1016/j.bbamem.2018.09.016.

Dai, J., & Sheetz, M. P. (1999). Membrane tether formation from blebbing cells. *Biophysical Journal*, 77(6), 3363–3370.

Daillant, J., Bellet-Amalric, E., Braslau, A., Charitat, T., Fragneto, G., Graner, F., et al. (2005). Structure and fluctuations of a single floating lipid bilayer. *Proceedings of the National Academy of Sciences of the United States of America*, *102*(33), 11639.

Davis, J. H., Clair, J. J., & Juhasz, J. (2009). Phase equilibria in DOPC/DPPC-d62/cholesterol mixtures. *Biophysical Journal*, *96*(2), 521–539. https://doi.org/10.1016/j.bpj.2008.09.042.

Deamer, D. W. (2010). From "banghasomes" to liposomes: A memoir of Alec Bangham, 1921-2010. *The FASEB Journal: Official Publication of the Federation of American Societies for Experimental Biology*, *24*(5), 1308–1310. https://doi.org/10.1096/fj.10-0503.

Dieluweit, S., Csiszar, A., Rubner, W., Fleischhauer, J., Houben, S., & Merkel, R. (2010). Mechanical properties of bare and protein-coated giant unilamellar phospholipid vesicles. A comparative study of micropipet aspiration and atomic force microscopy. *Langmuir*, *26*(13), 11041–11049. https://doi.org/10.1021/la1005242.

Dimova, R. (2019). Giant vesicles and their use in assays for assessing membrane phase state, curvature, mechanics, and electrical properties. *Annual Review of Biophysics*, *48*(1), 93–119. https://doi.org/10.1146/annurev-biophys-052118-115342.

Doktorova, M., Heberle, F. A., Eicher, B., Standaert, R. F., Katsaras, J., London, E., et al. (2018). Preparation of asymmetric phospholipid vesicles for use as cell membrane models. *Nature Protocols*, *13*(9), 2086–2101. https://doi.org/10.1038/s41596-018-0033-6.

Domenech, O., Redondo, L., Picas, L., Morros, A., Montero, M. T., & Hernandez-Borrell, J. (2007). Atomic force microscopy characterization of supported planar bilayers that mimic the mitochondrial inner membrane. *Journal of Molecular Recognition, 20*(6), 546–553. https://doi.org/10.1002/jmr.849.

Dufrêne, Y. F., Ando, T., Garcia, R., Alsteens, D., Martinez-Martin, D., Engel, A., et al. (2017). Imaging modes of atomic force microscopy for application in molecular and cell biology. *Nature Nanotechnology, 12*, 295. https://doi.org/10.1038/nnano.2017.45.

Dufrene, Y. F., Barger, W. R., Green, J. B. D., & Lee, G. U. (1997). Nanometer-scale surface properties of mixed phospholipid monolayers and bilayers. *Langmuir, 13*(18), 4779–4784. https://doi.org/10.1021/la970221r.

Dufrene, Y. F., Boland, T., Schneider, J. W., Barger, W. R., & Lee, G. U. (1998). Characterization of the physical properties of model biomembranes at the nanometer scale with the atomic force microscope. *Faraday Discussions, 111*, 79–94.

Dufrêne, Y. F., & Lee, G. U. (2000). Advances in the characterization of supported lipid films with the atomic force microscope. *Biochimica et Biophysica Acta (BBA)—Biomembranes, 1509*(1), 14–41. https://doi.org/10.1016/S0005-2736(00)00346-1.

Ehrig, J., Petrov, E. P., & Schwille, P. (2011). Phase separation and near-critical fluctuations in two-component lipid membranes: Monte Carlo simulations on experimentally relevant scales. *New Journal of Physics, 13*(4), 045019. https://doi.org/10.1088/1367-2630/13/4/045019.

El Kirat, K., Morandat, S., & Dufrene, Y. F. (2010). Nanoscale analysis of supported lipid bilayers using atomic force microscopy. *Biochimica et Biophysica Acta—Biomembranes, 1798*(4), 750–765. https://doi.org/10.1016/j.bbamem.2009.07.026.

Evans, E., Heinrich, V., Ludwig, F., & Rawicz, W. (2003). Dynamic tension spectroscopy and strength of biomembranes. *Biophysical Journal, 85*(4), 2342–2350.

Evans, E., & Ritchie, K. (1997). Dynamic strength of molecular adhesion bonds. *Biophysical Journal, 72*(4), 1541–1555.

Evans, E., & Yeung, A. (1994). Hidden dynamics in rapid changes of bilayer shape. *Chemistry and Physics of Lipids, 73*(1), 39–56. https://doi.org/10.1016/0009-3084(94)90173-2.

Fidorra, M., Heimburg, T., & Seeger, H. M. (2009). Melting of individual lipid components in binary lipid mixtures studied by FTIR spectroscopy, DSC and Monte Carlo simulations. *Biochimica et Biophysica Acta—Biomembranes, 1788*(3), 600–607. https://doi.org/10.1016/j.bbamem.2008.12.003.

Florin, E. L., Moy, V. T., & Gaub, H. E. (1994). Adhesion forces between individual ligand-receptor pairs. *Science, 264*(5157), 415–417. https://doi.org/10.1126/science.8153628.

Franz, V., Loi, S., Muller, H., Bamberg, E., & Butt, H. H. (2002). Tip penetration through lipid bilayers in atomic force microscopy. *Colloids and Surfaces, B: Biointerfaces, 23*(2–3), 191–200.

Garcia, R. (2020). Nanomechanical mapping of soft materials with the atomic force microscope: Methods, theory and applications. *Chemical Society Reviews, 49*(16), 5850–5884. https://doi.org/10.1039/D0CS00318B.

García-Arribas, A. B., Alonso, A., & Goñi, F. M. (2016). Cholesterol interactions with ceramide and sphingomyelin. *Chemistry and Physics of Lipids, 199*, 26–34. https://doi.org/10.1016/j.chemphyslip.2016.04.002.

García-Arribas, A. B., González-Ramírez, E. J., Sot, J., Areso, I., Alonso, A., & Goñi, F. M. (2017). Complex effects of 24:1 sphingolipids in membranes containing dioleoylphosphatidylcholine and cholesterol. *Langmuir, 33*(22), 5545–5554. https://doi.org/10.1021/acs.langmuir.7b00162.

Garcia-Manyes, S., Guell, A. G., Gorostiza, P., & Sanz, F. (2005). Nanomechanics of silicon surfaces with atomic force microscopy: An insight to the first stages of plastic deformation. *Journal of Chemical Physics, 123*(11), 114711. https://doi.org/10.1063/1.2035094.

Garcia-Manyes, S., Oncins, G., & Sanz, F. (2005a). Effect of ion-binding and chemical phospholipid structure on the nanomechanics of lipid bilayers studied by force spectroscopy. *Biophysical Journal, 89*, 1812–1826. https://doi.org/10.1529/biophysj.105.064030.

Garcia-Manyes, S., Oncins, G., & Sanz, F. (2005b). Effect of temperature on the nanomechanics of lipid bilayers studied by force spectroscopy. *Biophysical Journal, 89*(6), 4261–4274. https://doi.org/10.1529/biophysj.105.065581.

Garcia-Manyes, S., Redondo-Morata, L., Oncins, G., & Sanz, F. (2010). Nanomechanics of lipid bilayers: Heads or tails? *Journal of the American Chemical Society, 132*(37), 12874–12886. https://doi.org/10.1021/ja1002185.

Garcia-Manyes, S., & Sanz, F. (2010). Nanomechanics of lipid bilayers by force spectroscopy with AFM: A perspective. *Biochimica et Biophysica Acta—Biomembranes, 1798*(4), 741–749. https://doi.org/10.1016/j.bbamem.2009.12.019.

Giannotti, M. I., & Vancso, G. J. (2007). Interrogation of single synthetic polymer chains and polysaccharides by AFM-based force spectroscopy. *Chemphyschem, 8*(16), 2290–2307. https://doi.org/10.1002/cphc.200700175.

Giles, R., Cleveland, J. P., Manne, S., Hansma, P. K., Drake, B., Maivald, P., et al. (1993). Noncontact force microscopy in liquids. *Applied Physics Letters, 63*(5), 617–618. https://doi.org/10.1063/1.109967.

Gillissen, J. J. J., Jackman, J. A., Tabaei, S. R., Yoon, B. K., & Cho, N.-J. (2017). Quartz crystal microbalance model for quantitatively probing the deformation of adsorbed particles at low surface coverage. *Analytical Chemistry, 89*(21), 11711–11718. https://doi.org/10.1021/acs.analchem.7b03179.

Goñi, F. M. (2019). "Rafts": A nickname for putative transient nanodomains. *Chemistry and Physics of Lipids, 218*, 34–39. https://doi.org/10.1016/j.chemphyslip.2018.11.006.

González-Ramírez, E. J., Artetxe, I., García-Arribas, A. B., Goñi, F. M., & Alonso, A. (2019). Homogeneous and heterogeneous bilayers of ternary lipid compositions containing equimolar ceramide and cholesterol. *Langmuir, 35*(15), 5305–5315. https://doi.org/10.1021/acs.langmuir.9b00324.

Gumí-Audenis, B., Costa, L., Carlá, F., Comin, F., Sanz, F., & Giannotti, M. I. (2016). Structure and nanomechanics of model membranes by atomic force microscopy and spectroscopy: Insights into the role of cholesterol and sphingolipids. *Membranes, 6*(4), 58. https://doi.org/10.3390/membranes6040058.

Gumí-Audenis, B., Costa, L., Ferrer-Tasies, L., Ratera, I., Ventosa, N., Sanz, F., et al. (2018). Pulling lipid tubes from supported bilayers unveils the underlying substrate contribution to the membrane mechanics. *Nanoscale, 10*, 14763–14770. https://doi.org/10.1039/C8NR03249A.

Gumí-Audenis, B., Costa, L., Redondo-Morata, L., Milhiet, P.-E., Sanz, F., Felici, R., et al. (2018). In-plane molecular organization of hydrated single lipid bilayers: DPPC: Cholesterol. *Nanoscale, 10*(1), 87–92. https://doi.org/10.1039/c7nr07510c.

Gumí-Audenis, B., Sanz, F., & Giannotti, M. I. (2015). Impact of galactosylceramides on the nanomechanical properties of lipid bilayer models: An AFM-force spectroscopy study. *Soft Matter, 11*(27), 5447–5454. https://doi.org/10.1039/C5SM01252J.

Guyomarc'h, F., Zou, S., Chen, M., Milhiet, P.-E., Godefroy, C., Vié, V., et al. (2014). Milk sphingomyelin domains in biomimetic membranes and the role of cholesterol: Morphology and nanomechanical properties investigated using AFM and force spectroscopy. *Langmuir, 30*(22), 6516–6524. https://doi.org/10.1021/la501640y.

Hardy, G. J., Nayak, R., & Zauscher, S. (2013). Model cell membranes: Techniques to form complex biomimetic supported lipid bilayers via vesicle fusion. *Current Opinion in Colloid & Interface Science, 18*(5), 448–458. https://doi.org/10.1016/j.cocis.2013.06.004.

Hasan, I. Y., & Mechler, A. (2015). Viscoelastic changes measured in partially suspended single bilayer membranes. *Soft Matter*, *11*(27), 5571–5579. https://doi.org/10.1039/C5SM00278H.

Hassinger, J. E., Oster, G., Drubin, D. G., & Rangamani, P. (2017). Design principles for robust vesiculation in clathrin-mediated endocytosis. *Proceedings of the National Academy of Sciences of the United States of America*, *114*(7), E1118–E1127. https://doi.org/10.1073/pnas.1617705114.

Heimburg, T. (1998). Mechanical aspects of membrane thermodynamics. Estimation of the mechanical properties of lipid membranes close to the chain melting transition from calorimetry. *Biochimica et Biophysica Acta—Biomembranes*, *1415*(1), 147–162. https://doi.org/10.1016/s0005-2736(98)00189-8.

Heimburg, T. (2000). A model for the lipid pretransition: Coupling of ripple formation with the chain-melting transition. *Biophysical Journal*, *78*(3), 1154–1165. http://www.sciencedirect.com/science/article/pii/S0006349500766732.

Hill, T. L. (2013). *Thermodynamics of small systems*. **Reprint Dover Publications Inc.

Hochmuth, F. M., Shao, J. Y., Dai, J., & Sheetz, M. P. (1996). Deformation and flow of membrane into tethers extracted from neuronal growth cones. *Biophysical Journal*, *70*(1), 358–369.

Hohner, A. O., David, M. P. C., & Rädler, J. O. (2010). Controlled solvent-exchange deposition of phospholipid membranes onto solid surfaces. *Biointerphases*, *5*(1), 1–8. https://doi.org/10.1116/1.3319326.

Hugel, T., & Seitz, M. (2001). The study of molecular interactions by AFM force spectroscopy. *Macromolecular Rapid Communications*, *22*(13), 989–1016. https://doi.org/10.1002/1521-3927(20010901)22:13<989::AID-MARC989>3.0.CO;2-D.

Hung, W.-C., Lee, M.-T., Chen, F.-Y., & Huang, H. W. (2007). The condensing effect of cholesterol in lipid bilayers. *Biophysical Journal*, *92*(11), 3960–3967. https://doi.org/10.1529/biophysj.106.099234.

Hurley, J. H., Boura, E., Carlson, L.-A., & Różycki, B. (2010). Membrane budding. *Cell*, *143*(6), 875–887. https://doi.org/10.1016/j.cell.2010.11.030.

Ira, Zou, S., Ramirez, D. M. C., Vanderlip, S., Ogilvie, W., Jakubek, Z. J., et al. (2009). Enzymatic generation of ceramide induces membrane restructuring: Correlated AFM and fluorescence imaging of supported bilayers. *Journal of Structural Biology*, *168*(1), 78–89. https://doi.org/10.1016/j.jsb.2009.03.014.

Jiménez-Rojo, N., García-Arribas, A. B., Sot, J., Alonso, A., & Goñi, F. M. (2014). Lipid bilayers containing sphingomyelins and ceramides of varying N-acyl lengths: A glimpse into sphingolipid complexity. *Biochimica et Biophysica Acta—Biomembranes*, *1838*(1 Pt. B), 456–464. https://doi.org/10.1016/j.bbamem.2013.10.010.

Jing, Y., Trefna, H., Persson, M., Kasemo, B., & Svedhem, S. (2013). Formation of supported lipid bilayers on silica: Relation to lipid phase transition temperature and liposome size. *Soft Matter*, *10*(1), 187–195. https://doi.org/10.1039/C3SM50947H.

Johnston, L. J. (2007). Nanoscale imaging of domains in supported lipid membranes. *Langmuir: The ACS Journal of Surfaces and Colloids*, *23*(11), 5886–5895. https://doi.org/10.1021/la070108t.

Kaasgaard, T., Leidy, C., Crowe, J. H., Mouritsen, O. G., & Jørgensen, K. (2003). Temperature-controlled structure and kinetics of ripple phases in one- and two-component supported lipid bilayers. *Biophysical Journal*, *85*(1), 350–360. https://doi.org/10.1016/S0006-3495(03)74479-8.

Kahya, N., Scherfeld, D., Bacia, K., & Schwille, P. (2004). Lipid domain formation and dynamics in giant unilamellar vesicles explored by fluorescence correlation spectroscopy. *Journal of Structural Biology*, *147*(1), 77–89. https://doi.org/10.1016/j.jsb.2003.09.021.

Kechagia, J. Z., Ivaska, J., & Roca-Cusachs, P. (2019). Integrins as biomechanical sensors of the microenvironment. *Nature Reviews Molecular Cell Biology*, *20*(8), 457–473. https://doi.org/10.1038/s41580-019-0134-2.

Keller, C. A., & Kasemo, B. (1998). Surface specific kinetics of lipid vesicle adsorption measured with a quartz crystal microbalance. *Biophysical Journal*, *75*(3), 1397–1402.

Keller, D., Larsen, N. B., Moller, I. M., & Mouritsen, O. G. (2005). Decoupled phase transitions and grain-boundary melting in supported phospholipid bilayers. *Physical Review Letters*, *94*(2), 025701. https://doi.org/10.1103/PhysRevLett.94.025701.

Kodera, N., Yamamoto, D., Ishikawa, R., & Ando, T. (2010). Video imaging of walking myosin V by high-speed atomic force microscopy. *Nature*, *468*(7320), 72–76. https://doi.org/10.1038/nature09450.

Kurniawan, J., Ventrici de Souza, J. F., Dang, A. T., Liu, G., & Kuhl, T. L. (2018). Preparation and characterization of solid-supported lipid bilayers formed by Langmuir–Blodgett deposition: A tutorial. *Langmuir*, *34*(51), 15622–15639. https://doi.org/10.1021/acs.langmuir.8b03504.

Lee, G. U., Chrisey, L. A., & Colton, R. J. (1994). Direct measurement of the forces between complementary strands of DNA. *Science*, *266*(5186), 771–773. https://doi.org/10.1126/science.7973628.

Leonenko, Z. V., Finot, E., Ma, H., Dahms, T. E. S., & Cramb, D. T. (2004). Investigation of temperature-induced phase transitions in DOPC and DPPC phospholipid bilayers using temperature-controlled scanning force microscopy. *Biophysical Journal*, *86*(6), 3783–3793. https://doi.org/10.1529/biophysj.103.036681.

Lin, W.-C., Blanchette, C. D., & Longo, M. L. (2007). Fluid-phase chain unsaturation controlling domain microstructure and phase in ternary lipid bilayers containing GalCer and cholesterol. *Biophysical Journal*, *92*(8), 2831–2841. https://doi.org/10.1529/biophysj.106.095422.

Lin, D. C., Dimitriadis, E. K., & Horkay, F. (2007). Robust strategies for automated AFM force curve analysis—I. Non-adhesive indentation of soft, inhomogeneous materials. *Journal of Biomechanical Engineering*, *129*(3), 430–440. https://doi.org/10.1115/1.2720924.

Lind, T. K., & Cárdenas, M. (2016). Understanding the formation of supported lipid bilayers via vesicle fusion—A case that exemplifies the need for the complementary method approach (review). *Biointerphases*, *11*(2), 020801. https://doi.org/10.1116/1.4944830.

Lipowsky, R., & Seifert, U. (1991). Adhesion of vesicles and membranes. *Molecular Crystals and Liquid Crystals*, *202*(1), 17–25. https://doi.org/10.1080/00268949108035656.

Loi, S., Sun, G., Franz, V., & Butt, H. J. (2002). Rupture of molecular thin films observed in atomic force microscopy. II. Experiment. *Physical Review. E, Statistical, Nonlinear, and Soft Matter Physics*, *66*(3), 031602. https://doi.org/10.1103/PhysRevE.66.031602.

Losada-Pérez, P., Jiménez-Monroy, K. L., van Grinsven, B., Leys, J., Janssens, S. D., Peeters, M., et al. (2014). Phase transitions in lipid vesicles detected by a complementary set of methods: Heat-transfer measurements, adiabatic scanning calorimetry, and dissipation-mode quartz crystal microbalance. *Physica Status Solidi A*, *211*(6), 1377–1388. https://doi.org/10.1002/pssa.201431060.

Losada-Pérez, P., Khorshid, M., Yongabi, D., & Wagner, P. (2015). Effect of cholesterol on the phase behavior of solid-supported lipid vesicle layers. *The Journal of Physical Chemistry B*, *119*(15), 4985–4992. https://doi.org/10.1021/acs.jpcb.5b00712.

Losada-Pérez, P., Mertens, N., de Medio-Vasconcelos, B., Slenders, E., Leys, J., Peeters, M., et al. (2015). Phase transitions of binary lipid mixtures: A combined study by adiabatic scanning calorimetry and quartz crystal microbalance with dissipation monitoring. *Advances in Condensed Matter Physics*, *2015*, 479318. https://doi.org/10.1155/2015/479318.

Mabrey, S., & Sturtevant, J. M. (1976). Investigation of phase transitions of lipids and lipid mixtures by high sensitivity differential scanning calorimetry. *Proceedings of the National Academy of Sciences of the United States of America*, *73*(11), 3862–3866. https://doi.org/10.1073/pnas.73.11.3862.

Maeda, N., Senden, T. J., & di Meglio, J.-M. (2002). Micromanipulation of phospholipid bilayers by atomic force microscopy. *Biochimica et Biophysica Acta (BBA)—Biomembranes*, *1564*(1), 165–172.

Majewska, M., Mrdenovic, D., Pieta, I. S., Nowakowski, R., & Pieta, P. (2020). Nanomechanical characterization of single phospholipid bilayer in ripple phase with PF-QNM AFM. *Biochimica et Biophysica Acta (BBA) – Biomembranes*, 183347. https://doi.org/10.1016/j.bbamem.2020.183347.

Mannock, D. A., Lewis, R. N. A. H., & McElhaney, R. N. (2010). A calorimetric and spectroscopic comparison of the effects of ergosterol and cholesterol on the thermotropic phase behavior and organization of dipalmitoylphosphatidylcholine bilayer membranes. *Biochimica et Biophysica Acta (BBA)—Biomembranes*, *1798*(3), 376–388. https://doi.org/10.1016/j.bbamem.2009.09.002.

Marcus, W. D., & Hochmuth, R. M. (2002). Experimental studies of membrane tethers formed from human neutrophils. *Annals of Biomedical Engineering*, *30*(10), 1273–1280. https://doi.org/10.1114/1.1528614.

Marszalek, P. E., Li, H., & Fernandez, J. M. (2001). Fingerprinting polysaccharides with single-molecule atomic force microscopy. *Nature Biotechnology*, *19*(3), 258–262.

Marszalek, P. E., Lu, H., Li, H. B., Carrion-Vazquez, M., Oberhauser, A. F., Schulten, K., et al. (1999). Mechanical unfolding intermediates in titin modules. *Nature*, *402*(6757), 100–103.

Mennicke, U., & Salditt, T. (2002). Preparation of solid-supported lipid bilayers by spin-coating. *Langmuir*, *18*(21), 8172–8177. https://doi.org/10.1021/la025863f.

Mierzwa, B. E., Chiaruttini, N., Redondo-Morata, L., von Filseck, J. M., Koenig, J., Larios, J., et al. (2017). Dynamic subunit turnover in ESCRT-III assemblies is regulated by Vps4 to mediate membrane remodelling during cytokinesis. *Nature Cell Biology*, *19*(7), 787–798. https://doi.org/10.1038/ncb3559.

Mills, T. T., Huang, J., Feigenson, G. W., & Nagle, J. F. (2009). Effects of cholesterol and unsaturated DOPC lipid on chain packing of saturated gel-phase DPPC bilayers. *General Physiology and Biophysics*, *28*(2), 126–139.

Mingeot-Leclercq, M. P., Deleu, M., Brasseur, R., & Dufrene, Y. F. (2008). Atomic force microscopy of supported lipid bilayers. *Nature Protocols*, *3*(10), 1654–1659. https://doi.org/10.1038/nprot.2008.149.

Miyoshi, T., Lönnfors, M., Peter Slotte, J., & Kato, S. (2014). A detailed analysis of partial molecular volumes in DPPC/cholesterol binary bilayers. *Biochimica et Biophysica Acta (BBA)—Biomembranes*, *1838*(12), 3069–3077. https://doi.org/10.1016/j.bbamem.2014.07.004.

Morandat, S., Azouzi, S., Beauvais, E., Mastouri, A., & El Kirat, K. (2013). Atomic force microscopy of model lipid membranes. *Analytical and Bioanalytical Chemistry*, *405*(5), 1445–1461. https://doi.org/10.1007/s00216-012-6383-y.

Mouritsen, O. G., & Bloom, M. (1984). Mattress model of lipid-protein interactions in membranes. *Biophysical Journal*, *46*(2), 141–153.

Munguira, I., Casuso, I., Takahashi, H., Rico, F., Miyagi, A., Chami, M, et al. (2016). Glasslike membrane protein diffusion in a crowded membrane. *ACS Nano*, *10*(2), 2584–2590. https://doi.org/10.1021/acsnano.5b07595.

Nagle, J. F., & Tristram-Nagle, S. (2000). Structure of lipid bilayers. *Biochimica et Biophysica Acta—Biomembranes*, *1469*(3), 159–195. https://doi.org/10.1016/s0304-4157(00)00016-2.

Nawaz, S., Sánchez, P., Schmitt, S., Snaidero, N., Mitkovski, M., Velte, C., et al. (2015). Actin filament turnover drives leading edge growth during myelin sheath formation in the central nervous system. *Developmental Cell*, *34*(2), 139–151. https://doi.org/10.1016/j.devcel.2015.05.013.

Neupane, S., De Smet, Y., Renner, F. U., & Losada-Pérez, P. (2018). Quartz crystal microbalance with dissipation monitoring: A versatile tool to monitor phase transitions in biomimetic membranes. *Frontiers in Materials*, *5*, 46. https://doi.org/10.3389/fmats.2018.00046.

Noy, A., Vezenov, D. V., & Lieber, C. M. (1997). Chemical force microscopy. *Annual Review of Materials Science*, *27*(1), 381–421. https://doi.org/10.1146/annurev.matsci.27.1.381.

Olsson, A. L. J., Quevedo, I. R., He, D., Basnet, M., & Tufenkji, N. (2013). Using the quartz crystal microbalance with dissipation monitoring to evaluate the size of nanoparticles deposited on surfaces. *ACS Nano*, *7*(9), 7833–7843. https://doi.org/10.1021/nn402758w.

Pan, J., Mills, T. T., Tristram-Nagle, S., & Nagle, J. F. (2008). Cholesterol perturbs lipid bilayers nonuniversally. *Physical Review Letters*, *100*(19), 198103. https://doi.org/10.1103/PhysRevLett.100.198103.

Parot, P., Dufrêne, Y. F., Hinterdorfer, P., Le Grimellec, C., Navajas, D., Pellequer, J. L., et al. (2007). Past, present and future of atomic force microscopy in life sciences and medicine. *Journal of Molecular Recognition*, *20*(6), 418–431. https://doi.org/10.1002/jmr.857.

Peschel, A., Langhoff, A., Uhl, E., Dathathreyan, A., Haindl, S., Johannsmann, D., et al. (2016). Lipid phase behavior studied with a quartz crystal microbalance: A technique for biophysical studies with applications in screening. *The Journal of Chemical Physics*, *145*(20), 204904. https://doi.org/10.1063/1.4968215.

Picas, L., Montero, M. T., Morros, A., Cabanas, M. E., Seantier, B., Milhiet, P.-E., et al. (2009). Calcium-induced formation of subdomains in phosphatidylethanolamine-phosphatidylglycerol bilayers: A combined DSC, (31)P NMR, and AFM study. *Journal of Physical Chemistry B*, *113*(14), 4648–4655. https://doi.org/10.1021/jp8102468.

Picas, L., Rico, F., & Scheuring, S. (2012). Direct measurement of the mechanical properties of lipid phases in supported bilayers. *Biophysical Journal*, *102*(1), L1–L3. https://doi.org/10.1016/j.bpj.2011.11.4001.

Piggot, T. J., Allison, J. R., Sessions, R. B., & Essex, J. W. (2017). On the calculation of acyl chain order parameters from lipid simulations. *Journal of Chemical Theory and Computation*, *13*(11), 5683–5696. https://doi.org/10.1021/acs.jctc.7b00643.

Pinkwart, K., Schneider, F., Lukoseviciute, M., Sauka-Spengler, T., Lyman, E., Eggeling, C., et al. (2019). Nanoscale dynamics of cholesterol in the cell membrane. *Journal of Biological Chemistry*, *294*(34), 12599–12609. https://doi.org/10.1074/jbc.RA119.009683.

Pramanik, S. K., Seneca, S., Ethirajan, A., Neupane, S., Renner, F. U., & Losada-Pérez, P. (2016). Ionic strength dependent vesicle adsorption and phase behavior of anionic phospholipids on a gold substrate. *Biointerphases*, *11*(1), 019006. https://doi.org/10.1116/1.4939596.

Radmacher, M. (1997). Measuring the elastic properties of biological samples with the AFM. *IEEE Engineering in Medicine and Biology Magazine*, *16*(2), 47–57. https://doi.org/10.1109/51.582176.

Rascol, E., Devoisselle, J.-M., & Chopineau, J. (2016). The relevance of membrane models to understand nanoparticles–cell membrane interactions. *Nanoscale*, *8*(9), 4780–4798. https://doi.org/10.1039/C5NR07954C.

Rawicz, W., Olbrich, K. C., McIntosh, T., Needham, D., & Evans, E. (2000). Effect of chain length and unsaturation on elasticity of lipid bilayers. *Biophysical Journal, 79*(1), 328–339.

Redondo-Morata, L., Giannotti, M. I., & Sanz, F. (2012). Stability of lipid bilayers as model membranes: Atomic force microscopy and spectroscopy approach. In *Atomic force microscopy in liquid: Biological applications*. https://doi.org/10.1002/9783527649808.ch10.

Redondo-Morata, L., Giannotti, M. I., & Sanz, F. (2012a). AFM-based force-clamp monitors lipid bilayer failure kinetics. *Langmuir, 28*(15), 6403–6410. https://doi.org/10.1021/la3005147.

Redondo-Morata, L., Giannotti, M. I., & Sanz, F. (2012b). Influence of cholesterol on the phase transition of lipid bilayers: A temperature-controlled force spectroscopy study. *Langmuir, 28*(35), 12851–12860. https://doi.org/10.1021/la302620t.

Redondo-Morata, L., Giannotti, M. I., & Sanz, F. (2014). Structural impact of cations on lipid bilayer models: Nanomechanical properties by AFM-force spectroscopy. *Molecular Membrane Biology, 31*(1), 17–28. https://doi.org/10.3109/09687688.2013.868940.

Redondo-Morata, L., Oncins, G., & Sanz, F. (2012). Force spectroscopy reveals the effect of different ions in the nanomechanical behavior of phospholipid model membranes: The case of potassium cation. *Biophysical Journal, 102*(1), 66–74. https://doi.org/10.1016/j.bpj.2011.10.051.

Redondo-Morata, L., Sanford, R. L., Andersen, O. S., & Scheuring, S. (2016). Effect of statins on the nanomechanical properties of supported lipid bilayers. *Biophysical Journal, 111*(2), 363–372. https://doi.org/10.1016/j.bpj.2016.06.016.

Reimhult, E., Hook, F., & Kasemo, B. (2003). Intact vesicle adsorption and supported biomembrane formation from vesicles in solution: Influence of surface chemistry, vesicle size, temperature, and osmotic pressure. *Langmuir, 19*(5), 1681–1691. https://doi.org/10.1021/la0263920.

Relat-Goberna, J., Beedle Amy, E. M., & Garcia-Manyes, S. (2017). The nanomechanics of lipid multibilayer stacks exhibits complex dynamics. *Small, 13*(24), 1700147. https://doi.org/10.1002/smll.201700147.

Reviakine, I., Gallego, M., Johannsmann, D., & Tellechea, E. (2012). Adsorbed liposome deformation studied with quartz crystal microbalance. *The Journal of Chemical Physics, 136*(8), 084702. https://doi.org/10.1063/1.3687351.

Reviakine, I., Johannsmann, D., & Richter, R. P. (2011). Hearing what you cannot see and visualizing what you hear: Interpreting quartz crystal microbalance data from solvated interfaces. *Analytical Chemistry, 83*(23), 8838–8848. https://doi.org/10.1021/ac201778h.

Richter, R. P. (2006). Formation of solid-supported lipid bilayers: An integrated view. *Langmuir, 22*(8), 3497–3505. https://doi.org/10.1021/la052687c.

Richter, R. P., & Brisson, A. (2003). Characterization of lipid bilayers and protein assemblies supported on rough surfaces by atomic force microscopy. *Langmuir, 19*(5), 1632–1640. https://doi.org/10.1021/la026427w.

Richter, R. P., & Brisson, A. R. (2005). Following the formation of supported lipid bilayers on mica: A study combining AFM, QCM-D, and ellipsometry. *Biophysical Journal, 88*(5), 3422–3433. https://doi.org/10.1529/biophysj.104.053728.

Richter, R., Mukhopadhyay, A., & Brisson, A. (2003). Pathways of lipid vesicle deposition on solid surfaces: A combined QCM-D and AFM study. *Biophysical Journal, 85*(5), 3035–3047.

Rico, F., Gonzalez, L., Casuso, I., Puig-Vidal, M., & Scheuring, S. (2013). High-speed force spectroscopy unfolds titin at the velocity of molecular dynamics simulations. *Science, 342*(6159), 741–743. http://science.sciencemag.org/content/342/6159/741.abstract.

Rief, M., Oesterhelt, F., Heymann, B., & Gaub, H. E. (1997). Single molecule force spectroscopy on polysaccharides by atomic force microscopy. *Science*, *275*(5304), 1295–1297. https://doi.org/10.1126/science.275.5304.1295.

Rog, T., Pasenkiewicz-Gierula, M., Vattulainen, I., & Karttunen, M. (2009). Ordering effects of cholesterol and its analogues. *Biochimica et Biophysica Acta—Biomembranes*, *1788*(1), 97–121. https://doi.org/10.1016/j.bbamem.2008.08.022.

Roux, A. (2013). The physics of membrane tubes: Soft templates for studying cellular membranes. *Soft Matter*, *9*(29), 6726–6736.

Sauerbrey, G. (1959). Verwendung von Schwingquarzen zur Wägung dünner Schichten und zur Mikrowägung. *Zeitschrift für Physik*, *155*(2), 206–222. https://doi.org/10.1007/BF01337937.

Schmidtke, D. W., & Diamond, S. L. (2000). Direct observation of membrane tethers formed during neutrophil attachment to platelets or P-selectin under physiological flow. *The Journal of Cell Biology*, *149*(3), 719.

Schneider, J., Dufrene, Y. F., Barger, W. R., & Lee, G. U. (2000). Atomic force microscope image contrast mechanisms on supported lipid bilayers. *Biophysical Journal*, *79*(2), 1107–1118.

Sebaaly, C., Greige-Gerges, H., & Charcosset, C. (2019). Chapter 11—Lipid membrane models for biomembrane properties' investigation. In A. Basile, & C. Charcosset (Eds.), *Current trends and future developments on (bio-) membranes* (pp. 311–340). Elsevier. https://doi.org/10.1016/B978-0-12-813606-5.00011-7.

Seeger, H. M., Di Cerbo, A., Alessandrini, A., & Facci, P. (2010). Supported lipid bilayers on mica and silicon oxide: Comparison of the main phase transition behavior. *Journal of Physical Chemistry B*, *114*(27), 8926–8933. https://doi.org/10.1021/jp1026477.

Sessa, G., & Weissmann, G. (1968). Phospholipid spherules (liposomes) as a model for biological membranes. *Journal of Lipid Research*, *9*(3), 310–318.

Shao, J.-Y., Ting-Beall, H. P., & Hochmuth, R. M. (1998). Static and dynamic lengths of neutrophil microvilli. *Proceedings of the National Academy of Sciences of the United States of America*, *95*(12), 6797.

Sheetz, M. P. (2001). Cell control by membrane-cytoskeleton adhesion. *Nature Reviews Molecular Cell Biology*, *2*, 392.

Shinoda, K., Shinoda, W., & Mikami, M. (2007). Molecular dynamics simulation of an archaeal lipid bilayer with sodium chloride. *Physical Chemistry Chemical Physics*, *9*(5), 643–650. https://doi.org/10.1039/b611543h.

Simons, K., & Ikonen, E. (1997). Functional rafts in cell membranes. *Nature*, *387*(6633), 569–572. https://doi.org/10.1038/42408.

Singer, S. J., & Nicolson, G. L. (1972). Fluid mosaic model of structure of cell-membranes. *Science*, *175*(4023), 720. https://doi.org/10.1126/science.175.4023.720.

Siontorou, C. G., Nikoleli, G.-P., Nikolelis, D. P., & Karapetis, S. K. (2017). Artificial lipid membranes: Past, present, and future. *Membranes*, 7(3), 38. https://doi.org/10.3390/membranes7030038.

Steinkühler, J., Sezgin, E., Urbančič, I., Eggeling, C., & Dimova, R. (2019). Mechanical properties of plasma membrane vesicles correlate with lipid order, viscosity and cell density. *Communications Biology*, *2*(1), 1–8. https://doi.org/10.1038/s42003-019-0583-3.

Strulson, M. K., & Maurer, J. A. (2011). Microcontact printing for creation of patterned lipid bilayers on tetraethylene glycol self-assembled monolayers. *Langmuir: The ACS Journal of Surfaces and Colloids*, *27*(19), 12052–12057. https://doi.org/10.1021/la201839w.

Sullan, R. M. A., Li, J. K., Hao, C. C., Walker, G. C., & Zou, S. (2010). Cholesterol-dependent nanomechanical stability of phase-segregated multicomponent lipid bilayers. *Biophysical Journal*, *99*(2), 507–516. https://doi.org/10.1016/j.bpj.2010.04.044.

Sullan, R. M. A., Li, J. K., & Zou, S. (2009a). Direct correlation of structures and nanomechanical properties of multicomponent lipid bilayers. *Langmuir, 25*(13), 7471–7477. https://doi.org/10.1021/la900395w.

Sullan, R. M. A., Li, J. K., & Zou, S. (2009b). Quantification of the nanomechanical stability of ceramide-enriched domains. *Langmuir, 25*(22), 12874–12877. https://doi.org/10.1021/la903442s.

Sun, M., Graham, J. S., Hegedüs, B., Marga, F., Zhang, Y., Forgacs, G., et al. (2005). Multiple membrane tethers probed by atomic force microscopy. *Biophysical Journal, 89*(6), 4320–4329.

Tabaei, S. R., Gillissen, J. J. J., Vafaei, S., Groves, J. T., & Cho, N.-J. (2016). Size-dependent, stochastic nature of lipid exchange between nano-vesicles and model membranes. *Nanoscale, 8*(27), 13513–13520. https://doi.org/10.1039/C6NR03817D.

Takahashi, H., Miyagi, A., Redondo-Morata, L., & Scheuring, S. (2016). Temperature-controlled high-speed AFM: Real-time observation of ripple phase transitions. *Small, 12*(44), 6106–6113. https://doi.org/10.1002/smll.201601549.

Tellechea, E., Johannsmann, D., Steinmetz, N. F., Richter, R. P., & Reviakine, I. (2009). Model-independent analysis of QCM data on colloidal particle adsorption. *Langmuir, 25*(9), 5177–5184. https://doi.org/10.1021/la803912p.

Tero, R. (2012). Substrate effects on the formation process, structure and physicochemical properties of supported lipid bilayers. *Materials, 5*(12), 2658–2680. https://doi.org/10.3390/ma5122658.

Van Lehn, R. C., Ricci, M., Silva, P. H. J., Andreozzi, P., Reguera, J., Voïtchovsky, K., et al. (2014). Lipid tail protrusions mediate the insertion of nanoparticles into model cell membranes. *Nature Communications, 5*(1), 4482. https://doi.org/10.1038/ncomms5482.

van Meer, G., & de Kroon, A. I. P. M. (2011). Lipid map of the mammalian cell. *Journal of Cell Science, 124*(1), 5.

van Meer, G., Voelker, D. R., & Feigenson, G. W. (2008). Membrane lipids: Where they are and how they behave. *Nature Reviews Molecular Cell Biology, 9*(2), 112–124. https://doi.org/10.1038/nrm2330.

Veatch, S. L., & Keller, S. L. (2003). Separation of liquid phases in giant vesicles of ternary mixtures of phospholipids and cholesterol. *Biophysical Journal, 85*(5), 3074–3083.

Vereb, G., Szöllősi, J., Matkó, J., Nagy, P., Farkas, T., Vígh, L., et al. (2003). Dynamic, yet structured: The cell membrane three decades after the Singer–Nicolson model. *Proceedings of the National Academy of Sciences of the United States of America, 100*(14), 8053–8058. https://doi.org/10.1073/pnas.1332550100.

Vist, M. R., & Davis, J. H. (1990). Phase equilibria of cholesterol/dipalmitoyl-phosphatidylcholine
mixtures: Deuterium nuclear magnetic resonance and differential scanning calorimetry. *Biochemistry, 29*(2), 451–464. https://doi.org/10.1021/bi00454a021.

Vogel, V. (2006). Mechanotransduction involving multimodular proteins: Converting force into biochemical signals. *Annual Review of Biophysics and Biomolecular Structure, 35*(1), 459–488. https://doi.org/10.1146/annurev.biophys.35.040405.102013.

Voinova, M. V., Rodahl, M., Jonson, M., & Kasemo, B. (1999). Viscoelastic acoustic response of layered polymer films at fluid-solid interfaces: Continuum mechanics approach. *Physica Scripta, 59*(5), 391. https://doi.org/10.1238/Physica.Regular.059a00391.

Wallace, E. J., Hooper, N. M., & Olmsted, P. D. (2006). Effect of hydrophobic mismatch on phase behavior of lipid membranes. *Biophysical Journal, 90*(11), 4104–4118. https://doi.org/10.1529/biophysj.105.062778.

Wargenau, A., & Tufenkji, N. (2014). Direct detection of the gel–fluid phase transition of a single supported phospholipid bilayer using quartz crystal microbalance with dissipation monitoring. *Analytical Chemistry, 86*(16), 8017–8020. https://doi.org/10.1021/ac5019183.

Winkler, P. M., Campelo, F., Giannotti, M. I., & García-Parajo, M. F. (2020). Planar plasmonic antenna arrays resolve transient nanoscopic heterogeneities in biological membranes. In *Proc. SPIE, Single molecule spectroscopy and superresolution imaging XIII, 112460F*. https://doi.org/10.1117/12.2543726.

Winkler, P. M., Regmi, R., Flauraud, V., Brugger, J., Rigneault, H., Wenger, J., et al. (2017). Transient nanoscopic phase separation in biological lipid membranes resolved by planar plasmonic antennas. *ACS Nano, 11*(7), 7241–7250. https://doi.org/10.1021/acsnano.7b03177.

Yeagle, P. L. (1989). Lipid regulation of cell membrane structure and function. *The FASEB Journal, 3*(7), 1833–1842. https://doi.org/10.1096/fasebj.3.7.2469614.

Zou, S., & Johnston, L. J. (2010). Ceramide-enriched microdomains in planar membranes. *Current Opinion in Colloid & Interface Science, 15*(6), 489–498. https://doi.org/10.1016/j.cocis.2010.06.003.

CHAPTER TWO

Membrane modulatory effects of omega-3 fatty acids: Analysis of molecular level interactions

Manuela A.A. Ayee*, Brendan C. Bunker, and Jordan L. De Groot
Department of Engineering, Dordt University, Sioux Center, IA, United States
*Corresponding author: e-mail address: Manuela.Ayee@dordt.edu

Contents

Abstract

Bioactive omega-3 polyunsaturated fatty acids have been shown to reduce the risk of death in patients with cardiovascular disease and alleviate the symptoms of other inflammatory diseases. However, the mechanisms of action of these effects remain unclear. It has been postulated that omega-3 polyunsaturated fatty acids modify cell membranes by incorporation into the membrane and altering the signaling properties of cellular receptors. In this chapter, we explore the effects of omega-3 polyunsaturated fatty acids on cell membrane structure and function. We present a review of the current evidence for the health benefits of these compounds and explore the molecular mechanisms through which omega-3 polyunsaturated fatty acids interact with

Current Topics in Membranes, Volume 86
ISSN 1063-5823
https://doi.org/10.1016/bs.ctm.2020.08.001

membrane lipids and modulate bilayer structure. Using computational models of multicomponent phospholipid bilayers, we assess the consequences of incorporation of these fatty acids on membrane lipid packing, water permeation, and membrane structure.

1. Introduction

Cardiovascular disease is the leading cause of death worldwide, accounting for 23% of total deaths in the United States in 2017, according to the Centers for Disease Control and Prevention (Kochanek, Murphy, Xu, & Arias, 2019). Cardiovascular disease encompasses the class of disorders that affect the heart or blood vessels, including arrhythmias (Jin, Lyon, & Akar, 2008; Morotti & Grandi, 2019), congestive heart failure (Aquilani et al., 2017; Grassi, Quarti-Trevano, & Esler, 2019), hypertension (Drummond, Vinh, Guzik, & Sobey, 2019), and stroke (Vemuganti, 2019), with the underlying mechanisms of action dependent on the specific disease. A combination of lifestyle choices and risk factors increase the likelihood of developing cardiovascular disease. Among the most common predictors are dietary habits, physical inactivity, smoking, obesity, hypertension, dyslipidemia, and diabetes (Mozaffarian, Wilson, & Kannel, 2008).

Numerous studies have shown a strong association between the risk factors for cardiovascular disease and the development of non-alcoholic fatty liver disease (NAFLD), which results from an abnormal accumulation of fat deposits in hepatocytes (known as steatosis), leading to progressive liver damage (Bonora & Targher, 2012; Byrne & Targher, 2015; Chandrasekharan & Alazawi, 2019; Domanski, Park, & Harrison, 2012; Stahl et al., 2019). Common risk factors for the development of steatosis due to NAFLD are obesity, diabetes, and high blood triglyceride levels (Bayard, Holt, & Boroughs, 2006; Marchesini, Marzocchi, Agostini, & Bugianesi, 2005; Neuschwander-Tetri, 2005; Sanyal, 2002), factors also associated with increased risk of cardiovascular disease. Significantly increased plasma levels of the biomarkers for inflammation have been identified in NAFLD patients, with concentrations of these biomarkers dependent on the severity of liver disease (Ajmal et al., 2014; Foroughi et al., 2016). The presence of these biomarkers suggests that patients with NAFLD develop chronic inflammation, which is known to be associated with the development of cardiovascular disease (Pepys & Hirschfield, 2003).

The consumption of dietary fish oils is believed to have beneficial health effects due to their high concentrations of omega-3 polyunsaturated fatty acids (PUFAs). These fatty acids are unable to be synthesized by the human body but are essential for maintaining health. It is known that elevated plasma levels of these fatty acids lead to a decrease in the levels of the biomarkers for inflammation, implying that they exhibit anti-inflammatory effects. In this chapter, we will review and discuss the current state of knowledge on the health benefits of omega-3 PUFAs and the current hypothesized mechanisms of action of these fatty acids in alleviating cardiovascular disease. We will then highlight some recent results from our molecular dynamics simulations of the effects of omega-3 PUFAs on the structural properties of symmetric lipid bilayer membranes.

2. Health benefits of omega-3 polyunsaturated fatty acids

Dietary sources rich in omega-3 PUFAs include fatty fish, algae, and various seeds, oils, vegetables, nuts, and beans (Covington, 2004). Studies suggest an inverse relationship between the consumption of omega-3 PUFAs and negative cardiovascular outcomes for patients with cardiovascular disease (see Khan et al., 2019 for a recent comprehensive review). Moreover, a multiethnic study by de Oliveira Otto et al. recently showed that increased dietary consumption and elevated concentrations of circulating omega-3 PUFAs were inversely correlated with the incidence of cardiovascular disease (de Oliveira Otto et al., 2013). Supplementation has also been shown to improve endothelial function and health (Dangardt et al., 2010; Goodfellow, Bellamy, Ramsey, Jones, & Lewis, 2000), and to boost the mechanical efficiency of the heart (Peoples, McLennan, Howe, & Groeller, 2008; Pepe & McLennan, 2002). Dietary supplementation of omega-3 PUFAs has been shown to influence the key mechanisms involved in regulating hepatic fat content, leading to lower levels of accumulated liver fat deposits in patients with NAFLD (reviewed recently in Jump, Lytle, Depner, & Tripathy, 2018, Scorletti & Byrne, 2018), and to lower the levels of circulating plasma triglycerides (Harris & Bulchandani, 2006). Therefore, the possibility of using omega-3 PUFA supplementation as a treatment method for early stage NAFLD has been assessed in several recent clinical trials and meta-analyses, with the conclusion that this treatment regime leads to improvements in liver steatosis and plasma lipid concentrations (Spooner & Jump, 2019).

In addition to these cardiovascular and hepatic health benefits, studies indicate that dietary consumption of omega-3 PUFAs may have positive effects on neurological health. In experimental, biochemical, and epidemiological observations, these long-chain fatty acids are shown to play a crucial role in fetal and infant brain development and function when consumed during pregnancy and lactation (Bernardi, de Escobar, Ferreira, & Silveira, 2012). Furthermore, in studies of Alzheimer's disease, omega-3 PUFA supplementation may be beneficial during the early stages of this neurodegenerative disorder (Canhada, Castro, Perry, & Luft, 2018) and for the treatment and prevention of age-related cognitive decline (Cole, Ma, & Frautschy, 2009). Due to evidence of these fatty acids being critical to neurological health, clinical and research studies have been undertaken to determine their role in attention-deficit/hyperactivity disorder (ADHD) and other developmental disorders in children, such as dyslexia, autism, and developmental coordination disorder. The results from these studies suggest that dietary supplementation with omega-3 PUFAs may assuage some ADHD symptoms (Richardson, 2006). Much of this evidence of neuroprotective properties of omega-3 PUFAs does not distinguish between the unique effects of different fatty acids on each condition. However, a growing body of evidence is finding that the various forms of omega-3 PUFAs may have specific beneficial effects on different aspects of neurological health, with both individual and combined effects becoming apparent (Dyall, 2015).

Omega-3 PUFAs have also been studied in the context of nutritional therapy for cancer patients. Combined with chemotherapy, a few studies have shown that omega-3 PUFA supplementation may improve the response to treatment for patients diagnosed with cancer (Laviano, Rianda, Molfino, & Rossi Fanelli, 2013). Several other health conditions and their related symptoms have been shown to benefit to varying extents from dietary intake of omega-3 PUFAs, including treatment for two common inflammatory eye diseases: dry eye disease and age-related macular degeneration (Wang & Daggy, 2017), symptom reduction in patients with rheumatoid arthritis (Rajaei et al., 2016), attenuation of inflammatory processes in patients with inflammatory bowel disease (Barbalho, de Goulart, Quesada, Bechara, & de de Carvalho, 2016), and potential anti-inflammatory effects for cystic fibrosis sufferers (Oliver & Jahnke, 2011). In all these cases, the beneficial effects of omega-3 PUFAs are attributed to their anti-inflammatory properties. As noted by the authors of many of the studies and reviews listed in this section, the evidence of specific health benefits of omega-3 PUFAs tends to be somewhat inconclusive in some cases, with unclear mechanisms of action and therapeutic targets. Therefore, further studies are necessary to draw firmer conclusions, including both experimental and theoretical work to elucidate mechanisms, as well as larger scale, long term, randomized clinical trials. It is also noted that the relative safety and lack of serious adverse effects related to the consumption of fish oils provide a robust incentive for continued studies on their potential therapeutic effects across a wide range of debilitating health conditions.

3. Chemical structures of long-chain omega-3 polyunsaturated fatty acids

Structurally, fatty acids consist of a carboxylic acid (C(=O)OH) attached to an aliphatic hydrocarbon chain. This chain can be either saturated or unsaturated (containing double bonded carbon atoms), referring to the ability of the hydrocarbon chain to resist addition reactions such as oxidation or hydrogenation. Fatty acids can be classified according to the length and degree of unsaturation of their hydrocarbon chains. Unsaturated fatty acids contain one (monounsaturated) or more (polyunsaturated) carbon-carbon (C=C) double bonds. Specifically, as illustrated in Fig. 1A and B, for the chemical structures of omega-3 polyunsaturated fatty acids, the last double bond is located three atoms away from the terminal (omega) carbon atom. Conversely, for the related, and equally important omega-6 polyunsaturated fatty acids, which are proinflammatory and consumed in substantially higher amounts compared to omega-3 PUFAs (Brigham et al., 2019; DiNicolantonio & O'Keefe, 2018; Wall, Ross, Fitzgerald, & Stanton, 2010), the final double bond is located on the sixth carbon atom from the terminal methyl group (Fig. 1C).

The typical diet in the Western world contains a much greater ratio of omega-6 PUFAs to omega-3 PUFAs. Both omega-3 and omega-6 PUFAs are derived from essential fatty acids, which are required for the proper maintenance of good health but are unable to be synthesized by the human body (Singh, 2005). They must therefore be obtained from dietary sources. For mammals, the two essential fatty acids are α-linolenic acid (ALA, an omega-3 PUFA) and linoleic acid (LA, an omega-6 PUFA). These are both known as short-chain fatty acids with 18 carbon atoms per chain and multiple double bonds. Upon consumption, enzymes competitively metabolize these two essential fatty acids by adding more double bonds and

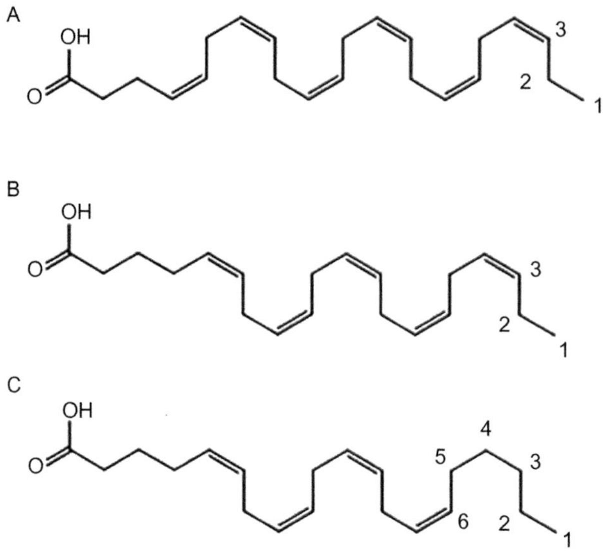

Fig. 1 Chemical structures of dietary omega-3 and omega-6 long-chain polyunsaturated fatty acids. (A) Docosahexaenoic acid (DHA) 22:6(n-3), with 22 carbon atoms and 6 double bonds, the last of which resides three carbons from terminal end (omega-3). (B) Eicosapentaenoic acid (EPA) 20:5(n-3) (omega-3). (C) Arachidonic acid 20:4(n-6) (omega-6) (ACD/ChemSketch, 2020).

elongating the aliphatic hydrocarbon chains, a process that occurs mainly in the liver. Anti-inflammatory effects result from the metabolism of the omega-3 PUFA, ALA, while the short-chain omega-6 PUFA, LA, metabolizes into long-chain fatty acids such as arachidonic acid (with a 20 carbon atom chain), which is a precursor of proinflammatory eicosanoids (Wall et al., 2010). Due to the competitive nature of these metabolic processes, an increased consumption of omega-6 PUFAs directly leads to decreased metabolism of omega-3 PUFAs. When consumed, the omega-3 ALA is metabolized into long-chain omega-3 PUFAs, including docosahexaenoic acid (DHA, with a 22-carbon atom chain, Fig. 1A) and eicosapentaenoic acid (EPA, with a 20 carbon atom chain, Fig. 1B). Both EPA and DHA can also be obtained directly from dietary sources such as fatty fish (Tur, Bibiloni, Sureda, & Pons, 2012). Once consumed through the diet and metabolized, PUFAs are transported around the body in the blood plasma and are able to partition into cellular membranes to varying degrees (Browning et al., 2012; Fielding, 2011).

Both EPA and DHA are known to have inflammation lowering properties, however, the specific mechanisms of this action, as well as the reasons for their effectiveness at reducing inflammation, continue to be elusive. Many hypothesized mechanisms regarding the various observed effects of omega-3 PUFAs have been postulated, although researchers agree that more studies need to be conducted to further provide clarity. In the following sections, we will focus on some proposed hypotheses specific to the observed effects of omega-3 PUFAs on cardiovascular health.

4. Hypothesized mechanisms of action of omega-3 PUFAs on cardiovascular health

In this section, we will review several hypothesized mechanisms of action that have been postulated in recent years regarding the observed beneficial health effects of omega-3 PUFAs on cardiovascular health. A variety of experimental approaches have been used to obtain the results and arrive at the conclusions on which these hypotheses were based. These include (1) human randomized controlled trials (RCTs), which are required by regulatory agencies in order to determine the effectiveness and safety of any new treatment (Kabisch, Ruckes, Seibert-Grafe, & Blettner, 2011), (2) human observational studies conducted on large sample populations, often without medical interventions, (3) *in vivo* experiments that are conducted in humans and living animals that attempt to provide mechanistic explanations for the effects documented in observational studies and clinical trials, and finally, (4) *in vitro* studies performed with cultured cells from either human or animal sources, which isolate these cells from their normal biological context to provide opportunities for more convenient and detailed mechanistic analyses.

4.1 Effects on atherosclerosis

Chronic inflammation has been implicated in all stages of the development and progression of atherosclerosis, demonstrated by increases in the circulating levels of biomarkers of inflammation, such as high-sensitivity C-reactive protein, with disease progression (Libby, Ridker, & Maseri, 2002; Ridker, Hennekens, Buring, & Rifai, 2000). This sustained inflammation alters lipid metabolism leading to dyslipidemia, which includes increased levels of circulating low-density lipoproteins (LDL) and triglycerides, among others (Tsoupras, Lordan, & Zabetakis, 2018). Moreover, oxidatively modified LDL (oxLDL) is known to be a major proinflammatory and proatherogenic factor, and has been recently shown to induce endothelial stiffness and disrupt membrane structure, resulting in changes in the biomechanical properties of endothelial cells (Ayee et al., 2017; Shentu et al., 2012). For patients with atherosclerosis, thinning of the fibrous cap covering the lipid cores of atherosclerotic plaques may result in plaque rupture, blood clot formation, and coronary thrombosis (Bentzon, Otsuka, Virmani, & Falk, 2014; Carr, Farb, Pearce, Virmani, & Yao, 1996; Libby et al., 2002). Plaques become more susceptible to rupture due to inflammation and the actions of oxidized lipids. Studies indicate that supplementation with long-chain omega-3 PUFAs may stabilize and possibly remodel vulnerable atherosclerotic plaques (reviewed recently by DiNoclantonio & O'Keefe, 2020). This may occur due to an omega-3 PUFA-induced decrease in the levels of systemic inflammatory biomarkers (discussed further in the section on *anti-inflammatory effects*). In clinical trials on dyslipidemic patients, a combined treatment regime with EPA and statin therapy led to an increased thickness of plaque fibrous caps and reduced macrophage accumulation, both possibly leading to stabilization of plaques that were vulnerable to rupture (Nishio et al., 2014). Omega-3 PUFAs also affect the deposition of LDL and other lipoproteins by macrophages into the arterial wall by modulating the levels of arterial lipoprotein lipase (Chang & Deckelbaum, 2013).

4.2 Effects on endothelial function

Studies show that omega-3 PUFAs may improve vasodilation (widening blood vessels and decreasing blood pressure) and inflammation related to endothelial dysfunction, a condition that can lead to the development of atherosclerosis, kidney disease, metabolic syndrome, and other disorders. The most significant improvements in endothelial function were observed in trials that combined both EPA and DHA supplementation (Zehr & Walker, 2018). In patients with metabolic syndrome, endothelial function (evaluated using flow mediated dilation of the brachial artery) and arterial elastic properties (evaluated with pulse wave velocity measurements to determine aortic stiffness) were ameliorated after daily administration of omega-3 PUFAs for 12 weeks (Tousoulis et al., 2014). In chronic kidney disease, omega-3 PUFAs were shown to improve endothelial dysfunction by increasing the expression and function of endothelial nitric oxide synthase (eNOS), an enzyme that synthesizes nitric oxide (NO), which is important for vasodilation (Zanetti, Gortan Cappellari, Barbetta, Semolic, & Barazzoni, 2017).

4.3 Regulation of platelet function

The role of omega-3 PUFAs on the regulation of platelet function is controversial, with some studies associating supplementation with reduced platelet aggregation in high-risk patients (reviewed by Gao et al., 2012), while other studies saw no effects on platelet function in healthy volunteers (Bagge, Schott, & Kander, 2018), or patients with combined, well-controlled diabetes and atherosclerosis (Poreba et al., 2017). It is known, however, that omega-3 PUFAs are important components of the lipid membranes of platelets, and that dietary supplementation of these fatty acids alters this lipid composition. These changes in composition may potentially influence the functions of platelet membranes as well as their interactions with blood vessel walls, thereby having consequences for thrombosis and cardiovascular disease (Abeywardena & Head, 2001; Adili, Hawley, & Holinstat, 2019; Mann, O'Connell, Baldwin, Singh, & Meyer, 2010). Most researchers agree that further studies need to be done to clarify discrepancies in results and to elucidate the mechanistic basis for platelet function regulation by omega-3 PUFAs.

4.4 Antiarrhythmic effects

The relationship between omega-3 PUFAs and atrial fibrillation is unclear, although some studies suggest that they aid in improving autonomic tone and possibly reducing the risk of arrhythmias (Anand, Alkadri, Lavie, & Milani, 2008; Monahan, Wilson, & Ray, 2004; Wu Jason et al., 2012). Many of the risk factors for atrial fibrillation onset and progression coincide with risk factors for cardiovascular disease (Blum et al., 2019; Morillo, Banerjee, Perel, Wood, & Jouven, 2017), therefore mitigating these factors may prevent onset and diminish severity of both disorders. In a recent review on the topic of omega-3 PUFAs and atrial fibrillation, Martino et al. (2016) concluded that a lack of uniformity between studies and the differing responses of study subjects to supplementation may be to blame for the disparate findings thus far. In terms of mechanism, it has been postulated that the antiarrhythmic effects of omega-3 PUFAs may be attributed to their capability to modulate cardiac ion channels, although it remains uncertain if they have direct effects on these channels (Moreno et al., 2012).

4.5 Effects on triglyceride levels

High levels of triglycerides in the plasma are associated with an increased risk for cardiovascular disease and NAFLD (Harchaoui, Visser, Kastelein, Stroes, & Dallinga-Thie, 2009; Stahl et al., 2019). Under normal conditions, the liver maintains the levels of lipids and triglycerides in the body. However, in patients with NAFLD, excess triglycerides accumulate in the liver, resulting in steatosis (Alves-Bezerra & Cohen, 2017).

Omega-3 PUFAs have been shown to reduce serum levels of triglycerides through decreasing the secretion of triglyceride-rich very low-density lipoproteins (VLDL) from the liver (Harris & Bulchandani, 2006). Although the molecular mechanisms are still unclear, studies have postulated that omega-3 PUFAs inhibit hepatic enzymes that synthesize triglycerides *de novo* and reduce its availability, leading to decreased secretion of VLDL (Green et al., 2020; Skulas-Ray et al., 2019).

4.6 Anti-inflammatory effects

Although omega-3 PUFAs have been shown to exhibit anti-inflammatory effects, as discussed in previous sections of this chapter, the precise mechanisms by which they regulate the immune system are complex and remain to be fully elucidated. Numerous recent reviews detail various potential mechanisms of omega-3 anti-inflammatory action, including, but not limited to: signal modulation in immune cells (Gutiérrez, Svahn, & Johansson, 2019), reduction of inflammatory biomarker levels (Baker et al., 2020; Natto, Yaghmoor, Alshaeri, & Van Dyke, 2019), decreasing circulating levels of very low-density lipoproteins (VLDL) (Sepidarkish et al., 2019), reducing membrane content of proinflammatory omega-6 PUFAs (Calder, 2012; Kromhout, Yasuda, Geleijinse, & Shimokawa, 2011), and incorporation into cellular membranes leading to localized disruption in membrane packing (Calder, 2012). The action of incorporation of omega-3 PUFAs into cellular membrane bilayers and the corresponding consequences for membrane structure and lipid packing is the focus of the next section of this chapter.

5. Membrane modulatory effects of omega-3 PUFAs

As mentioned previously, omega-3 PUFAs form an essential part of many cellular membranes, and dietary supplementation enables PUFAs to further partition into these lipid membranes (Miles et al., 2004). This action changes the lipid composition and potentially modifies the membrane biomechanical properties and function, leading to altered signaling of cellular receptors. In a study performed with HepG2 liver cancer cells, the omega-3 PUFAs DHA and EPA were found to have an inhibitory effect on the interleukin-6 (IL-6)-induced expression of C-reactive protein (CRP) (Wang, Hsieh, Chen, & Chiang, 2013). As an acute phase protein with elevated plasma levels during inflammation, CRP is an inflammatory biomarker and is primarily produced by liver hepatocytes in response to signaling by the cytokine, IL-6. CRP has further been shown to function as a marker to predict cardiovascular disease (Abd et al., 2011; Bisoendial, Boekholdt, Vergeer, Stroes, & Kastelein, 2010). In the liver, IL-6 is able to complex with the membrane-bound interleukin-6 receptor (IL-6R), which is expressed on hepatocytes and other liver cells (Schmidt-Arras & Rose-John, 2016). This IL-6/IL-6R complex induces downstream signaling, including the production of acute phase proteins such as CRP.

The molecular effects of omega-3 PUFAs on membrane properties remain unclear, therefore we recently employed coarse-grained molecular dynamics computer simulations to examine changes in the biophysical properties of model cell membranes in response to varying physiological compositions of omega-3 PUFAs. These simulations were used to assess the consequences of incorporation of these fatty acids on membrane packing, water permeation, and lipid dynamics.

5.1 Computational methods

We employed molecular dynamics simulations to determine the impact of omega-3 PUFAs on the properties of multicomponent lipid bilayers. We created four simulation systems with membrane concentrations of omega-3 PUFAs ranging from 1.6% to 6.0%. These concentrations are within reasonable agreement with physiological compositions of omega-3 PUFAs in both native and supplemented forms measured using various methods and in different cell types (Balogun, Albert, Ford, Brown, & Cheema, 2013; Browning et al., 2012; Puri et al., 2007). Indeed, in an immune function study, dietary EPA was found to be incorporated into cellular lipids in a dose-dependent manner (Rees et al., 2006). We decided on a composition of 1.6% omega-3 PUFAs as a baseline value, with each subsequent increase in concentration representing additional supplementation levels, as described in Table 1.

Table 1 Omega-3 PUFA concentrations in our multicomponent bilayer simulation systems.

Simulation system	Description	Omega-3 PUFA percentage
Native	Approximate physiological concentration	1.6
Supp1	Low supplementation	2.5
Supp2	Medium supplementation	3.2
Supp3	High supplementation	6.0

We created multicomponent bilayer systems containing phospholipids, sphingomyelin and 30% cholesterol. Of the remaining 70% lipid concentration, 30% was 1-palmitoyl-2-oleoylphosphatidylcholine (POPC), 26% N-stearoyl-D-erythro-sphingosylphosphorylcholine (sphingomyelin), 26% 1,2-bis(diphenylphosphino)ethane (DPPE), and 18% dipalmitoylphosphatidylcholine (DPPC). Into these multicomponent systems, we inserted the omega-3 PUFA concentrations listed in Table 1. We also created an additional system without incorporated omega-3 PUFAs to serve as a control. The components of our systems were chosen to reflect the major lipid constituents in liver plasma membranes (Ray, Skipski, Barclay, Essner, & Archibald, 1969).

Coarse-grained topologies for each membrane component were used based on the MARTINI force field (Marrink, De Vries, & Mark, 2004; Marrink, Risselada, Yefimov, Tieleman, & De Vries, 2007). In this force field, coarse-grained interaction sites for each molecule are created by grouping approximately four heavy atoms together and assigning properties to each site that preserve the physicochemical characteristics of the original molecule. This force field is parameterized against atomistic simulations and experimental data. Since four heavy atoms are generally grouped together to form one interaction site in the MARTINI force field, the structural differences between the omega-3 PUFAs DHA (22 carbon atoms) and EPA (20 carbon atoms) are indistinguishable. We therefore used the MARTINI lipid topology for DHA to represent combined membrane concentrations of both fatty acids. Molecular dynamics simulations were performed using the GROMACS simulation package (Abraham et al., 2015). Graphics were rendered using the Visual Molecular Dynamics program (Humphrey, Dalke, & Schulten, 1996). Each simulation system consisted of a central, multicomponent lipid bilayer surrounded on both sides by coarse-grained water particles, each of which corresponded to four real water molecules. Each system was built using the INSert membrANE (Insane.py) python script (Wassenaar, Ingólfsson, Böckmann, Tieleman, & Marrink, 2015). Snapshots of equilibrated simulations are shown in Fig. 2, corresponding to the native (1.6% omega-3 PUFA) simulation system (Fig. 2A) and high (6%) supplementation system (Fig. 2B). All the membrane lipids are represented as transparent spheres for clarity, except the omega-3 PUFAs.

5.2 Effects on bilayer membrane properties

In our simulations, we found that increasing concentrations of incorporated omega-3 PUFAs induced a dose-dependent lateral expansion of the bilayer

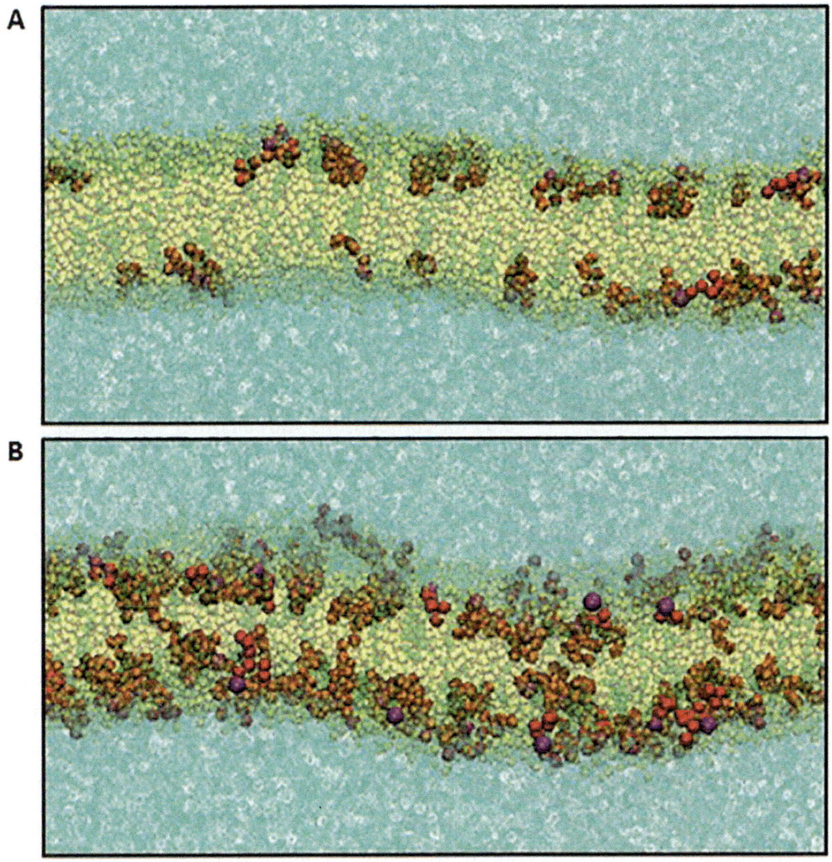

Fig. 2 Snapshots of simulation cells with lipid bilayers in water containing (A) a native concentration of omega-3 PUFAs (1.6%) and (B) a supplemented concentration (6%). The carboxylic acid group of the omega-3 PUFA molecules are represented with purple coarse-grained spheres and the hydrocarbon chains in red. Cholesterol molecules are depicted as green transparent spheres and POPC, DPPC, DPPE, and sphingomyelin as yellow transparent spheres. Water is depicted in cyan solvent representation.

membrane at constant surface pressure, as shown in Fig. 3A. This increase in lateral area is consistent with previous studies that show membranes containing phospholipids with DHA tails experience increased area per lipid values (Huber, Rajamoorthi, Kurze, Beyer, & Brown, 2002). A 5.4% increase in area per lipid was observed between the system containing no omega-3 PUFAs (control) and that containing 6% supplementation (linear increase from 46.2 to 48.7 Å). This increase in bilayer cross-sectional area was accompanied by overall bilayer thinning in the direction normal to the water/lipid interface, for most systems. The thickness of the bilayer was measured as the distance between the POPC head groups in each leaflet.

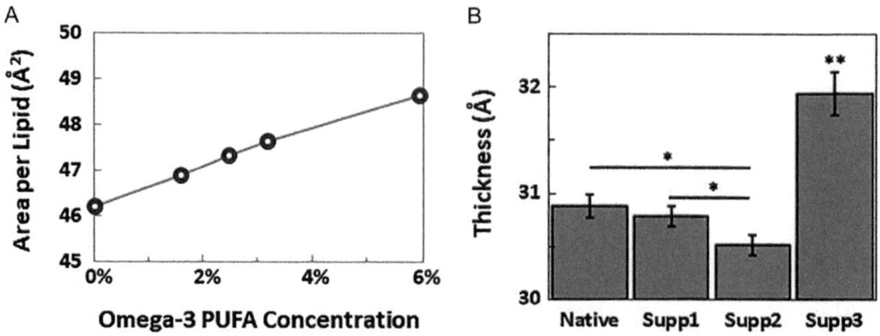

Fig. 3 (A) Lateral area per lipid values plotted against omega-3 PUFA concentrations. (B) Bilayer thickness values for systems with increasing omega-3 PUFA concentrations (Thickness data are presented as mean ± SEM and *P*-values calculated using unpaired *t*-test. $^*P < 0.05$; $^{**}P < 0.05$ for comparisons with all other conditions).

However, as shown in Fig. 3B, for the 6% supplementation system, a significant 3.5% increase in membrane thickness was observed when compared to the native (1.6%) system.

The lateral expansion of the membrane impacted the membrane lipid hydrocarbon chains (tails) that formed the hydrophobic core of the bilayer, resulting in changes in packing and tail conformation. The degree of membrane packing was determined by assessing the orientation and alignment of the lipid tails with respect to the normal to the water/lipid interface, using the orientational order parameter (Vermeer, De Groot, Réat, Milon, & Czaplicki, 2007), P_2 shown in Eq. (1). The order parameter is defined as an ensemble and time-averaged measure of alignment for each coarse-grained inter-particle pseudobond, with the time-varying θ being the angle between the pseudobond and the bilayer normal.

$$P_2 = -\left\langle \frac{3\cos^2\theta - 1}{2} \right\rangle \tag{1}$$

Decreasing P_2 values signify a decrease in lipid packing, or an increase in membrane fluidity. The presence of increasing concentrations of omega-3 PUFAs in our simulation systems resulted in a decrease in the order parameters of the lipid tail pseudobonds of all phospholipids and sphingomyelin present. This is shown in Fig. 4B (along with a coarse-grained representation of the phospholipid (Fig. 4A)) as time and ensemble averaged order parameters for each of the four pseudobonds in the sn-2 tail of POPC, specifically. Similar trends were observed for the sn-1 tails of POPC and the tails of the other membrane lipids (data not shown). Notably, the

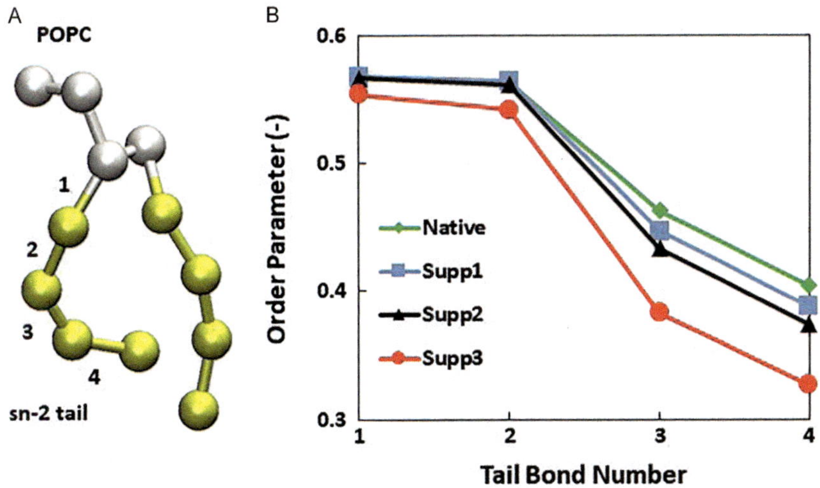

Fig. 4 Computational analysis of lipid tail order in omega-3 PUFA-containing bilayers. (A) Coarse-grained structure of POPC with head group and glycerol backbone represented in white and tails shown in yellow. Pseudobond numbers for the sn-2 tail are indexed. (B) Orientational order parameter profiles for pseudobonds between coarse-grained sites in the sn-2 tails of POPC (indexed in A) for the simulation systems containing increasing concentrations of omega-3 PUFAs (Data are shown as time and ensemble averaged mean $\pm$ SEM and *P*-values calculated using unpaired *t*-test $P < 0.01$. Error bars are small enough to fit with the marker for each point).

disordering effect is most pronounced for the simulation system containing 6% omega-3 PUFAs, indicating a pronounced increase in fluidity of the bilayer at that concentration. Increased disorder may also have consequences on membrane properties such as permeability and diffusion of different components.

The increase in membrane lateral area and disruption of lipid packing was accompanied by changes in the orientation of omega-3 PUFAs within the bilayer. This was quantified by time and ensemble averaged tilt angles with respect to the bilayer normal, shown in Fig. 5A. Significant increases in the average tilt angles of omega-3 PUFAs were observed with increasing concentrations within the bilayer membrane, ranging from $116.1 \pm 0.3°$ for the native system containing 1.6% omega-3 PUFAs, to $118.9 \pm 0.3°$ for the 6% supplementation system (an increase in average tilt angle of 3% from that of the native system). The schematic in Fig. 5B depicts the likely localization of omega-3 PUFAs within a lipid bilayer based on the calculated average tilt angle for the 6% supplementation system, and their potential effects on membrane packing.

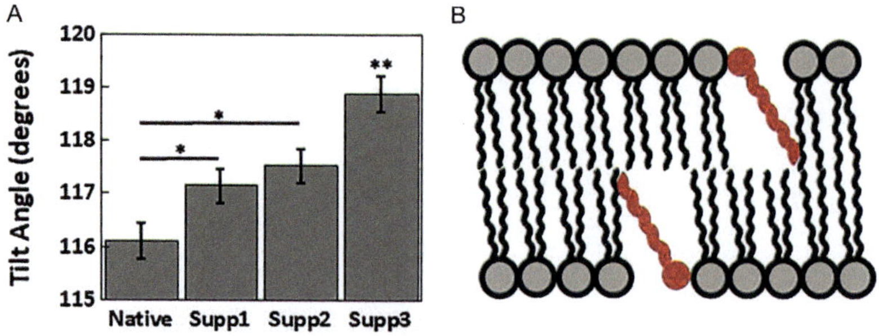

Fig. 5 (A) Average tail tilt angles of omega-3 PUFA molecules for simulation systems containing increasing concentrations of omega-3 PUFAs. (B) Schematic of the localization of omega-3 PUFA molecules within lipid bilayers based on the calculated average tilt angles in (A), and their effects on membrane packing (Data are presented as mean ± SEM and *P*-values calculated using unpaired *t*-test. *$P<0.05$; **$P<0.05$ for comparisons with all other conditions).

The presence of 6% omega-3 PUFAs in our simulations systems increased the hydration of the simulated bilayer membrane. Simulation snapshots in Fig. 6A focus in on the coarse-grained water particles near the lipid/water interface, with all the lipids hidden for clarity. The approximate positions of the lipid head groups in each leaflet are indicated by arrows below each snapshot. From these images, we observe that the lipid/water interface in the 6% supplementation system is less well defined than that of the 1.6% native system. We also observe an increase in water entry into the membrane toward the hydrophobic core, represented by the decrease in distance between the water/lipid interfaces of the two leaflets in the 6% (supp3) system. This decrease in distance cannot be accounted for by membrane thinning since we showed a significant increase in bilayer thickness for the 6% (supp3) system in Fig. 3B. This water permeation phenomenon is further highlighted in the normalized water density profiles shown in Fig. 6B, depicting the interior of one leaflet of the bilayer for each system. Here, we see that incorporation of 6% omega-3 PUFAs results in a detectable increase in water density inside the hydrophobic core of the bilayer, as compared to the native 1.6% system.

One of the most interesting results of the multicomponent bilayer simulations we performed was the effect of the high supplementation (6%) level of omega-3 PUFAs on cholesterol distribution within the membrane. In Fig. 7, we present two-dimensional time-averaged density maps of cholesterol in native (1.6%) and supp3 (6%) simulation systems, based on the center of mass of each cholesterol molecule. In the native system,

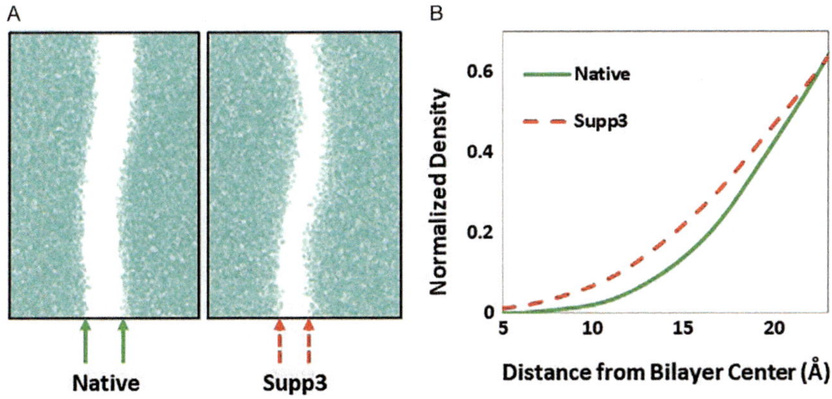

Fig. 6 Computational analysis of water permeation into omega-3 PUFA-containing bilayers. (A) Simulation snapshots of water penetration into bilayer containing 1.6% (native) and 6% (supp3) omega-3 PUFAs. Coarse-grained water particles are depicted in cyan solvent representation. Approximate lipid head group positions are indicated by arrows. (B) Water density profiles for one leaflet of native and supp3 bilayers containing 1.6% and 6% omega-3 PUFAs, respectively. The bilayer center is located at 0 Å.

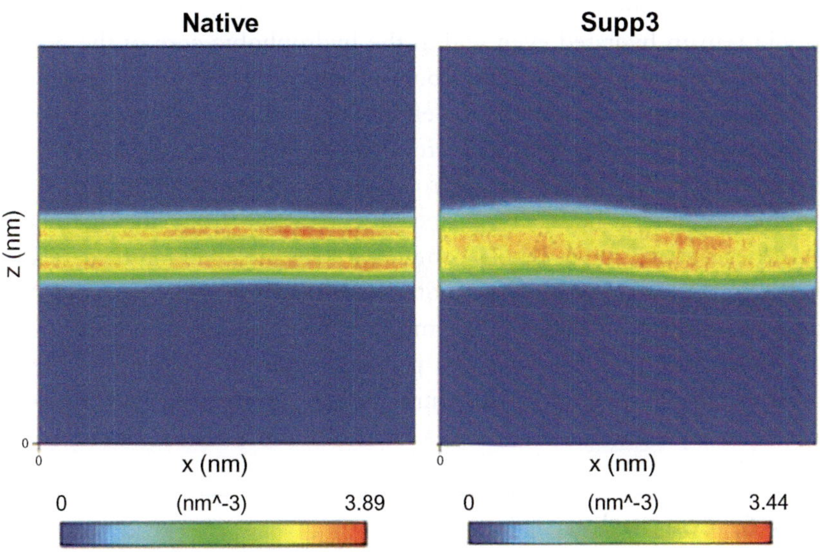

Fig. 7 Impact of omega-3 PUFA supplementation on cholesterol density. Two-dimensional density maps along the plane of the bilayer normal (XZ) for cholesterol in native (1.6%) and supp3 (6%) omega-3 PUFA simulation systems.

we observe, based on the two distinct regions of high density (red color), that cholesterol is distributed as expected between the two leaflets. In the 6% supplemented system, on the other hand, the distribution of cholesterol density is more diffuse throughout the core of the membrane, indicating that cholesterol molecules have greater mobility within the hydrophobic core of the bilayer.

Cholesterol is well known to reside just below the phospholipid head groups in lipid membranes, with the hydrated hydroxyl group of cholesterol forming hydrogen bonds with the carboxyl oxygen atoms of the phospholipid glycerol backbones (Léonard et al., 2001). Cholesterol is also well known to decrease membrane fluidity and increase lipid ordering (Ayee & Levitan, 2016), with cholesterol depletion causing decreased lipid packing. Our simulation results suggest that our high supplementation level of omega-3 PUFAs (6% incorporation in the bilayer membrane) causes cholesterol to redistribute within the bilayer, thereby reducing its ordering effect, allowing lipid tails to move more freely within the hydrophobic core, and increasing overall membrane fluidity. This cholesterol relocation may induce the increased water permeation we observed in our simulations of 6% supplementation, since the hydrophilic hydroxyl group of cholesterol would remain hydrated even within the hydrophobic core of the bilayer. The effects of cholesterol redistribution that we observed in our model membranes are reminiscent of the effects induced by cholesterol depletion on membrane fluidity and lipid order (Ayee & Levitan, 2016). Previous research has indicated that cholesterol depletion may play a role in shedding of the membrane-bound IL-6 receptor (IL-6R) from the membrane (Matthews et al., 2003). IL-6R complexes with the cytokine IL-6, leading to the production of acute phase proteins such as CRP. Further studies need to be performed to elucidate the mechanisms through which changes in membrane structure, fluidity, and packing induced by omega-3 PUFAs influence IL-6R shedding and downregulate the production CRP and other acute phase proteins.

6. Conclusion

In this chapter we have presented a review of current evidence for beneficial health effects of long-chain omega-3 polyunsaturated fatty acids and explored some molecular level effects of these fatty acids on cell bilayer membrane structure and function. We include a summary of some of the current hypothesized mechanisms of action of these fatty acids in alleviating cardiovascular disease. Our molecular level analyses of the membrane modulatory effects of omega-3 PUFAs indicate that supplementation of these fatty acids, which leads to membrane incorporation, may result in significant perturbation of the lipid packing, water permeation, and overall membrane structure, in a dose-dependent fashion. All these observations lead us to conclude that the lipid bilayer membrane becomes increasingly fluidized with the addition of omega-3 PUFAs. Further studies are needed to investigate the exact mechanisms through which omega-3 PUFA-remodeled and fluidized membranes regulate signaling pathways and lead to improved health outcomes.

Acknowledgment

We are very grateful to Dr. Nicolas A. Barbera for the critical reading of this manuscript.

References

Abd, T. T., Eapen, D. J., Bajpai, A., Goyal, A., Dollar, A., & Sperling, L. (2011). The role of C-reactive protein as a risk predictor of coronary atherosclerosis: Implications from the JUPITER trial. *Current Atherosclerosis Reports*, *13*(2), 154–161.

Abeywardena, M. Y., & Head, R. J. (2001). Longchain n-3 polyunsaturated fatty acids and blood vessel function. *Cardiovascular Research*, *52*(3), 361–371.

Abraham, M. J., Murtola, T., Schulz, R., Páll, S., Smith, J. C., Hess, B., et al. (2015). GROMACS: High performance molecular simulations through multi-level parallelism from laptops to supercomputers. *SoftwareX*, *1*, 19–25.

ACD/ChemSketch. (2020). *Computer software*. Advanced Chemistry Development, Inc. (Version 2019.2.1). www.acdlabs.com.

Adili, R., Hawley, M., & Holinstat, M. (2019). Regulation of platelet function and thrombosis by omega-3 and omega-6 polyunsaturated fatty acids. *Postaglandins & Other Lipid Mediators*, *139*, 10–18.

Ajmal, M. R., Yaccha, M., Malik, M. A., Rabbani, M., Ahmad, I., Isalm, N., et al. (2014). Prevalence of nonalcoholic fatty liver disease (NAFLD) in patients of cardiovascular diseases and its association with hs-CRP and TNF-α. *Indian Heart Journal*, *66*(6), 574–579.

Alves-Bezerra, M., & Cohen, D. E. (2017). Triglyceride metabolism in the liver. *Comprehensive Physiology*, *8*(1), 1–8. https://doi.org/10.1002/cphy.c170012.

Anand, R. G., Alkadri, M., Lavie, C. J., & Milani, R. V. (2008). The role of fish oil in arrhythmia prevention. *Journal of Cardiopulmonary Rehabilitation and Prevention*, *28*(2), 92–98. https://doi.org/10.1097/01.HCR.0000314202.09676.f0.

Aquilani, R., La Rovere, M. T., Corbellini, D., Pasini, E., Verri, M., Barbieri, A., et al. (2017). Plasma amino acid abnormalities in chronic heart failure. Mechanisms, potential risks and targets in human myocardium metabolism. *Nutrients*, *9*(11), 1251.

Ayee, M. A. A., LeMaster, E., Shentu, T. P., Singh, D. K., Barbera, N., Soni, D., et al. (2017). Molecular-scale biophysical modulation of an endothelial membrane by oxidized phospholipids. *Biophysical Journal*, *112*(2), 325–338.

Ayee, M. A., & Levitan, I. (2016). Paradoxical impact of cholesterol on lipid packing and cell stiffness. *Frontiers in Bioscience (Landmark Edition)*, *21*, 1245–1259.

Bagge, A., Schott, U., & Kander, T. (2018). High-dose omega-3 fatty acids have no effect on platelet aggregation on coagulation measured with static and flow-based aggregation instruments and Sonoclot; an observational study in healthy volunteers. *Scandinavian Journal of Clinical and Laboratory Investigation*, *78*(7–8), 539–545.

Baker, E. J., Valenzuela, C. A., De Souza, C. O., Yaqoob, P., Miles, E. A., & Calder, P. C. (2020). Comparative anti-inflammatory effects of plant-and marine-derived omega-3 fatty acids explored in an endothelial cell line. *Biochimica et Biophysica Acta (BBA)-Molecular and Cell Biology of Lipids*, *1865*(6), 158662.

Balogun, K. A., Albert, C. J., Ford, D. A., Brown, R. J., & Cheema, S. K. (2013). Dietary omega-3 polyunsaturated fatty acids alter the fatty acid composition of hepatic and plasma bioactive lipids in C57BL/6 mice: A lipidomic approach. *PLoS One*, *8*(11), e82399.

Barbalho, S. M., de Goulart, R. A., Quesada, K., Bechara, M. D., & de de Carvalho, A. C. A. (2016). Inflammatory bowel disease: Can omega-3 fatty acids really help? *Annals of Gastroenterology*, *29*(1), 37–43.

Bayard, M., Holt, J. D., & Boroughs, E. (2006). Nonalcoholic fatty liver disease. *American Family Physician*, *73*(11), 1961–1968.

Bentzon, J. F., Otsuka, F., Virmani, R., & Falk, E. (2014). Mechanisms of plaque formation and rupture. *Circulation Research*, *114*(12), 1852–1866.

Bernardi, J. R., de Escobar, R. S., Ferreira, C. F., & Silveira, P. P. (2012). Fetal and neonatal levels of omega-3: Effects on neurodevelopment, nutrition, and growth. *The Scientific World Journal*, *2012*, 1–8. https://doi.org/10.1100/2012/202473. 202473.

Bisoendial, R. J., Boekholdt, S. M., Vergeer, M., Stroes, E. S., & Kastelein, J. J. (2010). C-reactive protein is a mediator of cardiovascular disease. *European Heart Journal*, *31*(17), 2087–2091.

Blum, S., Aeschbacher, S., Meyre, P., Zwimpfer, L., Reichlin, T., Beer, J. H., et al. (2019). Incidence and predictors of atrial fibrillation progression. *Journal of the American Heart Association*, *8*(20), e012554. https://doi.org/10.1161/JAHA.119.012554.

Bonora, E., & Targher, G. (2012). Increased risk of cardiovascular disease and chronic kidney disease in NAFLD. *Nature Reviews Gastroenterology & Hepatology*, *9*(7), 372.

Brigham, E. P., Woo, H., McCormack, M., Rice, J., Koehler, K., Vulcain, T., et al. (2019). Omega-3 and omega-6 intake modifies asthma severity and response to indoor air pollution in children. *American Journal of Respiratory and Critical Care Medicine*, *199*(12), 1478–1486.

Browning, L. M., Walker, C. G., Mander, A. P., West, A. L., Madden, J., Gambell, J. M., et al. (2012). Incorporation of eicosapentaenoic acid and docosahexaenoic acids into lipid pools when given as supplements providing doses equivalent to typical intakes of oily fish. *The American Journal of Clinical Nutrition*, *96*(4), 748–758.

Byrne, C. D., & Targher, G. (2015). NAFLD: A multisystem disease. *Journal of Hepatology*, *62*(1), S47–S64.

Calder, P. C. (2012). Omega-3 polyunsaturated fatty acids and inflammatory processes: Nutrition or pharmacology? *British Journal of Clinical Pharmacology*, *75*(3), 645–662.

Canhada, S., Castro, K., Perry, I. S., & Luft, V. C. (2018). Omega-3 fatty acids' supplementation in Alzheimer's disease: A systematic review. *Nutritional Neuroscience*, *21*(8), 529–538. https://doi.org/10.1080/1028415X.2017.1321813.

Carr, S., Farb, A., Pearce, W., Virmani, R., & Yao, J. S. T. (1996). Atherosclerotic plaque rupture in symptomatic carotid artery stenosis. *Journal of Vascular Surgery*, *23*(5), 755–766.

Chandrasekharan, K., & Alazawi, W. (2019). Genetics of non-alcoholic fatty liver and cardiovascular disease: Implications for therapy? *Frontiers in Pharmacology*, *10*, 1413.

Chang, C. L., & Deckelbaum, R. J. (2013). Omega-3 fatty acids: Mechanisms underlying "protective effects" in atherosclerosis. *Current Opinion in Lipidology*, *24*(4), 345–350. https://doi.org/10.1097/MOL.0b013e3283616364.

Cole, G. M., Ma, Q.-L., & Frautschy, S. A. (2009). Omega-3 fatty acids and dementia. *Prostaglandins, Leukotrienes and Essential Fatty Acids*, *81*(2), 213–221. https://doi.org/10.1016/j.plefa.2009.05.015.

Covington, M. B. (2004). Omega-3 fatty acids. *American Family Physician*, *70*(1), 133–140.

Dangardt, F., Osika, W., Chen, Y., Nilsson, U., Gan, L.-M., Gronowitz, E., et al. (2010). Omega-3 fatty acid supplementation improves vascular function and reduces inflammation in obese adolescents. *Atherosclerosis*, *212*(2), 580–585.

de Oliveira Otto, M. C., Wu, J. H. Y., Baylin, A., Vaidya, D., Rich, S. S., Tsai, M. Y., et al. (2013). Circulating and dietary omega-3 and omega-6 polyunsaturated fatty acids and incidence of CVD in the multi-ethnic study of atherosclerosis. *Journal of the American Heart Association*, *2*(6), e000506.

DiNicolantonio, J. J., & O'Keefe, J. H. (2018). Importance of maintaining a low omega–6/omega–3 ratio for reducing inflammation. *Open Heart*, *5*(2), e000946.

DiNoclantonio, J. J., & O'Keefe, J. H. (2020). The benefits of omega-3 fats for stabilizing and remodeling atherosclerosis. *Missouri Medicine*, *117*(1), 65–69.

Domanski, J. P., Park, S. J., & Harrison, S. A. (2012). Cardiovascular disease and nonalcoholic fatty liver disease: Does histologic severity matter? *Journal of Clinical Gastroenterology*, *46*(5), 427–430.

Drummond, G. R., Vinh, A., Guzik, T. J., & Sobey, C. G. (2019). Immune mechanisms of hypertension. *Nature Reviews Immunology*, *19*(8), 517–532.

Dyall, S. C. (2015). Long-chain omega-3 fatty acids and the brain: A review of the independent and shared effects of EPA, DPA and DHA. *Frontiers in Aging Neuroscience*, 7(52), 1–15. https://doi.org/10.3389/fnagi.2015.00052.

Fielding, B. (2011). Tracing the fate of dietary fatty acids: Metabolic studies of postprandial lipaemia in human subjects. *Proceedings of the Nutrition Society*, *70*(3), 342–350.

Foroughi, M., Maghsoudi, Z., Khayyatzadeh, S., Ghiasvand, R., Askari, G., & Iraj, B. (2016). Relationship between non-alcoholic fatty liver disease and inflammation in patients with non-alcoholic fatty liver. *Advanced Biomedical Research*, *5*(1), 28. https://doi.org/10.4103/2277-9175.176368.

Gao, L., Cao, J., Mao, Q., Lu, X., Zhou, X., & Fan, L. (2012). Influence of omega-3 polyunsaturated fatty acid—Supplementation of platelet aggregation in humans: A meta-analysis of randomized controlled trials. *Atherosclerosis*, *226*(2), 328–334.

Goodfellow, J., Bellamy, M. F., Ramsey, M. W., Jones, C. J., & Lewis, M. J. (2000). Dietary supplementation with marine omega-3 fatty acids improve systemic large artery endothelial function in subjects with hypercholesterolemia. *Journal of the American College of Cardiology*, *35*(2), 265–270.

Grassi, G., Quarti-Trevano, F., & Esler, M. D. (2019). Sympathetic activation in congestive heart failure: An updated overview. *Heart Failure Reviews*, *2019*, 1–10.

Green, C. J., Pramfalk, C., Charlton, C. A., Gunn, P. J., Cornfield, T., Pavlides, M., et al. (2020). Hepatic de novo lipogenesis is suppressed and fat oxidation is increased by omega-3 fatty acids at the expense of glucose metabolism. *BMJ Open Diabetes Research & Care*, *8*(1), e000871. https://doi.org/10.1136/bmjdrc-2019-000871.

Gutiérrez, S., Svahn, S. L., & Johansson, M. E. (2019). Effects of omega-3 fatty acids on immune cells. *International Journal of Molecular Sciences*, *20*(20), 5028.

Harchaoui, K. E. L., Visser, M. E., Kastelein, J. J. P., Stroes, E. S., & Dallinga-Thie, G. M. (2009). Triglycerides and cardiovascular risk. *Current Cardiology Reviews*, *5*(3), 216–222. https://doi.org/10.2174/157340309788970315.

Harris, W. S., & Bulchandani, D. (2006). Why do omega-3 fatty acids lower serum triglycerides? *Current Opinion in Lipidology*, *17*(4), 387–393.

Huber, T., Rajamoorthi, K., Kurze, V. F., Beyer, K., & Brown, M. F. (2002). Structure of docosahexaenoic acid-containing phospholipid bilayers as studied by 2H NMR and molecular dynamics simulations. *Journal of the American Chemical Society*, *124*(2), 298–309.

Humphrey, W., Dalke, A., & Schulten, K. (1996). VMD: Visual molecular dynamics. *Journal of Molecular Graphics*, *14*(1), 33–38.

Jin, H., Lyon, A. R., & Akar, F. G. (2008). Arrhythmia mechanisms in the failing heart. *Pacing and Clinical Electrophysiology*, *31*(8), 1048–1056.

Jump, D. B., Lytle, K. A., Depner, C. M., & Tripathy, S. (2018). Omega-3 polyunsaturated fatty acids as a treatment strategy for nonalcoholic fatty liver disease. *Pharmacology & Therapeutics*, *181*, 108–125.

Kabisch, M., Ruckes, C., Seibert-Grafe, M., & Blettner, M. (2011). Randomized controlled trials. *Deutsches Ärzteblatt International*, *108*(39), 663–668. https://doi.org/10.3238/arztebl.2011.0663.

Khan, S. U., Khan, M. U., Riaz, H., Valavoor, S., Zhao, D., Vaughan, L., et al. (2019). Effects of nutritional supplements and dietary interventions on cardiovascular outcomes: An umbrella review and evidence map. *Annals of Internal Medicine*, *171*(3), 190–198.

Kochanek, K. D., Murphy, S. L., Xu, J., & Arias, E. (2019). Deaths: Final data for 2017 (cdc:79486). Vol. 68. (9). https://stacks.cdc.gov/view/cdc/79486.

Kromhout, D., Yasuda, S., Geleijinse, J. M., & Shimokawa, H. (2011). Fish oil and omega-3 fatty acids in cardiovascular disease: Do they really work? *European Heart Journal*, *33*(4), 436–443.

Laviano, A., Rianda, S., Molfino, A., & Rossi Fanelli, F. (2013). Omega-3 fatty acids in cancer. *Current Opinion in Clinical Nutrition and Metabolic Care*, *16*(2), 156–161. https://doi.org/10.1097/MCO.0b013e32835d2d99.

Léonard, A., Escrive, C., Laguerre, M., Pebay-Peyroula, E., Néri, W., Pott, T., et al. (2001). Location of cholesterol in DMPC membranes. A comparative study by neutron diffraction and molecular mechanics simulation. *Langmuir*, *17*(6), 2019–2030.

Libby, P., Ridker, P. M., & Maseri, A. (2002). Inflammation and atherosclerosis. *Circulation*, *105*(9), 1135–1143. https://doi.org/10.1161/hc0902.104353.

Mann, N. J., O'Connell, S. L., Baldwin, K. M., Singh, I., & Meyer, B. J. (2010). Effects of seal oil and tuna-fish oil on platelet parameters and plasma lipid levels in healthy subjects. *Lipids*, *45*(8), 669–681.

Marchesini, G., Marzocchi, R., Agostini, F., & Bugianesi, E. (2005). Nonalcoholic fatty liver disease and the metabolic syndrome. *Current Opinion in Lipidology*, *16*(4), 421–427. https://journals.lww.com/co-lipidology/Fulltext/2005/08000/Nonalcoholic_fatty_liver_disease_and_the_metabolic.6.aspx.

Marrink, S. J., De Vries, A. H., & Mark, A. E. (2004). Coarse grained model for semiquantitative lipid simulations. *The Journal of Physical Chemistry B*, *108*(2), 750–760.

Marrink, S. J., Risselada, H. J., Yefimov, S., Tieleman, D. P., & De Vries, A. H. (2007). The MARTINI force field: Coarse grained model for biomolecular simulations. *The Journal of Physical Chemistry B*, *111*(27), 7812–7824.

Martino, A., Pezzi, L., Magnano, R., Salustri, E., Penco, M., & Calo', L. (2016). Omega 3 and atrial fibrillation: Where are we? *World Journal of Cardiology*, *8*(2), 114–119. https://doi.org/10.4330/wjc.v8.i2.114.

Matthews, V., Schuster, B., Schütze, S., Bussmeyer, I., Ludwig, A., Hundhausen, C., et al. (2003). Cellular cholesterol depletion triggers shedding of the human interleukin-6 receptor by ADAM10 and ADAM17 (TACE). *Journal of Biological Chemistry*, *278*(40), 38829–38839.

Miles, E. A., Banerjee, T., Dooper, M. M., M'Rabet, L., Graus, Y. M., & Calder, P. C. (2004). The influence of different combinations of γ-linolenic acid, stearidonic acid and EPA on immune function in healthy young male subjects. *British Journal of Nutrition*, *91*(6), 893–903.

Monahan, K. D., Wilson, T. E., & Ray, C. A. (2004). Omega-3 fatty acid supplementation augments sympathetic nerve activity responses to physiological stressors in humans. *Hypertension*, *44*(5), 732–738. https://doi.org/10.1161/01.HYP.0000145292.38579.f4.

Moreno, C., Macías, Á., Prieto, Á., De La Cruz, A., González, T., & Valenzuela, C. (2012). Effects of n − 3 polyunsaturated fatty acids on cardiac ion channels. *Frontiers in Physiology*, *3*, 245.

Morillo, C. A., Banerjee, A., Perel, P., Wood, D., & Jouven, X. (2017). Atrial fibrillation: The current epidemic. *Journal of Geriatric Cardiology: JGC*, *14*(3), 195–203. https://doi.org/10.11909/j.issn.1671-5411.2017.03.011.

Morotti, S., & Grandi, E. (2019). Quantitative systems models illuminate arrhythmia mechanisms in heart failure: Role of the Na+-Ca2+-Ca2+/calmodulin-dependent protein kinase II-reactive oxygen species feedback. *Wiley Interdisciplinary Reviews: Systems Biology and Medicine*, *11*(2), e1434.

Mozaffarian, D., Wilson, P. W., & Kannel, W. B. (2008). Beyond established and novel risk factors: Lifestyle risk factors for cardiovascular disease. *Circulation*, *117*(23), 3031–3038.

Natto, Z. S., Yaghmoor, W., Alshaeri, H. K., & Van Dyke, T. E. (2019). Omega-3 fatty acids effects on inflammatory biomarkers and lipid profiles among diabetic and cardiovascular disease patients: A systematic review and meta-analysis. *Scientific Reports*, *9*(1), 1–10.

Neuschwander-Tetri, B. A. (2005). Nonalcoholic steatohepatitis and the metabolic syndrome. *The American Journal of the Medical Sciences*, *330*(6), 326–335.

Nishio, R., Shinke, T., Otake, H., Nakagawa, M., Nagoshi, R., Inoue, T., et al. (2014). Stabilizing effect of combined eicosapentaenoic acid and statin therapy on coronary thin-cap fibroatheroma. *Atherosclerosis*, *234*(1), 114–119.

Oliver, C., & Jahnke, N. (2011). Omega-3 fatty acids for cystic fibrosis. *The Cochrane Database of Systematic Reviews*, *8*, CD002201. https://doi.org/10.1002/14651858.CD002201.pub3.

Peoples, G. E., McLennan, P. L., Howe, P. R., & Groeller, H. (2008). Fish oil reduces heart rate and oxygen consumption during exercise. *Journal of Cardiovascular Pharmacology*, *52*(6), 540–547.

Pepe, S., & McLennan, P. L. (2002). Cardiac membrane fatty acid composition modulates myocardial oxygen consumption and postischemic recovery of contractile function. *Circulation*, *105*(19), 2303–2308.

Pepys, M. B., & Hirschfield, G. M. (2003). C-reactive protein: A critical update. *The Journal of Clinical Investigation*, *111*(12), 1805–1812.

Poreba, M., Mostowik, M., Siniarski, A., Golebiowska-Wiatrak, R., Piotr Malinowski, K., Haberka, M., et al. (2017). Treatment with high-dose n-3 PUFAs has no effect on platelet function, coagulation, metabolic status or inflammation in patients with atherosclerosis and type 2 diabetes. *Cardiovascular Diabetology*, *16*(50), 50.

Puri, P., Baillie, R. A., Wiest, M. M., Mirshahi, F., Choudhury, J., Cheung, O., et al. (2007). A lipidomic analysis of nonalcoholic fatty liver disease. *Hepatology*, *46*(4), 1081–1090.

Rajaei, E., Mowla, K., Ghorbani, A., Bahadoram, S., Bahadoram, M., & Dargahi-Malamir, M. (2016). The effect of omega-3 fatty acids in patients with active rheumatoid arthritis receiving DMARDs therapy: Double-blind randomized controlled trial. *Global Journal of Health Science*, *8*(7), 18–25. https://doi.org/10.5539/gjhs.v8n7p18.

Ray, T. K., Skipski, V. P., Barclay, M., Essner, E., & Archibald, F. M. (1969). Lipid composition of rat liver plasma membranes. *Journal of Biological Chemistry*, *244*(20), 5528–5536.

Rees, D., Miles, E. A., Banerjee, T., Wells, S. J., Roynette, C. E., Wahle, K. W., et al. (2006). Dose-related effects of eicosapentaenoic acid on innate immune function in healthy humans: A comparison of young and older men. *The American Journal of Clinical Nutrition*, *83*(2), 331–342.

Richardson, A. J. (2006). Omega-3 fatty acids in ADHD and related neurodevelopmental disorders. *International Review of Psychiatry (Abingdon, England)*, *18*(2), 155–172. https://doi.org/10.1080/09540260600583031.

Ridker, P. M., Hennekens, C. H., Buring, J. E., & Rifai, N. (2000). C-reactive protein and other markers of inflammation in the prediction of cardiovascular disease in women. *The New England Journal of Medicine, 342*(12), 836–843. https://doi.org/10.1056/NEJM200003233421202.

Sanyal, A. J. (2002). AGA technical review on nonalcoholic fatty liver disease. *Gastroenterology, 123*(5), 1705–1725.

Schmidt-Arras, D., & Rose-John, S. (2016). IL-6 pathway in the liver: From physiopathology to therapy. *Journal of Hepatology, 64*(6), 1403–1415.

Scorletti, E., & Byrne, C. D. (2018). Omega-3 fatty acids and non-alcoholic fatty liver disease: Evidence of efficacy and mechanism of action. *Molecular Aspects of Medicine, 64*, 135–146.

Sepidarkish, M., Morvaridzadeh, M., Akbari-Fakhrabadi, M., Almasi-Hashiani, A., Rezaeinejad, M., & Heshmati, J. (2019). Effect of omega-3 fatty acid plus vitamin E Co-Supplementation on lipid profile: A systematic review and meta-analysis. *Diabetes and Metabolic Syndrome: Clinical Research and Reviews, 13*(2), 1649–1656.

Shentu, T. P., Singh, D. K., Oh, M.-J., Sun, S., Sadaat, L., Makino, A., et al. (2012). The role of oxysterols in control of endothelial stiffness. *Journal of Lipid Research, 53*(7), 1348–1358.

Singh, M. (2005). Essential fatty acids, DHA, and human brain. *Indian Journal of Pediatrics, 72*(3), 239–242.

Skulas-Ray, A. C., Wilson, P. W. F., Harris, W. S., Brinton, E. A., Kris-Etherton, P. M., Richter, C. K., et al. (2019). Omega-3 fatty acids for the management of hypertriglyceridemia: A science advisory from the American Heart Association. *Circulation, 140*(12), e673–e691. https://doi.org/10.1161/CIR.0000000000000709.

Spooner, M. H., & Jump, D. B. (2019). Omega-3 fatty acids and nonalcoholic fatty liver disease in adults and children: Where do we stand? *Current Opinion in Clinical Nutrition and Metabolic Care, 22*(2), 103–110.

Stahl, E. P., Dhindsa, D. S., Lee, S. K., Sandesara, P. B., Chalasani, N. P., & Sperling, L. S. (2019). Nonalcoholic fatty liver disease and the heart: JACC state-of-the-art review. *Journal of the American College of Cardiology, 73*(8), 948–963.

Tousoulis, D., Plastiras, A., Siasos, G., Oikonomou, E., Verveniotis, A., Kokkou, E., et al. (2014). Omega-3 PUFAs improved endothelial function and arterial stiffness with a parallel antiinflammatory effect in adults with metabolic syndrome. *Atherosclerosis, 232*(1), 10–16. https://doi.org/10.1016/j.atherosclerosis.2013.10.014.

Tsoupras, A., Lordan, R., & Zabetakis, I. (2018). Inflammation, not cholesterol, is a cause of chronic disease. *Nutrients, 10*(5), 604. PubMed https://doi.org/10.3390/nu10050604.

Tur, J., Bibiloni, M., Sureda, A., & Pons, A. (2012). Dietary sources of omega 3 fatty acids: Public health risks and benefits. *British Journal of Nutrition, 107*(S2), S23–S52.

Vemuganti, R. (2019). Stroke: Molecular mechanisms and therapies. *Neuromolecular Medicine, 21*(4), 323–324. https://doi.org/10.1007/s12017-019-08580-4.

Vermeer, L. S., De Groot, B. L., Réat, V., Milon, A., & Czaplicki, J. (2007). Acyl chain order parameter profiles in phospholipid bilayers: Computation from molecular dynamics simulations and comparison with 2 H NMR experiments. *European Biophysics Journal, 36*(8), 919–931.

Wall, R., Ross, R. P., Fitzgerald, G. F., & Stanton, C. (2010). Fatty acids from fish: The anti-inflammatory potential of long-chain omega-3 fatty acids. *Nutrition Reviews, 68*(5), 280–289.

Wang, H., & Daggy, B. P. (2017). The role of fish oil in inflammatory eye diseases. *Biomedicine Hub, 2*(1), 1–12. https://doi.org/10.1159/000455818.

Wang, T.-M., Hsieh, S.-C., Chen, J.-W., & Chiang, A.-N. (2013). Docosahexaenoic acid and eicosapentaenoic acid reduce C-reactive protein expression and STAT3 activation in IL-6-treated HepG2 cells. *Molecular and Cellular Biochemistry, 377*(1–2), 97–106.

Wassenaar, T. A., Ingólfsson, H. I., Böckmann, R. A., Tieleman, D. P., & Marrink, S. J. (2015). Computational lipidomics with insane: A versatile tool for generating custom membranes for molecular simulations. *Journal of Chemical Theory and Computation*, *11*(5), 2144–2155.

Wu Jason, H. Y., Lemaitre, R. N., King, I. B., Xiaoling, S., Sacks, F. M., Rimm, E. B., et al. (2012). Association of plasma phospholipid long-chain omega-3 fatty acids with incident atrial fibrillation in older adults. *Circulation*, *125*(9), 1084–1093. https://doi.org/10.1161/CIRCULATIONAHA.111.062653.

Zanetti, M., Gortan Cappellari, G., Barbetta, D., Semolic, A., & Barazzoni, R. (2017). Omega 3 polyunsaturated fatty acids improve endothelial dysfunction in chronic renal failure: Role of eNOS activation and of oxidative stress. *Nutrients*, *9*(8), 895. https://doi.org/10.3390/nu9080895.

Zehr, K. R., & Walker, M. K. (2018). Omega-3 polyunsaturated fatty acids improve endothelial function in humans at risk for atherosclerosis: A review. *Prostaglandins & Other Lipid Mediators*, *134*, 131–140. https://doi.org/10.1016/j.prostaglandins.2017.07.005.

CHAPTER THREE

Cell membrane mechanics and mechanosensory transduction

Boris Martinac[a,b,*], Yury A. Nikolaev[c], Giulia Silvani[a], Navid Bavi[d], Valentin Romanov[a], Yoshitaka Nakayama[a], Adam D. Martinac[e], Paul Rohde[a], Omid Bavi[f], and Charles D. Cox[a,b]

[a]Molecular Cardiology and Biophysics Division, Victor Chang Cardiac Research Institute, Sydney, Australia
[b]St. Vincent's Clinical School, University of New South Wales, Sydney, Australia
[c]Cellular & Molecular Physiology, Yale University, New Haven, CT, United States
[d]Institute for Biophysical Dynamics, University of Chicago, Chicago, IL, United States
[e]Neuroscience Research Australia and Prince of Wales Clinical School, UNSW, Kensington, Australia
[f]Department of Mechanical and Aerospace Engineering, Shiraz University of Technology, Shiraz, Iran
*Corresponding author: e-mail address: b.martinac@victorchang.edu.au

Contents

Abstract

The rapid progress in mechanobiology has brought together many scientific and engineering disciplines to work hand in hand toward better understanding of the role that mechanical force plays in functioning and evolution of different forms of life. New tools designed by engineers helped to develop new methods and techniques for investigation of mechanical properties of biological cells and tissues. This multidisciplinary approach made it clear that cell mechanics is tightly linked to intracellular signaling pathways, which directly regulate gene expression in response to mechanical stimuli originating outside or inside the cells. Mechanical stimuli act on mechanoreceptors

Current Topics in Membranes, Volume 86
ISSN 1063-5823
https://doi.org/10.1016/bs.ctm.2020.08.002

which convert these stimuli into intracellular signals. In this chapter, we review the current knowledge about cell mechanics and the role cell mechanics plays for the function of mechanosensitive ion channels as a special class of mechanoreceptors functioning as molecular transducers of mechanical stimuli on a millisecond timescale.

1. Introduction

Over the past decade of investigation, biomechanics research has entered an exciting era of exploration where mechanical forces are increasingly recognized as major regulators of both cell structure and function. The goal of current cell biomechanics research is to combine theoretical, experimental, and computational approaches to construct a realistic description of the cell mechanical behaviors. While some successes have been achieved in the development of tissue level continuum models (Ehlers, Krause, & Markert, 2011), it is still a challenge to study single cell mechanics, especially when considering the complexity of cell composition and architecture. Moreover, it remains unclear as to how the plasma membrane senses mechanical stimuli, transmits the stimuli to the cell interior or to other cells, and triggers subsequent biochemical and genetic response through mechanotransduction signaling pathways. This is further confounded by membrane electromotility resulting from repulsive forces generated by excess charge at the two polarizable monolayers of a cell membrane, which behaves as a whole like a mechanosensor due to flexoelectricity (Petrov, 2007). Voltage across the membrane allows for very fast stimuli applied locally to the bilayer. In addition, voltage-dependent flexing of the cell membrane becomes non-linear in ion channel-containing membranes, as shown for Shaker-transfected HEK cells (Beyder, 2007; Beyder & Sachs, 2009). As such, the understanding of how cells mechanically respond by deforming from physical loads (e.g. voltage, mechanical stimuli), is an important first step to further investigate how the transmission and distribution of these mechanical stimuli are eventually converted to biological and chemical responses in the cell (Hamill & Martinac, 2001).

In this review we first describe a number of experimental techniques and theoretical approaches adapted from soft matter physics and mechanical engineering used to probe cell mechanics and cell viscoelastic properties. We summarize experimental methods and techniques used to quantify membrane mechanical properties underlying cell deformation associated with mechanical stresses. Interdisciplinary research using these methods and techniques allows for light to be shed on the key role of membrane mechanisms in cell development, physiology, and disease. More specifically, to understand how the mechanical properties can be associated with cell components, we revisit cellular structure and composition by focusing on the cell membrane. We further describe mechanical and molecular features of the membrane and focus on mechanosensitive ion channels as specific mechanoreceptors, operating on a millisecond timescale, which link mechanical stimuli to intracellular signaling pathways. These mechanoreceptors enable cells to adjust to their mechanical environment by reprogramming the expression of cellular components supporting their physiological processes. Additionally, examples of currently known, well characterized, mechanosensitive ion channels are discussed, and the general principles of their gating mechanism are described.

2. Cell membrane mechanics

Cells are the structural and functional units of all living organisms and they are characterized by physical properties that allow them to withstand and respond to mechanical stimuli from the surrounding environment, whether isolated or part of a larger collective. Mechanical interaction between individual cells and the adjacent extracellular matrix (ECM) can determine the structural integrity of whole tissues (Wakatsuki, Kolodney, Zahalak, & Elson, 2000). Conversely, mechanical loads exerted at the tissue level and transmitted to individual cells can influence their physiological functions (Kim, Guilak, & Haider, 2008).

By forming a physical boundary between cells and their environment, the plasma membrane (PM) is a key interface mediating both cellular response to extra- and intracellular mechanical stimuli, and subsequent biochemical signaling. Its physical state and integrity are crucial for cell survival. The major function of the PM is to preserve cell integrity while enabling changes in cell shape and mechanical properties accommodating any variation in the environmental mechanical conditions. These changes may occur not only in response to cell processes such as division, migration or spreading but also to external mechanical forces present in many physiological or pathological scenarios (Fulda, Gorman, Hori, & Samali, 2010; Hoffman & Crocker, 2009). Mechanical stresses are also exerted in many therapeutical strategies as an alternative way to actively increase the sensitivity of the tissue for drug delivery studies (Peruzzi, Sinibaldi, Silvani, Ruocco, & Casciola, 2018; Stewart et al., 2016). For this purpose transient discontinuities in the PM can be introduced by mechanical means, such as local cavitation events using ultrasound-activated microbubbles that can temporary open inter-endothelial junction proteins at the cell-cell border (Silvani et al., 2019).

2.1 Mechanical force and molecular features of the plasma membrane

The plasma membrane has a complex anisotropic composition and spatial structure (Spector & Yorek, 1985). This allows it to perform a wide variety of general as well as specialized tasks. Although some PM characteristics may vary from cell to cell, certain general properties are the same for every cell, as illustrated in Fig. 1.

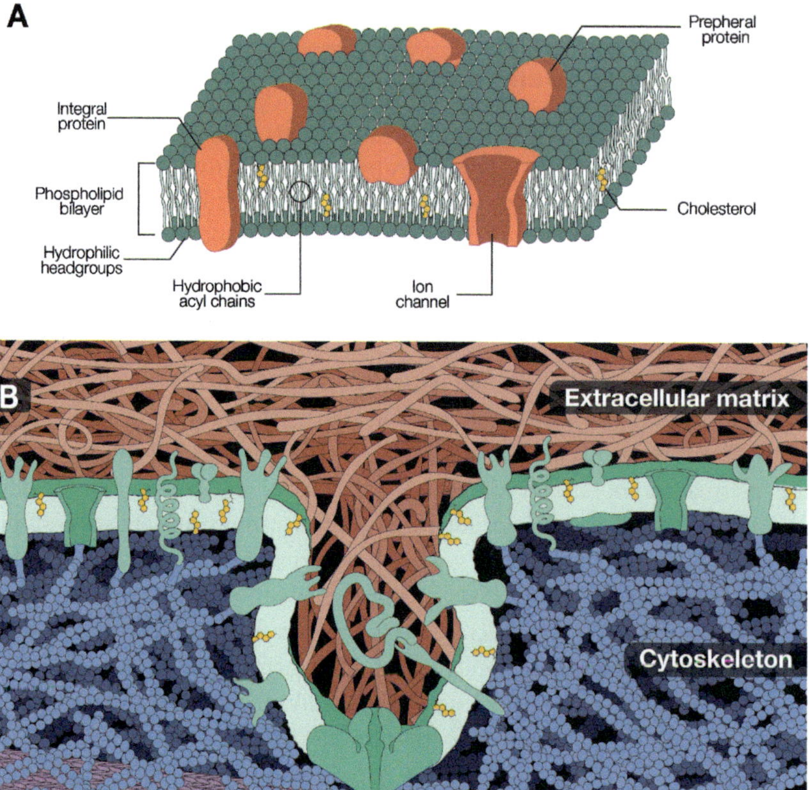

Fig. 1 Cell membrane. (A) Original fluid mosaic model of the cell membrane. (B) Currently accepted model of the cell membrane presented as a mosaic of different components, including the ECM, lipid bilayer, and the CSK. In many ways biological membranes behave as two-dimensional continua (Dai, Sheetz, Wan, & Morris, 1998).

At the microscopic scale, the PM can be thought of as a lipid bilayer with the primary function of selectively controlling the transport of solutes, water, and macromolecules between the cytoplasmatic and extracellular compartments. As amphipathic molecules, the membrane lipids are arranged with their hydrophilic heads exposed to the aqueous extra- and intracellular compartments, whereas the hydrophobic hydrocarbon tails are sandwiched in the interior of the bilayer held together via hydrophobic interactions (Alberts et al., 2002). The two monolayers (leaflets) constitute the bilayer which is typically 3.5–4.0 nm in thickness.

While we usually refer to the cellular membrane as the PM, it is worth mentioning that other membranes exist as intracellular membranes of subcellular organelles (e.g. endoplasmic reticulum, nucleus, Golgi apparatus, mitochondria) contributing to compartmentalization and overall function of the cell. It is now clear that different lipid species are not randomly distributed between membrane-bound subcellular organelles and that the chemical composition of the membrane of each organelle is intimately linked to its function (Van Meer, Voelker, & Feigenson, 2008). These entities play a limited role in terms of cellular deformation, but they do have a fundamental biological role during cell cycle.

The current, accepted model for the structure of the PM, called the fluid mosaic model (Fig. 1A), was first proposed by Singer and Nicolson (1972). According to this model, which is constantly evolving (Nicolson, 2014) but still provides a good basic description, the structure of the PM can be seen as a mosaic of components—primarily, phospholipids, cholesterol, and proteins—that are relatively free to move within the plane of the membrane, imbuing it with a highly fluid-like character (Espinosa, López-Montero, Monroy, & Langevin, 2011). Fluidity is a key feature of the cell membrane, nevertheless, experiments also showed that PM exhibit solid properties, e.g. elasticity (Evans & Hochmuth, 1978). Indeed, molecule diffusion in the PM is slower than in a pure lipid bilayer (Murase et al., 2004) not only because of its lateral organization, such as lipid-protein interactions, but also through the presence of other membrane associated molecular components.

Depending on the membrane's location and role of the cell, phospholipids can constitute up to 80% of the membrane composition with the remainder portion being proteins, both peripheral and integral. These proteins are randomly distributed and aggregated, forming a mosaic-like arrangement that plays a major role in assisting the transfer of molecules across the membrane. Peripheral proteins are localized on the PM surface forming temporary and weak connections. Conversely, integral membrane proteins are permanently connected to the cell membrane with a large portion allocated in the hydrophobic core of the membrane. Transmembrane proteins differ from the other proteins by the fact that they span the length of the lipid bilayer, often acting as pathways for ions and molecules. Among them, mechanosensitive ion channels form ion conduction pores, with gating (opening and closing), strongly altered by mechanical forces (Cox, Bavi, & Martinac, 2019; Martinac, 2004; Ranade, Syeda, & Patapoutian, 2015).

In addition to phospholipids and proteins, the membrane also contains glycolipids, sphingolipids, and cholesterol. While the number of glycolipids is small, constituting only about 2% of the total lipid content, cholesterol, tucked between the hydrophobic tails of the phospholipids, is a major membrane component, roughly 20% by weight depending on the type of cells (Veatch & Keller, 2002). It has the unique ability to increase lipid order and thickness of fluid membranes while keeping the phospholipid tails from coming into contact and in doing so maintaining their fluidity, flexibility, and diffusion rates (Mouritsen & Zuckermann, 2004).

Although our approach to modeling membrane mechanics relies on the modified, updated fluid mosaic model (Nicolson, 2014), the real PM is much more complex than the fluid mosaic representation would suggest. In addition to the composition of the lipid bilayer, the PM can be interpreted as a three-layered composite, whose architecture is also directly associated with two different molecular structures on both intra- and extracellular sides, as shown in Fig. 1B.

On the intracellular side, the membrane is strongly associated with the cortical cytoskeleton, which immediately underlies the membrane through different biochemical links (Kapus & Janmey, 2013) and continues toward the nucleus within the cytoplasm. The architecture of the cytoskeleton exhibits a dense and filamentous structure (i.e. microtubules, actin filaments, and intermediate filaments) providing a rigid structural support to maintain cell shape, although eventually reorganizing its composition during locomotion. This dynamic network of interlinking filaments also interacts with various transmembrane proteins often hampering their free diffusion either by steric interactions or direct chemical bonding (Csizmok, Follis, Kriwacki, & Forman-Kay, 2016). On the extracellular side is the glycocalyx, a highly hydrated fibrous meshwork of glycoproteins and glycolipids which has been shown to project out and covers most of the cellular membrane (Reitsma, Slaaf, Vink, Van Zandvoort, & Egbrink, 2007; Ushiyama, Kataoka, & Iijima, 2016). Although its function is not completely understood, it apparently plays an important role in cell recognition. It also functions as a compressible barrier providing cushioning and protection for the plasma membrane taking part in macromolecular transport across the endothelium (Yilmaz, Afsar, Ortiz, & Kanbay, 2019).

Another fundamental PM feature is its topography. The PM has a spontaneous curvature conferred by its composition and monolayer asymmetry with respect to its lipid composition. It contains micro and nanostructures, namely folds or membrane invagination (e.g. ruffles, microvilli and caveolae) actively maintained through mechanochemical feedback (Gauthier, Masters, & Sheetz, 2012; Jarsch, Daste, & Gallop, 2016) and is involved in molecular remodeling processes as a response to mechanical stresses, as explained later. Due to this complex composition and molecular structure, the cell membrane has a resting surface tension that may vary depending on the cell type (Pontes, Monzo, & Gauthier, 2017). Defined as the force per unit length across the membrane section, the lateral tension in the lipid bilayer is set by the nature of the lipids forming the bilayer itself and is thus influenced by external forces acting on the PM. The tension arises from a balance of forces, namely the hydrostatic or osmotic pressures from the cytosol, the forces exerted by the cytoskeleton, as well as adhesion forces between the surface and the cell and cell-cell interactions (Kozlov & Chernomordik, 2015; Lieber, Yehudai-Resheff, Barnhart, Theriot, & Keren, 2013). Values of resting tension for the PM are within a range of 0.01–0.04 mN/m (Gauthier et al., 2012).

If membrane tension is determined by a balance of mechanical forces then any variation from the environment would lead to a change in tension distribution, leading to a change in the mechanical behavior of the membrane. Recent reports have shown that tension is not equal over the cellular membrane leading to direct physiological consequences. Tension gradients have been shown to be important mechanical regulators for both mechanosensitive ion channel activation (Cox et al., 2019; Shi, Graber, Baumgart, Stone, & Cohen, 2018) or motile cell behavior (Lieber, Schweitzer, Kozlov, & Keren, 2015). In conclusion, tension plays a direct role in establishing mechanical properties of the cell.

2.1.1 Basic concepts in mechanobiology

The concept of force, in cell biology, can be related to the action exerted on individual cells or a cellular ensemble in the form of extensional, compressive (perpendicular to the surface), and shear stresses (parallel to the surface) (Le Roux, Anabel-Lise, Walani, Arroyo, & Roca-Cusachs, 2019) (Fig. 2A). Even if this concept is difficult to define, a wide range of mechanically driven

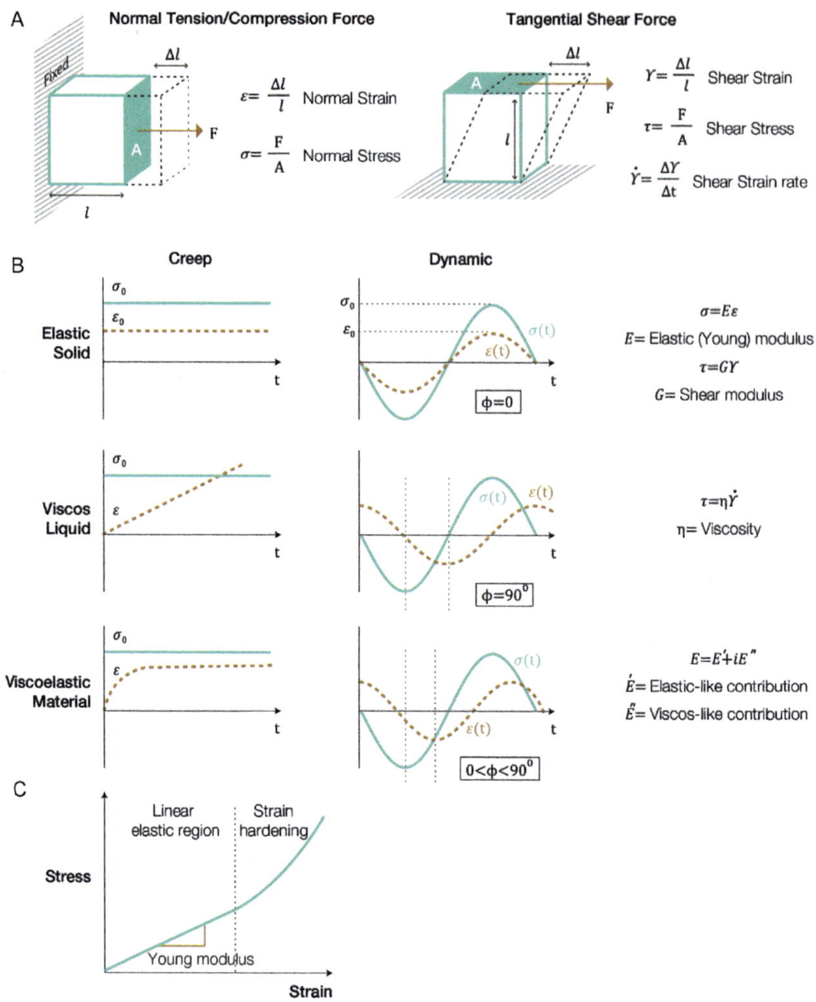

Fig. 2 Representation of the fundamental quantities used in mechanical characterization of a material. (A) Stress and strain are defined as force per unit area and deformation per unit length, respectively. The relationship between the stress and the strain defines the material mechanical properties. (B) Rheometry measurements allow for discrimination between elastic and viscous contributions of a material. (C) Diagram of a stress-strain relationship for a non-linear material. At small strains, the elastic modulus is calculated from the slope of the straight line (linear behavior). At larger strain, the material is stiffer and deforms less (non-linear behavior).

events exist that are crucial to the establishment of cellular function. For example, this may be seen in vascular endothelial cells when exposed to shear stress due to the movement of blood along the apical surface (Wang & Dimitrakopoulos, 2006). Further, elongation and compressive stresses can be generated by the lengthening and shortening of muscles as they pull on the surrounding tissue. Alveolar epithelial cells in the lung are cyclically loaded, undergoing stretching and relaxation (Yamamoto, Teramoto, Uetani, Igawa, & Shimizu, 2002) while cells in the intestinal tract are exposed to transmural pressure, shear flow, and cyclic strain (Gayer & Basson, 2009). Among other cells, red blood cells (RBCs) undergo stretching or compression due to osmotic gradients during circulation (Fung, 1981).

2.1.2 Structure and function of cell membrane

The cellular PM has evolved as an extremely complex interface able to deform its structure and trigger appropriate biochemical responses upon the application of a mechanical stress. Many biological processes, such as growth, differentiation, and transport (e.g. endocytosis, exocytosis or other membrane fusion events) are also accompanied by large local deformations of the membrane, reflected in the alteration of cell shape and structural integrity due to either extensional/compressive or bending forces (Huang & Ingber, 1999). This in turn may create tension gradients within the membrane. Upon an initial applied tensile stress, the PM can respond in several different ways, with the aim of supplying lipids to counteract tensional increases. Due to its peculiar composition, which is rich in lipid reserves in the form of folds and *endo*-membranes, upon tensile stress application, the apparent PM area increases while the real membrane area remains constant (Kosmalska et al., 2015). Caveolae are such structures (Sinha et al., 2011), playing a mechano-protective role within the PM. Exocytosis also contributes to PM area expansion by the fusion of intracellular fluid-filled vesicles, incorporating proteins and lipids into the PM (Vlahakis, Schroeder, Pagano, & Hubmayr, 2002) and by doing so, leads to changes in lateral tension. Only when these folds have been eliminated will the membrane area become stretched due to tensile stress, which is associated with a relatively rapid increase in extensional stiffness and resistance of the PM to extensional stresses (Chen, Helm, & Israelachvili, 1991; Dill & Flory, 1980). This results from the energy penalty associated with the exposure of the hydrophobic core of the PM to water as the spacing between individual lipid molecules increases. Like tensile stresses, compressive stresses also lead to lipid reorganization and formation of highly curved structures. This occurs because the two lipid layers can be independently modulated (Evans, 1974) allowing the hydrophilic head groups on the outer layer to move further apart while simultaneously decreasing intermolecular spacing in the inner layer. To better understand the mechanical behavior of the PM, an overview of the cellular mechanical properties is given in the following section.

2.1.3 Cellular rheology and rheometry

An ideal elastic material, such as a spring, deforms (strains) constantly upon loading (stress), maintaining its deformation until stress removal, and thus returning to its initial shape. The stress-strain curve during loading and unloading of the spring is the same and the ratio of stress to strain defines the elastic modulus of the spring. Depending on the stress type, different elastic moduli are defined. For instance, if measuring shear stress, then the ratio of stress to strain is the shear modulus. If it is elongation or compressional stress, then the ratio of stress to strain is generally called the Young's Modulus. In general, the greater the value of the modulus, the more resistant the material is to the particular form of deformation. Conversely, for an ideal viscous material, there is a linear relationship between the stress and strain rate. Viscous deformation is irreversible and creates permanent molecular structural deformation. In this case, the ratio of stress to the rate of strain defines the viscosity of a liquid and is a measure of the fluids resistance to flow.

Most biological materials show both viscous and elastic properties, exhibiting a non-linear stress-strain profile, under certain loading conditions (Fig. 2). Indeed, cells display an increased stress/strain slope under large deformations (strain-stiffening curve). On the other hand, under small deformations, the stress is linearly proportional to the strain (purely elastic region) with the elastic modulus directly calculated from the slope of the straight line. This peculiar behavior, termed viscoelasticity, arises from the complex composition of the cellular interior and represents the most intricate aspect of cell rheology (i.e. rheometry), which is the study of material flow in response to either an instantaneous applied stress (Fig. 2A) or the time-dependent characterization of material deformation of a viscoelastic system (Fig. 2B). Differentiating elastic from viscous contributions requires measurements on different time scales and at different stresses applied.

Several models have been proposed as a means of linking deformational information to a specific physiological response. The deformation of a viscoelastic material, such as a biological cell, leads to a multiphasic response, split between two phases: force loading and unloading. During the loading phase, the creep-response (the tendency of an initially semi-solid like object to deform) can be characterized by a power-law behavior. Upon load removal, the cell undergoes relaxation, the behavior of which can be characterized by an exponential function. Regardless of the phase, a cell demonstrates both elastic-solid and viscous fluid-like behavior, the magnitude of which depends on the measurement timescale and technique (Kollmannsberger & Fabry,

2011). Usually, the viscoelastic response has been modeled by a system of springs and dashpots (e.g. Burgers Model). Due to the force dissipative behavior of cells this makes difficult to establish any link to a characteristic relaxation time. Consequently, because of the breakdown of fitting such models to time-scales spanning several orders of magnitude, newer models have been proposed (Balland, Richert, & Gallet, 2005; Fabry et al., 2001). An in-depth review of these models is not within the scope of this chapter and we point the reader to these excellent reviews (Chen, 2014; Kollmannsberger & Fabry, 2011; Moeendarbary & Harris, 2014). Each model aims to measure an observed response (e.g. Youngs Modulus) and relate it within the context of the physiological response driving the behavior.

Unlike creep tests, defined as the application of a constant stress $\sigma = F/A$, dynamic rheometry tests are usually performed by applying an oscillatory stress over time while recording the resulting oscillatory deformation $\varepsilon(t)$ (Barnes, Hutton, & Walters, 1989) (Fig. 2B). In dynamic measurements performed on pure elastic-solids, deformation follows the applied oscillating stress, meaning that there is no phase shift between $\varepsilon(t)$ and $\sigma(t)$ signals. Similarly, creep tests will display instantaneous deformation, ε_0, proportional to the applied force. In both rheometer tests, the ratio between stress and strain is constant and corresponds to the elastic modulus $E = \sigma/\varepsilon$.

Different from pure elastic-solids, purely viscous-fluids will flow indefinitely when subjected to a creep test. The ratio at which the liquid flows under a given stress depends on its viscosity (Fig. 2C). Under dynamic loads, oscillating deformation is delayed compared to the applied oscillating stress, showing a 90-degree phase shift between $\varepsilon(t)$ and $\sigma(t)$ signals as shown in Fig. 2B. Viscoelastic materials exhibit an intermediary phase shift in the range of $0 < \varphi < 90$-degrees, depending on the elasticity (Phan-Thien & Mai-Duy, 2017). The response is then quantified through a complex modulus $E^* = E' + iE''$, allowing one to decouple the elastic-like contribution E' (the in-phase component of the response) from the viscous-like component E'' phase-log component. Since the viscoelastic response is time-dependent, the elastic or viscous-like behavior may dominate the response of such materials based on the experimental time scale. At high frequency, $E'' > E'$ the viscous behavior dominates, while at low frequency $E'' < E'$ the behavior is predominantly elastic, as observed for a creep test (Barnes et al., 1989).

Using different experimental techniques, biological membranes were found to have a high elastic modulus, ranging from hundreds of pascals to tens of kilo-pascals (Hochmuth, Mohandas, & Blackshear Jr., 1973; Hochmuth & Waugh, 1987). This is due to the resistance to stretch of the lipid bilayer which resists area expansion. Further, biological membranes exhibit low shear moduli on the order of $4–10 \times 10^{-3}$ N/m (Bausch, Ziemann, Boulbitch, Jacobson, & Sackmann, 1998; Espinosa et al., 2011), and highly variable bending moduli on the order of 10^{-19} Nm (Pontes et al., 2013), which is strongly influenced by protein-bilayer and cytoskeleton-bilayer interactions.

As complex, gel-like heterogeneous entities, cells contain hyperstructures of filament proteins or organelles that may contribute differently to cell elasticity and viscosity. The techniques described in this chapter provide an in-depth view of the connections between coupled systems within the cell, such as the contribution to cellular elasticity that arise due to PM-actin cortex interactions. However, to understand the intercellular origin of some of these forces that arise from specific proteins, FRET (Förster Resonance Energy Transfer) based-sensors can be employed (Guo, Sachs, & Meng, 2014). FRET sensors can be used to map the heterogeneous stress fields within living cells (Meng & Sachs, 2012) that arise from cell-substrate, cell-probe and cell-force interactions, including measurements of specific stress-induced protein activation arising from cell-dependent cytoskeletal prestress (Na et al., 2008) or probing force transmission through the cytoskeleton induced by fluidic shear (Rahimzadeh et al., 2011). While FRET sensors are capable of reporting multidimensional, intercellular stress fields, the techniques discussed in the following section focus on the extracellular domain. One of the major challenges of measuring the extracellular domain is correlating the measured mechanical properties with that of the local structural components in order to identify individual structural contributions. While a number of different instruments and methodologies have been developed to study these contributions, the reported values of cellular elasticity and viscosity vary strongly depending on the technology used. These variations may depend on cell culture conditions (e.g. temperature, pH and cell passage) as well as measurement protocols. In the following section we describe several important factors to be considered when choosing an instrument to measure the viscoelastic properties of cells.

2.2 Methods for determination of mechanical properties of cellular membranes

2.2.1 Choosing the right tool

Experiments on biological membranes started in the 1930s using sea urchin eggs (Cole, 1932). Later, the micropipette aspiration method was developed and used to measure the elastic properties of sea urchin eggs (Mitchison & Swann, 1954) and red blood cells (Rand & Burton, 1964). Unfortunately, this technique is mostly applicable to suspended cells and not to cells with complex structures, such as neuronal or adherent cells.

Advances in nanotechnology have led to the development of new, innovative solutions (Table 1). For example, developments in the field of microfabrication led to the creation of micrometer-sized cantilevers and other mechanical parts that offer greater spatial resolution for probing single cells with forces and displacements down to the pico-scale (Lang, Hegner, & Gerber, 2017). In general, techniques used for measuring the mechanical properties of cells can be subdivided into two categories as summarized in Fig. 3. The first category contains techniques that apply force to the whole cell while simultaneously recording its mechanical response (force-application techniques). The second category refers to techniques used for measurements of forces generated by cells during physiological processes, including development, contraction, and migration (force-sensing techniques) (Rodriguez, McGarry, & Sniadecki, 2013).

Techniques and methods used to apply controlled forces on cells include optical tweezers (laser traps) (Mio, Gong, Terray, & Marr, 2000; Williams, 2002), magnetic tweezers (Hosu, Jakab, Bánki, Tóth, & Forgacs, 2003; Kim & Saleh, 2009), atomic force microscopy (AFM) (Müller & Dufrene, 2010), micro-needle manipulation (Baldi, Fass, De Silva, Odde, & Ziaie, 2003), cell stretchers and flow rheometry (Gossett et al., 2012; Lu et al., 2004; Yost et al., 2000). To improve on experimental throughput and the span of force-loading rates, a new technique was developed called Acoustic Force Spectroscopy (AFS) (Sitters et al., 2015). Several techniques have also been developed to monitor the ability of cells to interact with and to generate forces on their local environment. These techniques include Tracking Force microscopy (TFM) (Style et al., 2014) and Micropillar Arrays (Schoen, Hu, Klotzsch, & Vogel, 2010). Another promising technique is Particle-Tracking Micro-rheology (PTM), used to measure the viscoelastic properties of the cell at high spatio-temporal resolution (Wirtz, 2009). Thanks to these sophisticated experimental techniques, our ability to study fundamental biological processes has been revolutionized over the last 20 years. However, due to various force directionalities, magnitudes, loading rates and modes of operation, different techniques stimulate different mechanical responses in cells. It is therefore important to evaluate the factors that can influence these measurements to choose the most appropriate experimental tool.

The choice of experimental tool mainly depends on the following key considerations, the length scale and compliance of the biological structure under investigation, as well as the specific information regarding that

Table 1 Techniques used for measurements of mechanical properties of biological cells.

Tool	Force application	Force sensing	Spatial sens. (nm)	Cell state	Force directionality
Micropipette aspiration	$1.1–10^3$ nN	–	1–100 μm	Suspended	Pull
Microneedle manipulation	1-pN	–	$1–10^3$ μm	Suspended	Push?
AFM	$10–10^7$ pN	$10–10^5$ pN	$1–10^5$ nm	Adherent	Push/pull
Optical tweezer	$0.01–10^3$ pN	1–100 pN	$10–10^5$ nm	Adherent	Push/pull
Optical stretcher	$10–10^3$ pN	–	$10–10^6$ nm	Suspended	Pull
Magnetic tweezer	$0.1–10^4$ pN	$10–10^3$ pN	0.1100 μm	Adherent	Pull
Magnetic twisting cytometry	1–100 pN	–	$1–10^3$ μm	Adherent	Angular
Acoustic force spectroscopy	$10^{-4}–10^3$ pN/s	$0.01–10^2$ pN		Adherent	Pull
Traction force microscopy	–	$10–10^6$ pN	–	Adherent	Push/pull
Micropillar array	–	0.01–1000 nN	$1–10^3$ μm	Adherent	Push/pull

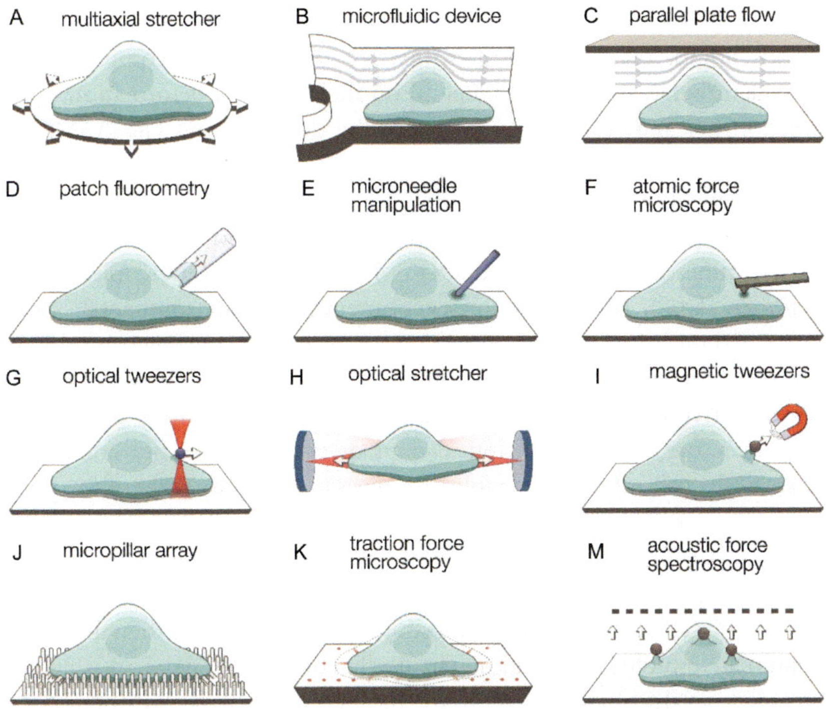

Fig. 3 Force-application and -sensing techniques. Techniques used to probe the rheological properties of cells by applying well-defined external loads and measuring the consequent deformations over time. Force-application AFM (*F*), micropillar array (*J*) and traction force microscopy (*K*) can also be used as force-sensing techniques for measuring forces exerted by cells on the surrounding substrate.

structure. Due to the size of cells (on the order of 10 μm) and their low elastic modulus, the mechanical measurement technique needs to be capable of applying and monitoring deformations and forces in the rage of micro to nanometers and nano- to piconewtons, respectively. Furthermore, for local measurements of cellular mechanical properties, the size of the probe must be much smaller than the cell being measured. Measurement techniques capable of monitoring and applying these loads at high spatial accuracy include atomic force microscopy (AFM) (Sirghi, Ponti, Broggi, & Rossi, 2008; Wojcikiewicz, Zhang, & Moy, 2004) and particle-tracking micro-rheology (PTM) (Tseng, Kole, & Wirtz, 2002).

The timescale of force-application measurements also determines the choice of the experimental tool. Given the intrinsically viscoelastic nature of a biological cell, whose response to mechanical force is neither time-independent, as in simple solids, nor scales linearly with time, as in a viscous liquid, the mechanical properties can exhibit a complex time-dependent behavior that may depend on the degree of deformation, force directionality and spatial distribution of the deformation. The AFM, and magnetic and optical tweezers can be operated within a closed-loop feedback system allowing for dynamic rheometry measurements and for capturing the complex time-dependent mechanical response of the cell over a wide range of timescales (Kollmannsberger & Fabry, 2007; Schitter, Allgöwer, & Stemmer, 2003). Moreover, the experimental and environmental conditions can significantly influence such measurements. Many mechanical measurements are performed under ambient conditions and as such, are not representative of the cell's native environment. For example, the increase in temperature when using optical traps or the choice of bio-compatible substrate for cell adhesion, capable of replicating the properties of the physiological environment are among common drawbacks. If the change in temperature may interfere with measurements this can be solved by using an environmentally controlled chamber equipped with a thermal controller. If the choice of a bio-compatible substrate may be a problem then the design of alternative biologically and mechanically compatible hydrogel is needed to provide a valuable alternative to the most commonly used synthetic polymer PDMS (Mata, Fleischman, & Roy, 2005). In addition, stiffness measurements of soft, thin objects using indentation techniques must be corrected to account for stiffness modulation due to the underlaying substrate. The thickness of the cell will determine the contribution from the substrate to the overall stiffness measurements with thinner sections of the cell appearing stiffer (Dimitriadis, Horkay, Maresca, Kachar, & Chadwick, 2002; Rahimzadeh et al., 2011). When quantifying cellular stiffness on gels, several factors must be considered and accounted for to eliminate cell-substrate stiffness artifacts, including the stiffness of the gel and its thickness (Gavara & Chadwick, 2012). Depending on gel thickness and stiffness, cells can feel the underlaying rigid support (below the gel) if the gel is thinner than 10–30 μm (Buxboim, Rajagopal, Brown, & Discher, 2010; Gavara & Chadwick, 2012).

2.3 Single cell-force spectroscopy

One of the most exciting developments in the field of biological physics is the ability to manipulate single cells or even single molecules. Single molecule measurements can provide extensive insights into the molecular mechanisms, interactions, and functions, complementing and often substituting bulk assays. Traditionally, these measurements are made with the following techniques: Atomic force microscopy (AFM), optical and magnetic tweezers, Particle-Tracking Micro-rheology (PTM), and Traction Force Microscopy (TFM). These techniques are summarized in the following section, specifically focusing on strengths, limitations, and practical considerations. The recently developed Acoustic Force Spectroscopy (AFS) technique will also be discussed.

2.3.1 Atomic force microscopy (AFM)

AFM is a sub-nanometer 3D surface scanning probe technique that has rapidly become a valuable tool for imaging live cell surfaces. It is used for characterizing cell structure and probing the global and local mechanical properties of cells (Kreplak, 2016; Rettler, Hoeppener, Sigusch, & Schubert, 2013). The key component of the AFM is the sharp tip, mounted on a micro-fabricated cantilever beam, usually made from silicon, which rasters (scans) over the sample surface by utilizing piezoelectric micro-positioners to interact with and eventually deform the sample. Lateral or vertical displacement of the cantilever is detected by a position-sensitive photodiode, which triggers the fast feedback loop system maintaining a constant relationship (e.g. distance) between tip and sample while communicating with the computer used to generate an image of the sample surface. Although the AFM is primarily an imaging tool, it can also be used for cell rheology, capable of applying piconewton forces. When used in force spectroscopy mode, the tip of the AFM cantilever is pressed against the cell surface while the force and the subsequent cellular deformation is monitored. Thus, elasticity can be computed from the applied force plotted against indentation, by considering tip geometry and an appropriate contact model (Wojcikiewicz et al., 2004). Moreover, using a feedback loop system, levels of strain and stress can be controlled over time. Creep experiments and oscillatory tests can be applied to measure time-dependent mechanical properties (Efremov, Okajima, & Raman, 2020).

AFM has also been used to investigate the mechanics of individual biomolecules (Engel & Gaub, 2008; Pleshakova, Bukharina, Archakov, & Ivanov, 2018), to characterize the mechanical properties of the cell nucleus (Hansma, Kasuya, & Oroudjev, 2004; Liu et al., 2014), and to measure cytoskeletal structure and organization (Pogoda et al., 2012) as well as to monitor changes in cell mechanics during differentiation (Collinsworth, Zhang, Kraus, & Truskey, 2002) and disease progression (Lam, Rosenbluth, &

Fletcher, 2007; Rosenbluth, Lam, & Fletcher, 2006). For such studies the AFM can be used in multiple configurations. One example is using AFM as a molecular force probe by moving the cantilever only in the vertical direction, perpendicular to the sample plane, allowing for measurements of inter- and intramolecular interaction forces (Rief, Oesterhelt, Heymann, & Gaub, 1997). In this configuration, the tip comes into direct contact with the sample, leading to attachment of the molecule of interest (DNA, RNA or proteins) to the tip, at which point the tip is retracted, bending the cantilever. As a result, elongation of the molecule as a function of extensional force can be recorded with high resolution. Force is generally calculated by using Hooke's Law for bending of a cantilever of known stiffness, which is typically in the order of 101–105 pN/nm (Neumeister & Ducker, 1994). This configuration offers another advantage by allowing to correlate sample topography with mechanical properties across the sample surface. It can be used to generate a force-distance curve based on indentation and retraction at specific points while scanning the cell's surface. One important advantage of the AFM technique is the simple and rapid sample preparation and the capability to perform measurements of biological samples under near-physiological conditions (Bustamante, Rivetti, & Keller, 1997). On the other hand, the main drawback and limitations of the AFM technique result from the large size and high stiffness of the cantilever, limiting the applicable force range. Moreover, considering the difficulty to discriminate specific to non-specific interactions of the tip with cell structures being investigated, significant improvements have been made by combining AFM with high resolution fluorescence microscopy (e.g. total internal reflection fluorescence microscopy—TIRF) (Sarkar, Robertson, & Fernandez, 2004).

2.3.2 Optical tweezers

Optical tweezers, also known as an optical trap, present another single-cell manipulation technique, which utilizes a focused laser beam with a high numerical aperture (NA) microscope objective to interact with objects. Any dielectric particles in proximity of the laser beam focus can be trapped, taking advantage of two optical forces directed toward the focus. The first force is the scattering force along the beam propagation trajectory. The second is the force resulting from the interaction of the dipole generated through polarization of a dielectric particle by the optical field and the optical pressure gradient near the focus of the laser. As long as the displacement is small (150 nm) the relationship between the force and the displacement is linear and the optical tweezers can be compared to a Hooke's spring with stiffness depending on laser power, particle polarization and the optical gradient.

Due to their high sensitivity (piconewton resolution) and the capacity to measure particle displacement with sub-nanometer accuracy and sub-millisecond resolution, optical tweezers are well suited to subcellular measurements, such as the pulling of membrane tethers (Ayala et al., 2016; Williams, 2002). In these experiments a small particle attached to the cell membrane is pulled away from the cell. The applied force and resulting particle displacement are interpreted in terms of the mechanical response of the membrane. Using an external movable mirror, an oscillatory force can also be generated to perform dynamic rheometry. The resulting phase shift amplitude of the bead motion is related to the viscoelastic behavior of the membrane.

For stable, consistent operation, several technical requirements should be considered. First, a 1.2 NA objective should be used to achieve the smallest focal spot producing the largest optical gradient leading to a Gaussian output intensity profile. For this reason, water or oil-immersion objectives are often used. Although oil-immersion objectives offer the highest NA, they introduce spherical aberrations that degrade optical tweezer performance in bulk solution (Fällman & Axner, 2003). On the other hand, water immersion objectives do not suffer from spherical aberrations and permit trapping deep in solution. Second, high precision optical trapping also requires power stability. Power fluctuations may result in force fluctuations, increasing measurement noise. Third, working in the near-infrared wavelengths (800–1100 nm) is necessary to avoid biological damage.

Optical tweezers are a versatile tool for biological investigation; however, several limitations need to be considered. First, operating at small distances has limited linearity between force and displacement. Second, there is a limit on the amount of force that can be applied using light. With this technique, the range of applied forces is 0.1–100 pN, where the lower limit (0.1 pN) is due to the minimum force needed for trap stability. Third, optical perturbations may affect the distribution of the optical field gradient, decreasing the performance of the optical tweezers. For this reason, optically homogeneous and highly purified samples are needed. This can also help to increase selectivity, given that any dielectric particle near the focus can be simultaneously trapped. Also, an increase in laser power, typically 109–1012 $W\,cm^{-1}$, may induce phototoxicity and thermal stress that may damage cell structure and influence cell mechanical properties.

In addition to being used as a tool for probing whole-cell mechanical properties of red blood cells (Dao, Lim, & Suresh, 2003; Zhu, Avsievich, Popov, & Meglinski, 2020), fibroblasts (Park, Koch, Richard, Käs, & Shih, 2005), tumor cells (Guck et al., 2005) and other cell types (Huang, Anvari, Torres, LeBaron, & Athanasiou, 2003; Titushkin & Cho, 2006; Walker et al., 1999), optical tweezers have also been used to determine the mechanical properties of subcellular structures such as DNA and proteins (Allemand, Bensimon, & Croquette, 2003; Bustamante, Bryant, & Smith, 2003). Additionally, this technique has been used to investigate the forces generated by molecular motors (Knight, Veigel, Chambers, & Molloy, 2001), those created by individual RNA polymerase molecules during transcription (Abbondanzieri, Greenleaf, Shaevitz, Landick, & Block, 2005), and those developed during protein unfolding/folding (Cecconi, Shank, Bustamante & Marqusee, 2005) or binding/unbinding (Litvinov, Bennett, Weisel, & Shuman, 2005).

2.3.3 Magnetic tweezers

Magnetic force spectroscopy is based on manipulating magnetic beads through an externally applied magnetic field. It consists of a permanent magnet arranged above an inverted microscope stage on which the sample is placed, providing an upward pulling force on magnetically functionalized beads embedded or attached to a molecule of interest (Tanase, Biais, & Sheetz, 2007; Vilfan, Lipfert, Koster, Lemay, & Dekker, 2009). Essentially, magnetic tweezers function as optical tweezers, controlling the movement of the bead, attached to the molecule of interest, by the interaction of its induced magnetic dipole with the field gradient of the magnet. The generated force is proportional to the gradient of the magnetic field and directed toward the area with the strongest field. The corresponding displacement of the magnetic bead is used to obtain cellular properties. Again, as in optical tweezers, the steep gradient of the magnetic field ensures linearity between force and displacement only for short distances, limiting the applicability of this force on particles near the magnet. However, larger magnets yield higher magnetic field strengths and a shallower field gradient, resulting in smaller force variation.

Magnetic tweezers come in a variety of configurations. While a single magnet can be used to pull on a magnetic particle, a minimum of two magnets, each wound with a separate field coils and arranged at predetermined angles, are required to generate torque (Lipfert, Lee, Ordu, Kerssemakers, & Dekker, 2014; Tokarev et al., 2012). Beads can also be manipulated in 3D, requiring a stronger magnetic field generated by the placement of several magnets around the bead (De Vlaminck, Henighan, Van Loenhout, Burnham, & Dekker, 2012). Advantages of this technique include no sample heating or photodamage. Magnetic manipulation is extremely selective for magnetic beads and is generally not influenced by sample preparation protocols. A drawback of magnetic tweezers is their low temporal and spatial resolution. However, this can be improved with the addition of a high-speed camera, reaching Angström-levels (Lansdorp, Tabrizi, Dittmore, & Saleh, 2013).

Magnetic tweezers were initially developed to measure the mechanical properties of cells but have since been employed for manipulation of individual cellular components, such as DNA (De Vlaminck et al., 2012; Yan, Skoko, & Marko, 2004) lipid bilayers (Heinrich & Waugh, 1996), and cell receptors (Mannix et al., 2008). Magnetic tweezers have also been used to determine the tensile strength of ligand/receptor bonds (Danilowicz, Greenfield, & Prentiss, 2005) and to investigate the effect of applied force on focal adhesion proteins (Matthews et al., 2004) or cell-cell junctions (Kris, Kamm, & Sieminski, 2008).

2.3.4 Traction force microscopy (TFM)

Traction force microscopy measures traction forces generated by adhering cells during physiological process and is often chosen for the simplicity and flexibility of its implementation (Style et al., 2014). These forces, typically on the order of a few nanonewtons, are not measured directly but calculated from the observed deformation using different detection mechanisms such as micropillar array deflection (Schoen et al., 2010) or embedded beads displacement in soft gels (Plotnikov, Sabass, Schwarz, & Waterman, 2014). The gels consist of a high density of fluorescent nanoparticles randomly distributed in a gel-like, elastic material, tracked by a microscope to create a vector field as an approximation of the overall deformation (Sabass, Gardel, Waterman, & Schwarz, 2008). The particle examined is embedded within an elastic, homogeneous, and isotropic material, thus the relationship between the displacement field and traction field can be determined from its mechanical properties.

This technique has several advantages. Since the cells of interest can be seeded on top or within the elastic material, this technique can be used with both adherent and non-adherent cell types. Further, the properties of the gel can be chemically altered to expose the cells to environments with different stiffnesses and compositions. The variety of different bead types and sizes allow for more accurate measurements over the region upon which the force is applied.

There are limitations associated with traction force microscopy. New technologies have addressed some of these drawbacks with the combination of super-resolution optical microscopy, which helped to increase the resolution of the displacement of fluorescent nanoparticles in 3D (Legant et al., 2010; Sabass et al., 2008). This however introduced additional limitations, such as longer acquisition times. Also, applied forces are strongly influenced by the size of the nanoparticles and temperature fluctuations. Moreover, a random dispersion of fluorescence markers may limit the post-processing of TFM samples for the localization of investigated protein activity by immunostaining. To overcome these intrinsic limitations, alternative substrates have been developed taking advantage of advanced microfabrication technologies (Fu et al., 2010). For example, micrometer-sized elastomeric pillars, generated by soft lithography, can act as springs, bending laterally when a cell adherent to its top exerts traction on the elastomeric pillar.

Traction force microscopy has been used to measure the forces produced by a vast number of different cell types (Curtze, Dembo, Miron, & Jones, 2004; Doyle & Lee, 2005; Iwadate & Yumura, 2008; Tolić-Nørrelykke & Wang, 2005) and to determine the contribution of active cytoskeletal contraction to these traction forces (Lombardi, Knecht, Dembo, & Lee, 2007) as well as to elucidate the effects of pathological processes on these forces (Marganski, De Biase, Burgess, & Dembo, 2003; Rösel et al., 2008). It has also been used to measure forces at focal adhesions (Stricker, Sabass, Schwarz, & Gardel, 2010) and to determine the traction forces during migration and development (Delanoë-Ayari et al., 2008).

2.3.5 Particle-tracking micro-rheology (PTM)

Cellular and subcellular processes depend on the mechanical deformability of the cytoplasm. The previously mentioned techniques function by applying external forces directly on the cell surface, measuring the response of the combined subcellular structure with little information regarding the contribution of each individual component. A new rheological method, called particle-tracking micro-rheology (PTM), allows for localized mechanical measurements inside the cytoplasm by tracking thermally driven motion of embedded fluorescence particles (Wirtz, 2009). These beads, typically less than 1 μm in diameter, are injected and rapidly dispersed into the cytoplasm of living cells. Subsequently, the recorded movements of the bead, tracked by fluorescence microscopy, are analyzed in terms of cytoplasmic viscoelasticity (Tseng et al., 2002).

In a purely viscous environment, particles move randomly due to thermal fluctuation. This behavior is the characteristic of Brownian motion (Massiera, Van Citters, Biancaniello, & Crocker, 2007) and the particle diffusion constant can be calculated to derive the viscosity of the medium using Stokes-Einstein equation (Mason, 2000). In contrast, in a purely elastic environment such as dense, cross-linked network of polymers, the embedded particles are stuck within the network and thermal agitation is not strong enough to induce substantial motion. In biological media, which is described as viscoelastic, beads follow a more complex, time-dependent behavior, distinctly different from Brownian motion. Recent studies have shown that at short timescales the dominant response is mainly elastic whereas a more viscous behavior is typical over longer time periods (Tseng et al., 2002). However, the high variability of these viscoelastic responses depends mainly on the type of cell and/or the physiological conditions. This technique has been employed to elucidate the existence of functional links between the nucleus and the cytoskeleton (Hale, Sun, & Wirtz, 2009) and to investigate the influence of environmental perturbation on cell viscoelasticity (e.g. shear stress; Dangaria & Butler, 2007; Lee et al., 2006), including stretching (Trepat et al., 2007), temperature and pH (Picard & Donald, 2009).

2.3.6 Acoustic force spectroscopy (AFS)

Acoustic force spectroscopy (AFS), first described in 2014 (Sitters et al., 2015) is a relatively new method for probing cellular properties. Operationally, the AFS is straightforward to setup and use. At its core, the technique can be broken into two major components, one is the microfluidic chip and the other a piezoelectric element. The microfluidic channel, fabricated from glass, provides a well-defined environment which can be seeded with the objects of interest. The piezo element is an electrically sensitive structure that mechanically deforms upon the application of an electrical charge. By placing the piezo element adjacent to a microfluidic channel and applying an oscillatory voltage across the piezo element, a standing acoustic wave can be generated (Sitters et al., 2015). Any object placed within the flow channel will experience acoustic force, driving the object toward the acoustic node. The vertical (z-direction) displacement of a bead under the influence of an acoustic field will depend on several factors, including bead material, density, and size. As with magnetic tweezers, particle displacement is tracked using a digital camera.

Initially, the technique utilized an opaque piezo element, coupled with an inverted microscope to demonstrate simultaneous, real-time tracking of 2000 microspheres. Forces on these beads were shown to range from sub-piconewtons to hundreds of piconewtons with a response time below 1 ms (Sitters et al., 2015). In order to improve on particle-tracking accuracy, a semi-transparent piezo element was introduced (Kamsma, Creyghton, Sitters, Wuite, & Peterman, 2016). The force-loading capabilities of the AFS were also improved by the formation of a flatter force profile across the flow chamber. This was shown to work in combination with water and oil-immersion objectives, although with steep reduction in force magnitude (Kamsma et al., 2016).

The AFS offers several advantages over traditional measuring techniques but also introduces a few challenges. One of the core strengths of the technique is massive-parallel, real-time measuring of particle or cell interaction with the acoustic field. Depending on the object being trapped, the tracking resolution is sub-100 nm. The number of objects tracked in real-time depends on the field-of-view, determined by the optical setup. Depending on particle size-distribution, the acoustic force may vary by up to 30% within a 1.8 mm^2 field-of-view (Kamsma et al., 2018). Further, application of high driving voltages may lead to sample heating. However, at a maximum voltage of 42 V at 14.3 MHz applied for no more than a few seconds, the resultant sample heating was shown to be less than 2 °C (Kamsma et al., 2018).

While still an emerging technology, three distinctly different applications have been demonstrated. First, the AFS has been used in high throughput fashion to measure several DNA related events including monitoring of the torsionally constrained and unconstrained behavior of individual dsDNA molecules, the mechanical properties of dsDNA after protein-filament assembly (Sitters et al., 2015), and the real-time assembly of a virus particle around a DNA template (Van Rosmalen et al., 2020). Second, the AFS has been used to probe the viscoelastic properties of red blood cells (RBCs) as a function of various drug treatments, specifically demonstrating that RBC deformability increases due to lipid accumulation after exposure to RBC-derived extracellular vesicles (Sorkin et al., 2018). Lastly, the AFS was used to measure the binding kinetics of CD4+T lymphocytes to fibronectin as a model for CD4+-endothelium invasion. The AFS was able to monitor the binding behavior of hundreds of CD4+ cells in parallel with forces up to 1 nN (Kamsma et al., 2018).

2.4 Cell membrane properties of organisms with and without cell wall

Living cells and organisms are continuously exposed to environmental mechanical forces ranging from gravity and osmotic pressure to wind, sound, and touch. During evolution of life, different strategies have been employed by various organisms. Bacteria, Archaea, fungi, and plants evolved a rigid cell wall, which protects their fragile plasma membrane from damage by excessive mechanical force. In contrast, animals have adopted a different strategy allowing them to possess a high degree of cell deformability and nevertheless keep their cell membrane mechanically protected (Gillespie & Walker, 2001; Hamill & Martinac, 2001).

In this section, we briefly describe the differences in structure, composition, and physical strength between the membranes of cell-walled and cell wall-less organisms.

2.4.1 Bacterial cell envelopes

The cell wall is an outer protective membrane present in many cells including plants, fungi, algae, and bacteria. The main functions of the cell wall are to provide structural support and protection for the cell. In bacteria, the cell wall plays a major role in protecting these microbes from sudden changes in environmental osmolarity (Cox et al., 2019). For example, the cell swelling due to water uptake upon a hypoosmotic shock could easily destroy a bacterial cell if it were not protected by the cell wall, which on average can resist an osmotic pressure increase of up to 30 atm (Cox, Bavi, & Martinac, 2018; Hamilton et al., 2015; Martinac & Kloda, 2003; Peyronnet, Tran, Girault, & Frachisse, 2014). This is because bacterial cell walls are up to several hundred thousand times stiffer compared to animal cell membranes. Young's elastic modulus, which in addition to shear modulus, bending stiffness, and viscosity is an important mechanical property of biological cell membranes, and varies between 50 and 150 and 100–200 MPa for the Gram-negative and Gram-positive bacterial cell walls, respectively (Tuson et al., 2012). In contrast, it varies only between a few and tens of kPa for animal cell membranes (Steltenkamp, Rommel, Wegener, & Janshoff, 2006). Based on these properties it has been estimated that bacterial cell walls can sustain approx. 90% of the cell turgor leaving the 10% share to be sustained by the cytoplasmic membrane, which mainly consists of the lipid bilayer (Cox et al., 2018). As discussed in the section on bacterial and archaeal MS channels, this amount of osmotic pressure exerted on the plasma membrane is within the force range required for activation of MscL and MscS, which have been shown to serve as osmotic safety valves in the physiology of bacterial cells (Levina et al., 1999).

2.4.2 Animal cell membranes

Animal cells do not have a cell wall. Instead, they possess excess membrane area in the form of ruffles, folds, and microvilli, protecting the fragile lipid bilayer from excessive strain (Martinac, 2014). The very soft membrane bilayer is supported by the extracellular matrix (ECM) serving as external scaffolding as well as by the cortical cytoskeleton (CSK) (Fig. 1B). The softness of animal cell membranes is characterized by a relatively small Young's modulus within the kPa range, which has enabled animals to evolve as highly mobile organisms moving slow or fast with ease from place to place. This membrane softness is also important for many membrane-transport processes, including endocytosis and exocytosis, which are enabled by large, local deformations of the membrane lipid bilayer (Dai & Sheetz, 1997).

The recent growing interest in mechanobiology research has led to rapid progress in the understanding of the importance of mechanical properties of biological cell membranes for a variety of fundamental cellular processes. Specific progress has been made by realizing that cell mechanics is linked to intracellular signaling, which directly controls and regulates gene expression (Cox et al., 2019). Consequently, many cellular processes are accompanied by morphological changes resulting from the intrinsic deformability of cell membranes induced by mechanical stimuli acting on them. An excellent example illustrating this is the response of the tissue cells to the stiffness of their substrate (Discher, Janmey, & Wang, 2005).

The recognition of substrate sensing as a fundamental property of most tissue cells represents a major contribution toward understanding the importance of cell membrane mechanics for the force-dependent evolution of life forms. Cell membrane and tissue elasticity correlate tightly with the tissue cell response to the stiffness of their microenvironment. For their growth and development, tissue cells depend on their adherence to a substrate, which can be rigid or soft. Tissue cells, such as fibroblasts or neurons, can thus sense the stiffness of the tissue matrix they are part of. Growth on soft materials, for example, is characteristic of cancer cells (Abidine et al., 2018). For the cellular mechanics and stiffness sensing adhesion complexes and the actin-myosin cytoskeleton are important players allowing the tissue cells to "interpret" correctly local matrix stiffness in order to develop and differentiate accordingly. Furthermore, external mechanical stimuli acting on cell membranes are focused and distributed throughout the tissue by ECM via the physical coupling between ECM molecules (e.g. fibronectin) and the CSK through integrins. By forming focal adhesions on the surface of animal cell membranes integrins allow mechanical forces to be focused on CSK components, which may mediate directly or indirectly transmission of forces to membrane proteins (Wang, Butler, & Ingber, 1993), including mechanosensitive ion channels, whose conformational changes mediate conversion of mechanical stimuli into electrical or chemical signaling regulating cellular dynamics (Hamill & Martinac, 2001).

3. Mechanosensitive ion channels

3.1 Major types of mechanosensitive ion channels

There are numerous physiological processes requiring MS ion channel involvement as primary molecular transducers of mechanical stimuli in living cells. The sense of touch, hearing and proprioception as well as osmotic gradients and gravitropism all require MS channel involvement at the origin of mechanosensory signaling cascades regulating the physiology of these processes (Árnadóttir & Chalfie, 2010; Cox, Bavi, & Martinac, 2017; Douguet & Honoré, 2019; Hamill & Martinac, 2001; Sachs, 2010) (Fig. 4). They are found in all three kingdoms of organisms on the evolutionary tree (Bacteria, Archaea and Eukarya) and are thus likely to be among the evolutionarily oldest molecules, which evolved to support life.

In prokaryotes Bacteria and Archaea, MscS- and MscL-like families of ion channels are the only types of the MS class of channels known to exist (Martinac, 2004). The main variety of MS channels found in Eukaryotes includes (i) the family of two-pore-domain (2P-domain), weakly inward-rectifying K^+ channels consisting of 15 members of which three, TREK-1, TREK-2 and TRAAK, are mechanosensitive (Honoré, 2007; Patel & Honoré, 2001), (ii) ENaC epithelial amiloride-sensitive Na^+ channel belonging to the MEC/DEG (for "mechanosensory abnormal/degenerins") family (Árnadóttir & Chalfie, 2010), (iii) Piezo1 and Piezo2, the members of the long-sought-after family of MS ion channels (Murthy, Dubin, & Patapoutian, 2017), (iv) the superfamily of TRP (for "transient receptor potential") ion channels, which are involved in many mechanotransductory cellular processes (Hamill & Martinac, 2001; Minke & Cook, 2002), although mechanosensitivity of the mammalian TRP channels has been somewhat controversial (Gottlieb et al., 2008; Nikolaev et al., 2019) as will be discussed

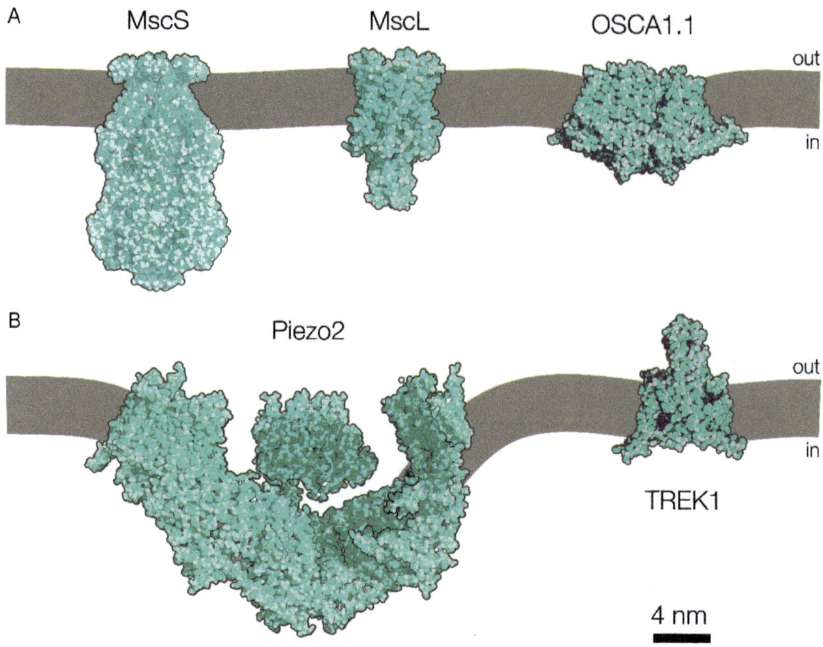

Fig. 4 Mechanosensitive channels found in organisms across the kingdoms of life forms.

later, and (v) the recently discovered OSCA/TMEM63 evolutionarily conserved family of mechanosensitive ion channels (Maity et al., 2019; Murthy et al., 2018). Each of these MS channel families is discussed next in more detail.

3.2 Bacterial and archaeal MS channels

Bacterial MS channels have been extensively studied in both Gram-negative and Gram-positive bacteria (Blount & Moe, 1999; Martinac, 2001; Zoratti & Ghazi, 1993). The most comprehensive characterization of bacterial MS channels has been completed in *E. coli* (Cox et al., 2018). The presence of MS channels in Archaea, the third domain of the universal phylogenetic tree (Pace, 1997; Woese, 1994), was first shown in the halophilic archaeon *Haloferax volcanii* (Le Dain et al., 1998), and later also in methanogenic archaeon *Methanococcus jannaschii* (Kloda & Martinac, 2001). The existence of MS channels in membranes of both prokaryotic organisms suggests that MS channels may have been among the first mechanoreceptors that evolved in primordial cells, which have enabled the bacterial and archaeal microbes throughout evolution to cope with environmental mechanical stimuli such as osmotic pressure (Martinac & Kloda, 2003).

3.2.1 MscL-like channels

MscL, the large conductance MS channel of *E. coli* was the first MS channel, whose gene was cloned and the structure was resolved by X-ray crystallography (Chang, Spencer, Lee, Barclay, & Rees, 1998; Sukharev, Blount, Martinac, Blattner, & Kung, 1994; Sukharev, Blount, Martinac, Kung, & Ching, 1997). MscL and his numerous homologous are the true osmosensors serving as osmotic valves by protecting Gram-negative and Gram-positive bacteria, the cyanobacteria from Synaechocystis sp. (Moe, Blount, & Kung, 1998; Sukharev et al., 1997), fungi (e.g. Neurospora) and fungus-like organisms (Phytophthora) as well as Archaea (e.g. Methanosarcina) (Martinac et al., 2014) from potentially deleterious effects of hypoosmotic shocks exceeding the pressure that can be sustained by the cell wall only. The 3D homopentameric structure of MscL-like channels (Chang et al., 1998; Li et al., 2015) has been highly conserved among all these species suggesting that these membrane channel proteins have successfully fulfilled their function as osmosensors for billions of years (Cox et al., 2018; Martinac et al., 2014).

MscL-like channels have the largest pore and highest activation threshold among all known mechanically gated channels. EcMscL has been characterized to require large tension threshold of 10–12 mN/m for its activation, which is close to lytic tension of the lipid bilayer (Nomura et al., 2012). Activation by amphipaths, such as LPC, has proven that MscL is gated by the "force-from-lipids" principle, which in essence corresponds to the changes of transbilayer pressure profile (Martinac et al., 2018; Teng, Loukin, Anishkin, & Kung, 2015) that will be dealt with in detail later in this chapter.

The 3D crystal structure of *Mycobacterium tuberculosis* MscL in a closed state shows a homopentameric channel with each monomer consisting of an N-terminal short amphipathic helix S1, two transmembrane helices (TM1 and TM2) connected by a periplasmic loop and a cytoplasmic C-terminal helix (Chang et al., 1998). The closed-to-open transition is driven by increased membrane tension and as initiation, the S1 helix is dragged away from the central pore on the cytoplasmic side, and simultaneously TM1 helix starts to tilt horizontally away from the periplasmic side (Bavi et al., 2016). This "tug-of-war" helix tilting between S1 and TM1 leads to the "iris-like" expansion resulting in a large open pore of 30 Å in diameter of the MscL channel (Cruickshank, Minchin, Le Dain, & Martinac, 1997; Wang et al., 2014). MtMscL requires even larger tension of 22 mN/m for its activation than EcMscL (12 mN/m) upon reconstitution into azolectin liposomes. Mechanosensitivity of MtMscL becomes significantly reduced by adding phosphatidylinositol (PI) to the liposome mix (Zhong & Blount, 2013). Ion mobility mass spectrometry demonstrated that lipids bind selectively to MscL, especially PI, suggesting that there is a specific lipid-channel interaction playing a role for MscL mechanosensitivity (Laganowsky et al., 2014). A recently proposed "lipid moves first"-hypothesis suggests that lipid molecules are penetrating into the nanopockets (NP) at the entrance and proximal regions of MscL channels in the closed state, and the lipid acyl chains in the NPs prevent the closed-to-open conformational change like the safety-pin of a grenade (Kapsalis et al., 2019). In the MscL case, lipid chains act as negative allosteric modulators and by moving out from the NPs due to increased membrane tension should result in a decrease of free energy required to open the channel. Nevertheless, modulation of MscL mechanosensitivity by specific lipids requires further study.

3.2.2 MscS-like channels

MscS was the first bacterial mechanosensitive channel from *E. coli* originally discovered as one of the six MscS-like channels identified in giant spheroplasts (Edwards et al., 2012; Martinac, Buechner, Delcour, Adler, & Kung, 1987). Unlike MscL-like channels, MscS-like channels have been implicated not only in osmoregulation of Bacteria and Archaea, but also in the functions in the eukaryotic organellar membranes, such as chloroplasts (Haswell & Meyerowitz, 2006), mitochondria (Lee, Kim, & Cohen, 2016) and endoplasmic reticulum (Nakayama, Yoshimura, & Iida, 2012). In cell-walled organisms, bacteria, archaea, fungi, algae, and plants, MscS-like channels are highly diverse in structure and function (Kloda & Martinac, 2002). In bacteria, for example, MscCG, the MscS-like channel of *C. glutamicum* functions as glutamate exporter (Börngen et al., 2010), suggesting that the MscCG channel family has been evolutionarily specialized (Cox, Nakayama, Nomura, & Martinac, 2015; Nakayama et al., 2018).

E. coli MscS has been studied as the model channel for all MscS-like channels because of its simplest structure. Like EcMscL, EcMscS was reconstituted into liposomes and characterized by the patch clamp technique showing that the channel requires moderate tension of 5–7 mN/m for its activation (Nomura et al., 2012; Sukharev, 2002). EcMscS is like EcMscL also activated by amphipaths, meaning that the gating of MscS follows the "force-from-lipids" principle (Martinac, Adler, & Kung, 1990; Teng et al., 2015). The structural dynamics directing the gating mechanism of MscS is significantly different from MscL. The crystal structure of *E. coli* MscS shows a homoheptameric channel in an inactivated state with each monomer consisting of three transmembrane helices (TM1, TM2, and TM3) (Bass, Strop, Barclay, & Rees, 2002). The pore-forming helix TM3 is bent at the kink G113, which leaves a void region between TM2 and TM3. Mass-spec and MD simulations suggested that lipid molecules are interdigitating into this void as a lipid nano-pocket (NP) (Pliotas et al., 2015). The "lipid moves first" model proposed for MscS gating suggests that upon removing the interdigitating lipids from NPs by membrane tension the channel can open by changing its conformation from closed to open. However, the recent cryo-EM EcMscS structure revealed that MscS transmembrane domain is positioned ~17 Å higher in the bilayer compared to its crystal structure (Reddy et al., 2019). Thus, it is unlikely that there are interdigitating lipids present in the NPs and the "lipid-moves-first" model will have to be revised in the future.

Mechanosensitivity of EcMscS is affected by membrane stiffness, whereas EcMscL is mainly affected by membrane thickness (Nomura et al., 2012; Xue et al., 2020). This is consistent with the previously proposed "Jack-in-the box" model (Malcolm, Blount, & Maurer, 2015). As discussed later, the MscS sensitivity to membrane stiffness results from the effect of the transbilayer pressure profile on the channel structural changes between the closed and open conformations (Ridone et al., 2018). Not surprisingly, the structure and lipid content of bacterial membranes are highly diverse (López-Lara & Geiger, 2017) given the variety of the environments in which they exist. In fact, *C. glutamicum* MscCG channels display different activation/deactivation profile in their native soft cell membrane compared to the harder cell membrane of *E. coli* (Nakayama et al., 2018). This membrane stiffness sensing mechanism may be a key to understand the gating mechanism of very diverse members of the MscS-like channel family (Cox et al., 2015).

E. coli cells possesses six MscS-like channels, which were most likely a result of two successive gene duplication events during evolution (Cox et al., 2015). Initially, there were three MscS-like channels identified in *E. coli*, which were named as MscM (M for mini; also referred to as YjeP) and MscS (S for small) based on their ion conductance, and MscK (named for its dependence on external K^+; also referred to as kefA). The other three MscS-like channels include YbdG (MscM-like), ybiO (its expression depends on the presence of external Na^+) and ynaI (Edwards et al., 2012; Martinac & Cox, 2017). Although all MscS-like proteins can function as MS channels in patch clamp experiments not all of them can protect bacterial cell integrity upon increase in cellular turgor caused by a hypo-osmotic shock. This is due to low levels of expression or integration into the membrane of YbiO, YnaI and YjeP (Edwards et al., 2012).

3.3 MS channels of eukaryotes

3.3.1 DEG/ENaC channels

The founding members of the DEG/ENaC subfamily of proteins mediate the gentle touch response in *Caenorhabditis elegans* (Árnadóttir & Chalfie, 2010). Mutations of DEG genes induced cell death (*degeneration*) confined to touch-receptor neurons, thus the family was named degenerins. The exact molecular make-up of the MS channel that mediates the mechanosensitive currents is still being mapped-out reminiscent of the auditory MS channel complex. It is clear, that not all DEG proteins (UNC, MEC, DEL, DEG) are ion channel subunits. The ion conducting pore seems to be formed by MEC-4 and MEC-10 proteins (MEC = ***mec***hanosensitive) which are likely tethered to a host of auxiliary scaffold proteins such as MEC-2 and MEC-6 (Cueva, Mulholland, & Goodman, 2007; Du, Guoqiang, William, & Chalfie, 1996; Emtage, Guoqiang, Hartwieg, & Chalfie, 2004).

The human genome contains genes which are enclosing orthologues of DEG proteins called epithelial Na^+ channel (ENaC) proteins. Along with proteins from *C. elegans* this family shares a common trimeric structure where each monomer consists of two transmembrane domains connected by a large extracellular region (Noreng, Bharadwaj, Posert, Yoshioka, & Baconguis, 2018). In addition to DEGs and ENaCs this family also includes a number of other ion channel subclasses including proton-gated acid sensing ion channels (ASICs) (Cristofori-Armstrong & Rash, 2017; Lin et al., 2016). Although these channels have been implied to play a role in mechanotransduction processes like blood pressure sensing and nociception, clear evidence that these channels are directly activated by mechanical stimuli is lacking (Ben-Shahar, 2011; Lin et al., 2016).

The ENaC channel is expressed in a variety of epithelial cells including the kidney, lung, and colon, where it is involved in maintenance of body salt and water homeostasis. In addition, ENaC channels have been linked to mechanotransduction processes in arterial baroreceptors and vascular smooth muscle cells. The ENaC channel has three subunits (α-, β-, and γ-ENaC), which show differential sensitivity to shear stress. Much like MEC channels ENaC is thought to be tethered to the extracellular matrix however, the mechanism seems to be different involving N-linked glycosylation. ENaC responds to shear force in endothelial cells and its shear force-sensing mechanism was shown to depend on direct contact with the extracellular matrix via two glycosylated asparagines. This direct connection of the alpha subunit of ENaC with the extracellular matrix is crucial for shear force-sensing role of ENaC in blood pressure regulation (Knoepp et al., 2020).

3.3.2 Two-pore domain K^+ channels (K2P)

Two-pore domain K^+ channels (K_{2P}) are widely distributed in excitable and non-excitable cells and show extensive polymodal regulation (i.e. mechanical force, pH, heat) (Douguet & Honoré, 2019; Patel et al., 1998). This channel family also has a unique topology relative to other K^+ channels characterized by two pore-forming domains. Currently, the K_{2P} subfamily includes 15 members of which only 3 show robust mechanical activation: ***T***wik-***re***lated ***K***$^+$ channels (TREK1; $K_{2P}2.1$ and TREK-2; $K_{2P}10.1$), and the ***T***wik-***r***elated ***a***rachidonic acid ***a***ctivated ***K***$^+$ (TRAAK; $K_{2P}4.1$).

The TREK-1 channel is regulated by a variety of physical and chemical stimuli including membrane tension (activation tension ~0.5–7 mN/m) (Brohawn, Zhenwei, & MacKinnon, 2014). Membrane patch excision or disruption of the actin cytoskeleton increased channel activity and reduced desensitization suggesting that membrane tension is likely to be directly transmitted to the channel via the lipid bilayer a fact confirmed by channel reconstitution. Importantly however, the expression of TREK-1 alters the cytoskeletal network and induces the formation of membrane protrusions rich in actin suggesting dynamic channel-cytoskeletal interaction (Lauritzen et al., 2005). Much like bacterial channels TREK-1 activity is modulated by amphipathic molecules characteristic of channels gated according to the force-from-lipids paradigm (Cox et al., 2019). Anionic amphipaths including trinitrophenol and PUFAs such as arachidonic acid stimulates the activity of the TREK-1 channel. In contrast, cationic amphipaths such as chlorpromazine inhibit channel activity. TREK-1/2 channels are also blocked by antidepressants and activated by volatile general anesthetic agents, the mechanism of which is still disputed.

3.3.3 Piezo channels

The Piezo channel family is comprised of two members, Piezo1 and Piezo2. Initial discovery of these channels by the Patapoutian lab reported in 2010 has led to intense research focus as both ion channels are broadly expressed and seem to be essential in a host of physiological processes (Coste et al., 2010). These proteins are pore-forming subunits of homotrimeric non-selective cationic mechanosensitive channels with characteristic sensitivity to generic MS channel blockers such as streptomycin, Gd^{3+}, Ruthenium red and the spider toxin GsMTx4 (Guo & MacKinnon, 2017; Saotome et al., 2018; Wang et al., 2019). A study in which heteromer formation between Piezo1 and Piezo2 was tested, Gnanasambandam, Bae, Ziegler, Sachs, and Gottlieb (2018) were unable to observe any features suggesting the presence of heteromeric channels in their experiments (Gnanasambandam et al., 2018). However, using "split protein" technique the same authors were able to show that the co-assembly of the N- and C-terminal halves of the human Piezo1 mutant channels resulted in functional "heteromeric" mutant channels (Bae, Suchyna, Ziegler, Sachs, & Gottlieb, 2016) similar to what was previously shown for the potassium TOK1 (Saldaña, Naranjo, Coria, Peña, & Vaca, 2002) and *E. coli* MscL channel (Park, Berrier, Martinac, & Ghazi, 2004). As a mark of Piezo1 and Piezo2 importance, global knockouts of either channel in mice results in lethality. For Piezo1 this happens during the embryonic stage (E9.5–10.5) and for Piezo2 this occurs up to 24h after birth (Martinac & Cox, 2017).

Piezo1 channels have emerged as key transducers of mechanical force in the cardiovascular system whereas Piezo2 is largely confined to sensory neurons. Piezo1 channels sense membrane forces (activation tension ~1–3mN/m) and is sensitive to lipid bilayer composition (Ridone, Vassalli, & Martinac, 2019; Syeda et al., 2016). This does not preclude modulation or even direct interactions with ECM or cytoskeletal proteins akin to ENaC and TMC proteins (Martinac & Cox, 2017). Indeed, Bavi, Richardson, Heu, Martinac, and Poole (2019) showed that forces arising from actin-mediated contractility acted synergistically with the force-from-lipids to regulate PIEZO1 activation by stimuli applied at contacts between cells and their surroundings (Bavi et al., 2019). In other words, depending on the physical environment of the cells, and possibly their physiological conditions as well, either force-from-lipids or force-from-filament mechanism may be more effective in the force transmission to Piezo channels. Piezo2 does not open in classical stretch assays, which seems to suggest a greater reliance of this channel on molecular tethers (Wu, Lewis, & Grandl, 2017).

3.3.4 TMC (transmembrane channel-like) proteins

TMC1 and TMC2 (transmembrane channel-like proteins 1 and 2) are linked to dominant and recessive non-syndromic hearing loss and are localized to mechanosensory transduction sites at the tips of stereocilia in the ear (Corey & Holt, 2016; Pan et al., 2018). Recent work suggests that both TMC1 and TMC2 are pore-forming subunits of the auditory mechanotransduction channel underlying the sense of hearing. While the exact composition of the pore is still the subject of intense research TMC proteins seem to be integral components. However, when reconstituted into liposomes, TMC1 and TMC2 proteins were recently shown to form mechanically activated channels, indicating that like MscL and Piezo1 channels they are inherently mechanosensitive (Jia et al., 2020). TMC1 has also been shown to interact directly with several other proteins essential for auditory transduction. These likely act as molecular tethers transmitting force directly to the channel-like DEG/ENaC channels.

3.3.5 New channel candidates

One exciting feature of channel-mediated mechanosensory transduction is that there are many primary mechanosensitive currents whose genetic origins are still unknown. Recently, two new candidates have emerged as potential pore-forming subunits of mechanosensitive channels. This includes Elkin1 (TMEM87a) (Patkunarajah et al., 2020) and TACAN (TMEM120A) (Beaulieu-Laroche et al., 2020). Elkin1 seems to be a component of mechanotransduction pathway independent of Piezo1 in melanoma cells. TACAN is a protein which plays a role in the sensation of noxious mechanical stimuli. It is likely that more novel mechanosensitive channels will come to light in the future.

3.3.6 Contribution of TRP ion channels to mechanotransduction

The transient receptor potential (TRP) family of ion channels are involved in a wide variety of mechanosensory processes, including touch sensation, pain, blood pressure regulation, bone loading and detection of cerebrospinal fluid flow (Christensen & Corey, 2007; Eijkelkamp, Quick, & Wood, 2013; Lin & Corey, 2005; Patel et al., 2010; Yin & Kuebler, 2010).

In Drosophila, mechanosensitive NOMPC ion channel (mechanoreceptor potential C) is involved in touch sensation and hearing (Yan et al., 2013). NOMPC has been shown to be gated by mechanical force by pulling the tethers (composed of 29 ankyrin repeats) associated with microtubules (Zhang et al., 2015). Such association with microtubules was shown to be required for mechano-gating of NOMPC ion channel. In contrast, in mammals, mechanosensitivity of TRP ion channels has been the subject of debate. It is accepted that some TRP channels such as TRPV4, respond to mechanical force (Loukin, Zhou, Zhenwei, Saimi, & Kung, 2010; Teng,

Loukin, Zhou, & Kung, 2013) however, whether TRP channels are primary mechanosensors that respond to membrane stretch was never demonstrated.

TRP ion channels, in mammals are divided into six subfamilies (33 members total): TRPC or canonical, TRPV for vanilloid, TRPM for melastatin, TRPP for polycystin, TRPML for mucolipin and TRPA for ankyrin (Clapham, 2003). TRP channels are polymodally gated and are not only involved in mechanosensory processes but also in various sensations in mammals, including taste, temperature, pain, pressure, and vision (Julius, 2013; Vriens, Nilius, & Voets, 2014).

A recent report contested whether mammalian TRP family of ion channels can directly be activated by mechanical stretch (Nikolaev et al., 2019). In this study, 11 TRP channels from 6 subfamilies were tested for their ability to be gated by membrane stretch using patch clamp technique. It was found that TRP ion channels overexpressed in cellular systems were insensitive to membrane stretch. Furthermore, the purified TRPC6 was shown to be insensitive to bilayer stretch in liposomes, an artificial lipid system free from any cellular components such as cytoskeleton. Other studies also do not support the mechanosensitivity of TRP channels (Constantine, 2016; Gottlieb et al., 2008; Servin-Vences, Moroni, Lewin, & Poole, 2017). This does not exclude the TRP channel involvement in cellular mechanotransduction processes but does indicate, that they do not function as the primary mechanotransducers. If any of the TRP channels could act as a primary mechanotransducer, then its mechanosensitivity would be dependent on additional transmembrane proteins or tethers that link them to structural scaffold proteins, like NOMPC that is linked to microtubules (Jin et al., 2017).

3.3.7 Modulation of NMDAR by mechanical force

N-methyl-D-aspartate receptor (NMDAR) is a ligand-gated cation channel requiring binding of glycine and glutamate as well as the voltage-dependent removal of Mg^{2+} from the channel pore for its activation (Johnson, Battle, & Martinac, 2019). 3D X-ray structure of NMDAR, solved at 4 Å resolution, shows that the channel is assembled as dimer of GluN1/GluN2 heterodimers (Karakas & Furukawa, 2014; Lee & Gouaux, 2011). Over 25 years ago NMDAR was also shown to exhibit mechanosensitive properties (Paoletti & Ascher, 1994). In addition to membrane stretch, NMDAR could be activated by amphipaths lysophosphatidylcholine (LPC) as well as arachidonic acid (AA) (Casado & Ascher, 1998). These findings were later confirmed by patch clamp recordings from purified NMDAR channel proteins reconstituted into liposomes, which indicate that NMDAR is inherently mechanosensitive like MscS, MscL, TREK-1, TREK-2, TRAAK, Piezo1, and OSCA channels (Brohawn et al., 2014; Cox et al., 2016; Kloda, Martinac, & Adams, 2007; Murthy et al., 2018; Syeda et al., 2016). This shows that lipid-protein interactions and thus, force-from-lipids are sufficient for the channel activation by mechanical force. More recently, NMDAR was shown to be activated by shear stress (Maneshi et al., 2017). All these studies provide a substantial evidence for NMDAR mechanosensitivity.

Furthermore, it has been reported that gating of NMDAR channels also depends on functionally important interactions of the channels with cytoskeletal proteins, such as actin, which can directly regulate NMDAR function (Rosenmund & Westbrook, 1993). Thus, both force-from-lipids and force-from-filament mechanisms can modulate mechanosensitivity of this ligand-gated channel showing that cytoskeletal tension may interact with bilayer tension (Cox et al., 2016) and that the force activating MS channels can be shared between the lipid bilayer and the CSK, as demonstrated for Piezo1 channels (Cox et al., 2019).

3.4 Interaction between MS channels and cell membrane components

3.4.1 Understanding MS channel gating according to the force-from-lipids principle

MS channels can be modulated by global biophysical properties of the lipid bilayer, which include thickness (order parameter), area per lipid, area compressibility and bending elastic moduli, as well as intrinsic monolayer curvature (Bavi et al., 2017). These global properties are, however, a function of the nanoscale bilayer properties such as degree of fatty acid chain unsaturation, head group geometry and composition. For example, changing the concentration of cholesterol in mixed PC membranes increases the bilayer thickness and bending elastic moduli without changing the area compressibility. This has a significant effect on activation threshold of MS channels (Evans, Rawicz, & Smith, 2013; Nomura et al., 2012). In addition, the variation of bilayer thickness as well as the degree of freedom of the acyl chain in the vicinity of MS channels can modulate the channel gating properties via altering the lipid-protein hydrophobic mismatch (Perozo, Kloda, Cortes, & Martinac, 2002). Also, the flexibility of phospholipids enables them to penetrate into functionally important fenestrations where the lipids can directly interact with the pore-lining region of the channels (Aryal, Abd-Wahab, Bucci, Sansom, & Tucker, 2014, 2015; Bavi, Vossoughi,

Naghdabadi, & Jamali, 2016). Furthermore, the level of fatty acid saturation introduces changes in the "transbilayer pressure profile" (Ridone et al., 2018), which is balanced between the single monolayers along the thickness of the bilayer due to anisotropic forces resulting from the "hydrophobic effect" that water exerts on the lipid bilayer (Chandler, 2005).

Changes in the force balance between the two monolayers of the lipid bilayer have been shown to activate and modulate the gating of MS channels (Bavi, Cox, Perozo, & Martinac, 2017; Bavi, Vossoughi, Naghdabadi, & Jamali, 2014; Martinac et al., 2018; Nomura et al., 2012). For example, the change in the transbilayer pressure profile (also referred to as the lateral pressure profile) has been proposed to generate enough force for the transition from "down" to "up" conformation of TREK-2 MS channels (Aryal et al., 2017; Clausen, Jarerattanachat, Carpenter, Sansom, & Tucker, 2017). Consequently, anisotropic forces associated with the transbilayer pressure profile and modulation of MS channels by changes in the force balance inside the lipid bilayer underlie the force-from-lipids principle as the unifying paradigm of most likely all mechanotransduction processes originating at the cell membrane (Martinac et al., 2018) (Table 2).

Table 2 Activation of inherently mechanosensitive ion channels reconstituted into lipid bilayers by membrane tension.

Evolutionary domain	MS channel	Force activation mechanism	Tension required for channel half-activation *T*1/2 (mN/m)	References
Bacteria	*E. coli* MscL	Bilayer tension	~12	Nomura et al. (2012)
Bacteria	*E. coli* MscS	Bilayer tension	~6	Nomura et al. (2012)
Eukaryote	TREK-1	Bilayer tension	~0.5–7	Brohawn et al. (2014)
Eukaryote	TRAAK	Bilayer tension	~0.5–7	Brohawn et al. (2014)
Eukaryote	Piezo1	Bilayer tension	~2–3	Lewis and Grandl (2015) and Cox et al. (2016)
Eukaryote	OSCA1.1/ OSCA1.2	Bilayer tension	~4.5–5	Murthy et al. (2018)

3.4.2 Transbilayer pressure profile

In a hydrated lipid bilayer, the transbilayer pressure profile comes from the repulsive lipid-lipid interactions (head groups and chains) and the cohesive tension in the contact area of the lipid tails with water molecules. The amphipathic nature of lipid molecules helps them to self-assemble such that they minimize the exposure of carbon acyl tail to water. The integrity and stability of this structure requires a balance between the repulsive interaction of head groups, entropic (steric) repulsion between the acyl chains, and the attractive interaction (surface tension) of the polar-apolar interface of water and chains (Fig. 5A). The surface tension around the glycerol backbone prevents water penetration to the hydrophobic core of the bilayer.

The free energy of a hydrated lipid bilayer, Δg, represents the sum of the lipid-lipid interactions, g_{int}, and the surface lipid-water interactions, g_{sur} as follows:

$$\Delta g = \Delta g_{int} + \Delta g_{sur} \tag{1}$$

Δg_{int} corresponds to the van der Waals and possible electrostatic potentials of head groups and chains. Δg_{sur} corresponds to the surface energy, γA_l where γ is the surface tension and A_l is the surface area per lipid molecule (Marsh, 2007). Surface tension, γ, can be calculated from the derivative of free energy with respect to the area at the temperature, T, of the system:

$$\gamma = -\left(\frac{\partial \Delta g}{\partial A_l}\right)_T \tag{2}$$

The transbilayer pressure profile, P, has units of force per unit area and can be calculated from the derivative of the surface tension with respect to the axis along the bilayer normal, z:

$$P(z) = \left(\frac{\partial \gamma}{\partial z}\right) \tag{3}$$

In the absence of any external force, the sum of the positive and negative pressures in the lipid bilayer is zero because the pressures compensate for each other and the membrane is at rest:

$$\int_0^t P(z)dz = 0 \tag{4}$$

where t is the thickness of the bilayer.

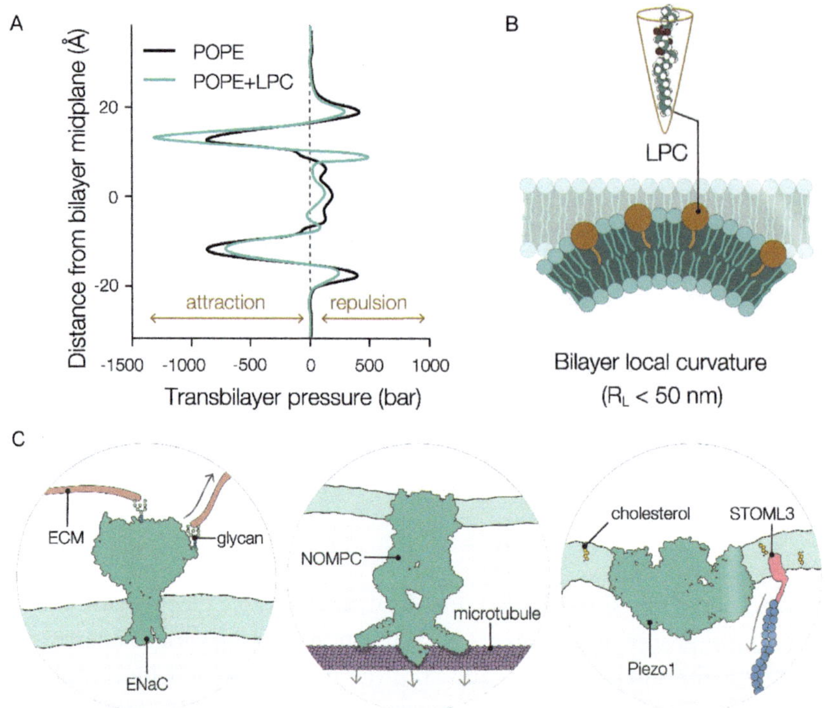

Fig. 5 Transbilayer pressure profile and bilayer curvature. (A) The effect of asymmetric incorporation of amphipathic molecules on the redistribution of the transbilayer pressure profile along the membrane thickness. (B) Schematic of the local curvature in a bilayer induced by adding LPC. (C) Examples of gating structures in activation of mechanosensitive ion channels by mechanical stimuli. Shear force activation of ENaC via ECM (left) (Knoepp et al., 2020), NompC activation by microtubules pulling on the channel ankyrin repeats (middle) (Zhang et al., 2015) and facilitation of the Piezo1 channel activation by STOML-3 interaction with membrane cholesterol (right) (Poole, Herget, Lapatsina, Ngo, & Lewin, 2014).

3.4.3 Parameters defining membrane mechanical properties

The defining parameters describing membrane mechanical properties include membrane surface tension (γ), local membrane curvature, bending modulus (K_b), area compressibility modulus (K_A), and Young's modulus (E). Here we provide a summary of how these parameters relate to each other in the context of membrane mechanics (Table 3). Note that these parameters depend on the properties of the membrane, which can vary from one lipid species to another. Especially, the presence of cholesterol or gel-liquid phase as well as a change in the acyl chain length or the linkage in the headgroup of

Table 3 Parameters defining membrane mechanical properties.

Parameter	**Symbol**	**Value**	**Unit**	**References**
Aqueous solution—lipid monolayer surface tension	γ	~50	mN/m	Gullingsrud and Schulten (2004)
Bending modulus	K_b	$\sim 10^{-20}$–10^{-19}	J	Bavi et al. (2014)
Area compressibility modulus	K_A	~16–250	mN/m	Bavi et al. (2014) and Picas, Rico, and Scheuring (2012)
Young's modulus	E	~4.5–19	MPa	Bavi et al. (2014) and Evans and Needham (1987)

a specific lipid can significantly affect the parameter values (Bavi et al., 2014; Evans & Needham, 1987; Picas et al., 2012).

The calculated values for $p(z)$ are large, varying between 400 and 1500 bar for different phospholipids (Bavi, Cox, Bavi, & Martinac, 2017; Cox et al., 2017, 2018; Gullingsrud & Schulten, 2004; Ollila, Róg, Karttunen, & Vattulainen, 2007). The resulting membrane surface tension is ~50 mN/m (Gullingsrud & Schulten, 2004) which can be calculated from the area under the transbilayer pressure profile curve (Fig. 5A) divided by bilayer thickness.

We should note that unlike the first moment of stress through the bilayer, the second moment, which is proportional to the spontaneous curvature of the lipid bilayer, could be non-zero even at rest:

$$\int_0^t zP(z)dz = K_b\chi \tag{5}$$

where χ is the equivalent local curvature (m^{-1}) and K_b is the bending modulus of the bilayer.

Eq. (5) describes the combined effect of changing the distribution of the pressure profile and the induced local curvature. For example, in regions of local curvature ($R_L < 50$ nm), which is in the range of experimentally determined values for caveolae, the transbilayer pressure profile changes of up to 20% can significantly change the overall membrane stress/strain which can modulate the transmembrane MS channels (Bavi et al., 2016). As shown in Fig. 5B, the change in local curvature of the membrane can result from the addition of amphipathic molecules such as lysophosphatidylcholine (LPC) to the lipid bilayer (Bavi, Cox, et al., 2016; Bavi, Nikolaev, Bavi, et al., 2017).

In addition to the bending modulus of the lipid bilayer, K_b, in Eq. (5), the area compressibility modulus, K_A, can be determined directly from the surface tension, γ:

$$K_A = A_l \left(\frac{\partial \gamma}{\partial A_l} \right)_T \tag{6}$$

Another way of determining the area compressibility modulus for tensionless membranes, using MD simulations, is by averaging the fluctuation of the surface area of a lipid molecule at a constant temperature, T:

$$K_A = \frac{k_B T \langle A \rangle}{\langle \delta A^2 \rangle} \tag{7}$$

where k_B is the Boltzmann constant, $\langle A \rangle$ is the average surface area, and $\langle \delta A^2 \rangle$ is the mean square fluctuation of the surface area of the lipid molecule.

Finally, the global mechanical properties of a membrane bilayer; bending, area compressibility, and Young modulus (E) (a measure of material elasticity) relate to each other as follows:

$$K_b = \frac{Et^3}{24} = \frac{K_A t^2}{24} \tag{8}$$

where t is thickness of the bilayer.

These characteristic properties of lipids provide quantitative criteria to verify if they can bear external forces or not, whereby the strength of external forces varies between different applications. For example, designing membrane desalination systems or efficient liposomal drug delivery systems using MscL channels as "smart" nanovalves needs sufficient information about the mechanical and thermal stability of both the lipid and the channel protein. In desalination platforms, bending and stretching stiffness of the bilayers under various mechanical loading greatly affect the desalination efficiency and structural stability of these systems (Bahari, 2020). In liposomal drug delivery systems, one should evaluate the mechanical strength of the lipid and the protein with regard to the traction force of attaching superparamagnetic particles to the N-terminus of the MscL channel and possible increase in temperature by the applied magnetic fields (Bavi et al., 2017; Martinac, Bavi, Bavi, & Martinac, 2017;

Nakayama et al., 2015). These applications emphasize the importance of calculations of the mechanical properties of the employed lipids and proteins.

3.4.4 CSK and ECM role in gating of MS ion channels

The current knowledge of basic principles of mechanosensory transduction processes involving MS channels includes the force-from-lipids principle as the essential physicochemical principle underlying MS channel gating by mechanical force (Cox et al., 2019; Teng et al., 2015). Originally conceived as an intuitive notion of close protein-lipid interactions between the MS channels and surrounding lipid bilayer (Martinac et al., 1990) the force-from-lipids principle has over the years found strong support thanks to discovery and functional characterization of new MS channels, including the Piezo and OSCA channel families (Coste et al., 2010; Murthy et al., 2018; Syeda et al., 2016). Nevertheless, given the complexity of animal and human cells it seems obvious that CSK and ECM should also contribute to the MS channel function as originally defined by the force-from-filament principle (Árnadóttir & Chalfie, 2010; Guharay & Sachs, 1984; Katta, Krieg, & Goodman, 2015; Martinac, 2014) (Fig. 5C). For many years, the two principles contradicted to each other and this dichotomy dominated discussions about the MS channel-mediated mechanotransduction. However, a dynamic force sharing between the lipid bilayer and scaffold proteins modulating inherent mechanosensitivity of Piezo1 channels, for example, Poole et al. (2014), Qi et al. (2015), and Retailleau et al. (2015), emerged as a meaningful resolve for this long-standing dichotomy. Another case in point is the NompC (TRPN) channel, which is activated by mechanical force pulling on microtubules coupled to a tether of 29 ankyrin repeats attached to the N-terminus of the channel (Retailleau et al., 2015; Zhang et al., 2015). Although NompC has not been shown to be gated directly by the force-from-lipids it has however, been shown to become no longer responsive to mechanical stimulation when the S4-S5 linker stops interacting directly with the lipid bilayer due to a single residue mutation within the linker (Jin et al., 2017). Consequently, the lipid bilayer alone seems essential and sufficient to mechanically gate Piezo or NompC channels (Cox et al., 2016; Jin et al., 2017; Syeda et al., 2016), whereas structural CSK proteins are required for modulation and fine-tuning of the channel gating depending on the cellular and tissue specific environment.

4. Summary and conclusions

As the interface enabling living cells to exchange nutrients and ions with their environment, the cell membrane is a major boundary transmitting various stimuli between intra- and extracellular compartments and converting them into intracellular signals. The focus of this chapter is on mechanical forces and cellular mechanosensory transduction enabled by complex interactions between different membrane components and a variety of membrane proteins. Divided in two major parts, the chapter first describes the mechanical properties of cell membranes and their components as well as the methods and techniques used to determine these properties, which is followed by focusing reader's attention on mechanosensitive ion channels by reviewing their diversity and mechanism of activation by mechanical forces.

In summary, the biophysical principles of mechanosensitive-channel-gating mechanisms can be reduced to two major paradigms known as "force-from-lipids" and "force-from-filament" principles. As reported in a number of recent publications and summarized here, both principles can be reduced to the anisotropic transbilayer pressure profile, which can change depending on the shape of a membrane protein, such as a mechanosensitive channel, and its specific interactions with the cytoskeleton and/or the extracellular matrix. Given that the lipid bilayer is much more compliant compared to a membrane protein a displacement of a mechanosensitive channel with respect to the lipid bilayer by the CSK/ECM pushing or pulling on the channel could change anisotropy of the transbilayer pressure profile and thus trigger the channel opening or closing. Consequently, the amount of force sharing between the lipid bilayer and other membrane components affecting structural dynamics of a mechanosensitive ion channel should depend on the channel interactions with various membrane components as dictated by its structure. This supports the notion of the force-from-lipids as the fundamental and unifying physicochemical principle underlying mechanosensory transduction at the cell membrane interface.

Further research in mechanobiology focusing on mechanosensory transduction may be expected to show that the "force sharing principle" applies not only to mechanosensitive ion channels but also to other types of membrane proteins, including ion exchangers and pumps or GPCRs. Given the substantial developments in the research on mechanotransduction over the last couple of years we may also look forward to identification of novel mechanosensitive channels, some of which were recently discovered and briefly mentioned in this chapter.

References

Abbondanzieri, E. A., Greenleaf, W. J., Shaevitz, J. W., Landick, R., & Block, S. M. (2005). Direct observation of base-pair stepping by RNA polymerase. *Nature, 438*, 460–465.

Abidine, Y., Constantinescu, A., Laurent, V. M., Rajan, V. S., Michel, R., Laplaud, V., et al. (2018). Mechanosensitivity of cancer cells in contact with soft substrates using AFM. *Biophysical Journal, 114*, 1165–1175.

Alberts, B., Johnson, A., Lewis, J., Raff, M., Roberts, K., & Walter, P. (2002). The lipid bilayer. In J. H. Wilson, & T. Hunt (Eds.), *Molecular biology of the cell* (4th ed.). New York: Garland Science.

Allemand, J.-F., Bensimon, D., & Croquette, V. (2003). Stretching DNA and RNA to probe their interactions with proteins. *Current Opinion in Structural Biology, 13*, 266–274.

Árnadóttir, J., & Chalfie, M. (2010). Eukaryotic mechanosensitive channels. *Annual Review of Biophysics, 39*, 111–137.

Aryal, P., Abd-Wahab, F., Bucci, G., Sansom, M. S. P., & Tucker, S. J. (2014). A hydrophobic barrier deep within the inner pore of the TWIK-1 K2P potassium channel. *Nature Communications, 5*, 4377.

Aryal, P., Abd-Wahab, F., Bucci, G., Sansom, M. S. P., & Tucker, S. J. (2015). Influence of lipids on the hydrophobic barrier within the pore of the TWIK-1 K2P channel. *Channels, 9*, 44–49.

Aryal, P., Jarerattanachat, V., Clausen, M. V., Schewe, M., McClenaghan, C., Argent, L., et al. (2017). Bilayer-mediated structural transitions control mechanosensitivity of the TREK-2 K2P channel. *Structure, 25*, 708–18. e2.

Ayala, Y. A., Pontes, B., Ether, D. S., Pires, L. B., Araujo, G. R., Frases, S., et al. (2016). Rheological properties of cells measured by optical tweezers. *BMC Biophysics, 9*, 5.

Bae, C., Suchyna, T. M., Ziegler, L., Sachs, F., & Gottlieb, P. A. (2016). Human PIEZO1 ion channel functions as a split protein. *PLoS One, 11*, e0151289.

Bahari, S. (2020). Osmotic water permeability of aquaporin-1 under radial tensile and compressive stimuli. *Desalination, 492*, 114524. https://doi.org/10.1016/j.desal.2020.114524.

Baldi, A., Fass, J. N., De Silva, M. N., Odde, D. J., & Ziaie, B. (2003). A micro-tool for mechanical manipulation of in vitro cell arrays. *Biomedical Microdevices, 5*, 291–295.

Balland, M., Richert, A., & Gallet, F. (2005). The dissipative contribution of myosin II in the cytoskeleton dynamics of myoblasts. *European Biophysics Journal, 34*, 255–261.

Barnes, H. A., Hutton, J. F., & Walters, K. (1989). *An introduction to rheology*. Elsevier.

Bass, R. B., Strop, P., Barclay, M., & Rees, D. C. (2002). Crystal structure of Escherichia coli MscS, a voltage-modulated and mechanosensitive channel. *Science, 298*, 1582–1587.

Bausch, A. R., Ziemann, F., Boulbitch, A. A., Jacobson, K., & Sackmann, E. (1998). Local measurements of viscoelastic parameters of adherent cell surfaces by magnetic bead microrheometry. *Biophysical Journal, 75*, 2038–2049.

Bavi, N., Bavi, O., Vossoughi, M., Naghdabadi, R., Hill, A. P., Martinac, B., et al. (2017). Nanomechanical properties of MscL α helices: A steered molecular dynamics study. *Channels (Austin, Tex.), 11*, 209–223.

Bavi, N., Cox, C. D., Bavi, O., & Martinac, B. (2017). Perturbation of bilayer surface tension differentially modulates mechanosensitive ion channels. *Biophysical Journal, 112*, 416a.

Bavi, N., Cox, C. D., Perozo, E., & Martinac, B. (2017). Toward a structural blueprint for bilayer-mediated channel mechanosensitivity. *Channels, 11*, 91–93.

Bavi, O., Cox, C. D., Vossoughi, M., Naghdabadi, R., Jamali, Y., & Martinac, B. (2016). Influence of global and local membrane curvature on mechanosensitive ion channels: a finite element approach. *Membranes, 6*, 14.

Bavi, N., Marien Cortes, D., Cox, C. D., Rohde, P. R., Liu, W., Deitmer, J. W., et al. (2016). The role of MscL amphipathic N terminus indicates a blueprint for bilayer-mediated gating of mechanosensitive channels. *Nature Communications, 7*, 1–13.

Bavi, N., Nikolaev, Y. A., Bavi, O., Ridone, P., Martinac, A. D., Nakayama, Y., et al. (2017). Principles of mechanosensing at the membrane interface. In R. Epand, & J. M. Ruysschaert (Eds.), *Springer Series in Biophysics: vol 19. The biophysics of cell membranes*. Singapore: Springer.

Bavi, N., Richardson, J., Heu, C., Martinac, B., & Poole, K. (2019). PIEZO1-mediated currents are modulated by substrate mechanics. *ACS Nano*, *13*, 13545–13559.

Bavi, O., Vossoughi, M., Naghdabadi, R., & Jamali, Y. (2014). The effect of local bending on gating of MscL using a representative volume element and finite element simulation. *Channels*, *8*, 344–349.

Bavi, O., Vossoughi, M., Naghdabadi, R., & Jamali, Y. (2016). The combined effect of hydrophobic mismatch and bilayer local bending on the regulation of mechanosensitive ion channels. *PLoS One*, *11*(3). https://doi.org/10.1371/journal.pone.0150578.

Beaulieu-Laroche, L., Christin, M., Donoghue, A., Agosti, F., Yousefpour, N., Petitjean, H., et al. (2020). TACAN is an ion channel involved in sensing mechanical pain. *Cell*, *180*, 956–967.e17.

Ben-Shahar, Y. (2011). Sensory functions for degenerin/epithelial sodium channels (DEG/ENaC). In *Advances in Genetics: 76* (pp. 1–26). Elsevier.

Beyder, A. (2007). *Electro-mechanical measurements on membranes of Shaker-transfected cells*. State University of New York at Buffalo.

Beyder, A., & Sachs, F. (2009). Electromechanical coupling in the membranes of Shaker-transfected HEK cells. *Proceedings of the National Academy of Sciences of the United States of America*, *106*, 6626–6631.

Blount, P., & Moe, P. C. (1999). Bacterial mechanosensitive channels: Integrating physiology, structure and function. *Trends in Microbiology*, 7, 420–424.

Börngen, K., Battle, A. R., Möker, N., Morbach, S., Marin, K., Martinac, B., et al. (2010). The properties and contribution of the Corynebacterium glutamicum MscS variant to fine-tuning of osmotic adaptation. *Biochimica et Biophysica Acta (BBA)-Biomembranes*, *1798*, 2141–2149.

Brohawn, S. G., Zhenwei, S., & MacKinnon, R. (2014). Mechanosensitivity is mediated directly by the lipid membrane in TRAAK and TREK1 K+ channels. *Proceedings of the National Academy of Sciences*, *111*, 3614–3619.

Bustamante, C., Bryant, Z., & Smith, S. B. (2003). Ten years of tension: Single-molecule DNA mechanics. *Nature*, *421*, 423–427.

Bustamante, C., Rivetti, C., & Keller, D. J. (1997). Scanning force microscopy under aqueous solutions. *Current Opinion in Structural Biology*, 7, 709–716.

Buxboim, A., Rajagopal, K., Brown, A.'. E. X., & Discher, D. E. (2010). How deeply cells feel: Methods for thin gels. *Journal of Physics: Condensed Matter*, *22*, 194116.

Casado, M., & Ascher, P. (1998). Opposite modulation of NMDA receptors by lysophospholipids and arachidonic acid: Common features with mechanosensitivity. *The Journal of Physiology*, *513*, 317–330.

Cecconi, C., Shank, E. A., Bustamante, C., & Marqusee, S. (2005). Direct observation of the three-state folding of a single protein molecule. *Science*, *309*, 2057–2060.

Chandler, D. (2005). Interfaces and the driving force of hydrophobic assembly. *Nature*, *437*, 640–647.

Chang, G., Spencer, R. H., Lee, A. T., Barclay, M. T., & Rees, D. C. (1998). Structure of the MscL homolog from Mycobacterium tuberculosis: A gated mechanosensitive ion channel. *Science*, *282*, 2220–2226.

Chen, J. (2014). Nanobiomechanics of living cells: A review. *Interface Focus*, *4*, 20130055.

Chen, Y. L., Helm, C. A., & Israelachvili, J. N. (1991). Measurements of the elastic properties of surfactant and lipid monolayers. *Langmuir*, 7, 2694–2699.

Christensen, A. P., & Corey, D. P. (2007). TRP channels in mechanosensation: Direct or indirect activation? *Nature Reviews Neuroscience*, *8*, 510–521.

Clapham, D. E. (2003). TRP channels as cellular sensors. *Nature, 426*, 517–524.
Clausen, M. V., Jarerattanachat, V., Carpenter, E. P., Sansom, M. S. P., & Tucker, S. J. (2017). Asymmetric mechanosensitivity in a eukaryotic ion channel. *Proceedings of the National Academy of Sciences, 114*, E8343–E8351.
Cole, K. S. (1932). Surface forces of the Arbacia egg. *Journal of Cellular and Comparative Physiology, 1*, 1–9.
Collinsworth, A. M., Zhang, S., Kraus, W. E., & Truskey, G. A. (2002). Apparent elastic modulus and hysteresis of skeletal muscle cells throughout differentiation. *American Journal of Physiology-Cell Physiology, 283*, C1219–C1227.
Constantine, M. (2016). *Structural and functional characterisation of heterologously expressed human transient receptor potential cation channel, subfamily M, member 4 (TRPM4)*. Australia: Victor Chang Cardiac Research Institute & Faculty of Medicine, UNSW.
Corey, D. P., & Holt, J. R. (2016). Are TMCs the mechanotransduction channels of vertebrate hair cells? *Journal of Neuroscience, 36*, 10921–10926.
Coste, B., Mathur, J., Schmidt, M., Earley, T. J., Ranade, S., Petrus, M. J., et al. (2010). Piezo1 and Piezo2 are essential components of distinct mechanically activated cation channels. *Science, 330*, 55–60.
Cox, C. D., Bae, C., Ziegler, L., Hartley, S., Nikolova-Krstevski, V., Rohde, P. R., et al. (2016). Removal of the mechanoprotective influence of the cytoskeleton reveals PIEZO1 is gated by bilayer tension. *Nature Communications, 7*, 1–13.
Cox, C. D., Bavi, N., & Martinac, B. (2017). Origin of the force: The force-from-lipids principle applied to piezo channels. In P. A. Gottlieb (Ed.), *79. Current topics in membranes* (pp. 59–96). Elsevier.
Cox, C. D., Bavi, N., & Martinac, B. (2018). Bacterial mechanosensors. *Annual Review of Physiology, 80*, 71–93.
Cox, C. D., Bavi, N., & Martinac, B. (2019). Biophysical principles of ion-channel-mediated mechanosensory transduction. *Cell Reports, 29*, 1–12.
Cox, C. D., Nakayama, Y., Nomura, T., & Martinac, B. (2015). The evolutionary 'tinkering' of MscS-like channels: Generation of structural and functional diversity. *Pflügers Archiv/European Journal of Physiology, 467*, 3–13.
Cristofori-Armstrong, B., & Rash, L. D. (2017). Acid-sensing ion channel (ASIC) structure and function: Insights from spider, snake and sea anemone venoms. *Neuropharmacology, 127*, 173–184.
Cruickshank, C. C., Minchin, R. F., Le Dain, A. C., & Martinac, B. (1997). Estimation of the pore size of the large-conductance mechanosensitive ion channel of Escherichia coli. *Biophysical Journal, 73*(4), 1925–1931.
Csizmok, V., Follis, A. V., Kriwacki, R. W., & Forman-Kay, J. D. (2016). Dynamic protein interaction networks and new structural paradigms in signaling. *Chemical Reviews, 116*, 6424–6462.
Cueva, J. G., Mulholland, A., & Goodman, M. B. (2007). Nanoscale organization of the MEC-4 DEG/ENaC sensory mechanotransduction channel in Caenorhabditis elegans touch receptor neurons. *Journal of Neuroscience, 27*, 14089–14098.
Curtze, S., Dembo, M., Miron, M., & Jones, D. B. (2004). Dynamic changes in traction forces with DC electric field in osteoblast-like cells. *Journal of Cell Science, 117*, 2721–2729.
Dai, J., & Sheetz, M. P. (1997). Cell membrane mechanics. In M. P. Sheetz (Ed.), *55. Methods in cell biology* (pp. 157–171). Elsevier.
Dai, J., Sheetz, M. P., Wan, X., & Morris, C. E. (1998). Membrane tension in swelling and shrinking molluscan neurons. *Journal of Neuroscience, 18*, 6681–6692.
Le Dain, A. C., Saint, N., Kloda, A., Ghazi, A., & Martinac, B. (1998). Mechanosensitive ion channels of the archaeon Haloferax volcanii. *Journal of Biological Chemistry, 273*, 12116–12119.

Dangaria, J. H., & Butler, P. J. (2007). Macrorheology and adaptive microrheology of endothelial cells subjected to fluid shear stress. *American Journal of Physiology-Cell Physiology, 293*, C1568–C1575.

Danilowicz, C., Greenfield, D., & Prentiss, M. (2005). Dissociation of ligand—receptor complexes using magnetic tweezers. *Analytical Chemistry*, 77, 3023–3028.

Dao, M., Lim, C. T., & Suresh, S. (2003). Mechanics of the human red blood cell deformed by optical tweezers. *Journal of the Mechanics and Physics of Solids, 51*, 2259–2280.

De Vlaminck, I., Henighan, T., Van Loenhout, M. T. J., Burnham, D. R., & Dekker, C. (2012). Magnetic forces and DNA mechanics in multiplexed magnetic tweezers. *PLoS One*, 7(8). https://doi.org/10.1371/journal.pone.0041432.

Delanoë-Ayari, H., Iwaya, S., Maeda, Y. T., Inose, J., Riviere, C., Sano, M., et al. (2008). Changes in the magnitude and distribution of forces at different dictyostelium developmental stages. *Cell Motility and the Cytoskeleton, 65*, 314–331.

Dill, K. A., & Flory, P. J. (1980). Interphases of chain molecules: Monolayers and lipid bilayer membranes. *Proceedings of the National Academy of Sciences*, 77, 3115–3119.

Dimitriadis, E. K., Horkay, F., Maresca, J., Kachar, B., & Chadwick, R. S. (2002). Determination of elastic moduli of thin layers of soft material using the atomic force microscope. *Biophysical Journal, 82*, 2798–2810.

Discher, D. E., Janmey, P., & Wang, Y.-l. (2005). Tissue cells feel and respond to the stiffness of their substrate. *Science, 310*, 1139–1143.

Douguet, D., & Honoré, E. (2019). Mammalian mechanoelectrical transduction: Structure and function of force-gated ion channels. *Cell, 179*, 340–354.

Doyle, A. D., & Lee, J. (2005). Cyclic changes in keratocyte speed and traction stress arise from Ca2+-dependent regulation of cell adhesiveness. *Journal of Cell Science, 118*, 369–379.

Du, H., Guoqiang, G., William, C. M., & Chalfie, M. (1996). Extracellular proteins needed for C. elegans mechanosensation. *Neuron, 16*, 183–194.

Edwards, M. D., Black, S., Rasmussen, T., Rasmussen, A., Stokes, N. R., Stephen, T.-L., et al. (2012). Characterization of three novel mechanosensitive channel activities in Escherichia coli. *Channels, 6*, 272–281.

Efremov, Y. M., Okajima, T., & Raman, A. (2020). Measuring viscoelasticity of soft biological samples using atomic force microscopy. *Soft Matter, 16*, 64–81.

Ehlers, W., Krause, R., & Markert, B. (2011). Modelling and remodelling of biological tissue in the framework of continuum biomechanics. *Proceedings in Applied Mathematics and Mechanics, 11*, 35–38.

Eijkelkamp, N., Quick, K., & Wood, J. N. (2013). Transient receptor potential channels and mechanosensation. *Annual Review of Neuroscience, 36*, 519–546.

Emtage, L., Guoqiang, G., Hartwieg, E., & Chalfie, M. (2004). Extracellular proteins organize the mechanosensory channel complex in C. elegans touch receptor neurons. *Neuron, 44*, 795–807.

Engel, A., & Gaub, H. E. (2008). Structure and mechanics of membrane proteins. *Annual Review of Biochemistry*, 77, 127–148.

Espinosa, G., López-Montero, I., Monroy, F., & Langevin, D. (2011). Shear rheology of lipid monolayers and insights on membrane fluidity. *Proceedings of the National Academy of Sciences, 108*, 6008–6013.

Evans, E. A. (1974). Bending resistance and chemically induced moments in membrane bilayers. *Biophysical Journal, 14*, 923–931.

Evans, E., & Needham, D. (1987). Physical properties of surfactant bilayer membranes: Thermal transitions, elasticity, rigidity, cohesion and colloidal interactions. *Journal of Physical Chemistry, 91*, 4219–4228.

Evans, E., Rawicz, W., & Smith, B. A. (2013). Concluding remarks back to the future: Mechanics and thermodynamics of lipid biomembranes. *Faraday Discussions, 161*, 591–611.

Evans, E. A., & Hochmuth, R. M. (1978). Mechanochemical properties of membranes. In F. Bronner, & A. Kleinzeller (Eds.), *Vol. 10. Current topics in membranes and transport* (pp. 1–64). Elsevier.

Fabry, B., Maksym, G. N., Butler, J. P., Glogauer, M., Navajas, D., & Fredberg, J. J. (2001). Scaling the microrheology of living cells. *Physical Review Letters*, *87*, 148102.

Fällman, E., & Axner, O. (2003). Influence of a glass-water interface on the on-axis trapping of micrometer-sized spherical objects by optical tweezers. *Applied Optics*, *42*, 3915–3926.

Fu, J., Wang, Y.-K., Yang, M. T., Desai, R. A., Yu, X., Liu, Z., et al. (2010). Mechanical regulation of cell function with geometrically modulated elastomeric substrates. *Nature Methods*, *7*, 733.

Fulda, S., Gorman, A. M., Hori, O., & Samali, A. (2010). Cellular stress responses: Cell survival and cell death. *International Journal of Cell Biology*, *2010*, 214074.

Fung, Y. C. (1981). Red blood cells and their deformability. In *Biomechanics*. New York, NY: Springer.

Gauthier, N. C., Masters, T. A., & Sheetz, M. P. (2012). Mechanical feedback between membrane tension and dynamics. *Trends in Cell Biology*, *22*, 527–535.

Gavara, N., & Chadwick, R. S. (2012). Determination of the elastic moduli of thin samples and adherent cells using conical atomic force microscope tips. *Nature Nanotechnology*, *7*, 733–736.

Gayer, C. P., & Basson, M. D. (2009). The effects of mechanical forces on intestinal physiology and pathology. *Cellular Signalling*, *21*, 1237–1244.

Gillespie, P. G., & Walker, R. G. (2001). Molecular basis of mechanosensory transduction. *Nature*, *413*, 194–202.

Gnanasambandam, R., Bae, C., Ziegler, L., Sachs, F., & Gottlieb, P. A. (2018). Functional analyses of heteromeric human PIEZO1 channels. *PLoS One*, *13*, e0207309.

Gossett, D. R., Henry, T. K., Lee, S. A., Ying, Y., Lindgren, A. G., Yang, O. O., et al. (2012). Hydrodynamic stretching of single cells for large population mechanical phenotyping. *Proceedings of the National Academy of Sciences*, *109*, 7630–7635.

Gottlieb, P., Folgering, J., Maroto, R., Raso, A., Wood, T. G., Kurosky, A., et al. (2008). Revisiting TRPC1 and TRPC6 mechanosensitivity. *Pflügers Archiv/European Journal of Physiology*, *455*, 1097–1103.

Guck, J., Schinkinger, S., Lincoln, B., Wottawah, F., Ebert, S., Romeyke, M., et al. (2005). Optical deformability as an inherent cell marker for testing malignant transformation and metastatic competence. *Biophysical Journal*, *88*, 3689–3698.

Guharay, F., & Sachs, F. (1984). Stretch-activated single ion channel currents in tissue-cultured embryonic chick skeletal muscle. *The Journal of Physiology*, *352*, 685–701.

Gullingsrud, J., & Schulten, K. (2004). Lipid bilayer pressure profiles and mechanosensitive channel gating. *Biophysical Journal*, *86*, 3496–3509.

Guo, Y. R., & MacKinnon, R. (2017). Structure-based membrane dome mechanism for piezo mechanosensitivity. *eLife*, *6*, e33660.

Guo, J., Sachs, F., & Meng, F. (2014). Fluorescence-based force/tension sensors: A novel tool to visualize mechanical forces in structural proteins in live cells. *Antioxidants & Redox Signaling*, *20*, 986–999.

Hale, C. M., Sun, S. X., & Wirtz, D. (2009). Resolving the role of actoymyosin contractility in cell microrheology. *PLoS One*, *4*(9). https://doi.org/10.1371/journal.pone.0007054.

Hamill, O. P., & Martinac, B. (2001). Molecular basis of mechanotransduction in living cells. *Physiological Reviews*, *81*, 685–740.

Hamilton, E. S., Jensen, G. S., Maksaev, G., Katims, A., Sherp, A. M., & Haswell, E. S. (2015). Mechanosensitive channel MSL8 regulates osmotic forces during pollen hydration and germination. *Science*, *350*, 438–441.

Hansma, H. G., Kasuya, K., & Oroudjev, E. (2004). Atomic force microscopy imaging and pulling of nucleic acids. *Current Opinion in Structural Biology*, *14*, 380–385.

Haswell, E. S., & Meyerowitz, E. M. (2006). MscS-like proteins control plastid size and shape in Arabidopsis thaliana. *Current Biology, 16*, 1–11.

Heinrich, V., & Waugh, R. E. (1996). A piconewton force transducer and its application to measurement of the bending stiffness of phospholipid membranes. *Annals of Biomedical Engineering, 24*, 595–605.

Hochmuth, R. M., Mohandas, N., & Blackshear, P. L., Jr. (1973). Measurement of the elastic modulus for red cell membrane using a fluid mechanical technique. *Biophysical Journal, 13*, 747–762.

Hochmuth, R. M., & Waugh, R. E. (1987). Erythrocyte membrane elasticity and viscosity. *Annual Review of Physiology, 49*, 209–219.

Hoffman, B. D., & Crocker, J. C. (2009). Cell mechanics: Dissecting the physical responses of cells to force. *Annual Review of Biomedical Engineering, 11*, 259–288.

Honoré, E. (2007). The neuronal background K 2P channels: Focus on TREK1. *Nature Reviews Neuroscience, 8*, 251–261.

Hosu, B. G., Jakab, K., Bánki, P., Tóth, F. I., & Forgacs, G. (2003). Magnetic tweezers for intracellular applications. *Review of Scientific Instruments, 74*, 4158–4163.

Huang, W., Anvari, B., Torres, J. H., LeBaron, R. G., & Athanasiou, K. A. (2003). Temporal effects of cell adhesion on mechanical characteristics of the single chondrocyte. *Journal of Orthopaedic Research, 21*, 88–95.

Huang, S., & Ingber, D. E. (1999). The structural and mechanical complexity of cell-growth control. *Nature Cell Biology, 1*, E131–E138.

Iwadate, Y., & Yumura, S. (2008). Actin-based propulsive forces and myosin-II-based contractile forces in migrating dictyostelium cells. *Journal of Cell Science, 121*, 1314–1324.

Jarsch, I. K., Daste, F., & Gallop, J. L. (2016). Membrane curvature in cell biology: An integration of molecular mechanisms. *Journal of Cell Biology, 214*, 375–387.

Jia, Y., Zhao, Y., Kusakizako, T., Wang, Y., Pan, C., Zhang, Y., et al. (2020). TMC1 and TMC2 proteins are pore-forming subunits of mechanosensitive ion channels. *Neuron, 105*, 310–21. e3.

Jin, P., Bulkley, D., Guo, Y., Zhang, W., Guo, Z., Huynh, W., et al. (2017). Electron cryo-microscopy structure of the mechanotransduction channel NOMPC. *Nature, 547*, 118–122.

Johnson, L., Battle, A., & Martinac, B. (2019). Remembering mechanosensitivity of NMDA receptors. *Frontiers in Cellular Neuroscience, 13*, 533. https://doi.org/10.3389/fncel.2019.00533.

Julius, D. (2013). TRP channels and pain. *Annual Review of Cell and Developmental Biology, 29*, 355–384.

Kamsma, D., Bochet, P., Oswald, F., Alblas, N., Goyard, S., Wuite, G. J. L., et al. (2018). Single-cell acoustic force spectroscopy: Resolving kinetics and strength of T cell adhesion to fibronectin. *Cell Reports, 24*, 3008–3016.

Kamsma, D., Creyghton, R., Sitters, G., Wuite, G. J. L., & Peterman, E. J. G. (2016). Tuning the music: Acoustic force spectroscopy (AFS) 2.0. *Methods, 105*, 26–33.

Kapsalis, C., Wang, B., El Mkami, H., Pitt, S. J., Schnell, J. R., Smith, T. K., et al. (2019). Allosteric activation of an ion channel triggered by modification of mechanosensitive nano-pockets. *Nature Communications, 10*, 4619.

Kapus, A., & Janmey, P. (2013). Plasma membrane—Cortical cytoskeleton interactions: A cell biology approach with biophysical considerations. *Comprehensive Physiology, 3*, 1231–1281.

Karakas, E., & Furukawa, H. (2014). Crystal structure of a heterotetrameric NMDA receptor ion channel. *Science, 344*, 992–997.

Katta, S., Krieg, M., & Goodman, M. B. (2015). Feeling force: Physical and physiological principles enabling sensory mechanotransduction. *Annual Review of Cell and Developmental Biology, 31*, 347–371.

Kim, E., Guilak, F., & Haider, M. A. (2008). The dynamic mechanical environment of the chondrocyte: A biphasic finite element model of cell-matrix interactions under cyclic compressive loading. *Journal of Biomechanical Engineering*, *130*(6), 061009.

Kim, K., & Saleh, O. A. (2009). A high-resolution magnetic tweezer for the single-molecule study of DNA-protein interactions. *Biophysical Journal*, *96*, 289a.

Kloda, A., & Martinac, B. (2001). Structural and functional differences between two homologous mechanosensitive channels of Methanococcus jannaschii. *The EMBO Journal*, *20*, 1888–1896.

Kloda, A., & Martinac, B. (2002). Common evolutionary origins of mechanosensitive ion channels in archaea, bacteria and cell-walled eukarya. *Archaea*, *1*, 35–44.

Kloda, A., Martinac, B., & Adams, D. H. (2007). Polymodal regulation of NMDA receptor-channels. *Channels*, *1*, 334–343.

Knight, A. E., Veigel, C., Chambers, C., & Molloy, J. E. (2001). Analysis of single-molecule mechanical recordings: Application to acto-myosin interactions. *Progress in Biophysics and Molecular Biology*, *77*, 45–72.

Knoepp, F., Ashley, Z., Barth, D., Baldin, J.-P., Jennings, M., Kazantseva, M., et al. (2020). Shear force sensing of epithelial Na+ channel (ENaC) relies on N-glycosylated asparagines in the palm and knuckle domains of αENaC. *Proceedings of the National Academy of Sciences*, *117*, 717–726.

Kollmannsberger, P., & Fabry, B. (2007). BaHigh-force magnetic tweezers with force feedback for biological applications. *Review of Scientific Instruments*, *78*, 114301.

Kollmannsberger, P., & Fabry, B. (2011). Linear and nonlinear rheology of living cells. *Annual Review of Materials Research*, *41*, 75–97.

Kosmalska, A. J., Casares, L., Elosegui-Artola, A., Thottacherry, J. J., Moreno-Vicente, R., González-Tarragó, V., et al. (2015). Physical principles of membrane remodelling during cell mechanoadaptation. *Nature Communications*, *6*, 1–11.

Kozlov, M. M., & Chernomordik, L. V. (2015). Membrane tension and membrane fusion. *Current Opinion in Structural Biology*, *33*, 61–67.

Kreplak, L. (2016). Introduction to atomic force microscopy (AFM) in biology. *Current Protocols in Protein Science*, *85*, 17.7.1–17.7.21.

Kris, A. S., Kamm, R. D., & Sieminski, A. L. (2008). VASP involvement in force-mediated adherens junction strengthening. *Biochemical and Biophysical Research Communications*, *375*, 134–138.

Laganowsky, A., Reading, E., Allison, T. M., Ulmschneider, M. B., Degiacomi, M. T., Baldwin, A. J., et al. (2014). Membrane proteins bind lipids selectively to modulate their structure and function. *Nature*, *510*, 172–175.

Lam, W. A., Rosenbluth, M. J., & Fletcher, D. A. (2007). Chemotherapy exposure increases leukemia cell stiffness. *Blood*, *109*, 3505–3508.

Lang, H. P., Hegner, M., & Gerber, C. (2017). Nanomechanical cantilever array sensors. In B. Bhushan (Ed.), *Springer handbook of nanotechnology*. Springer, Berlin, Heidelberg: Springer Handbooks.

Lansdorp, B. M., Tabrizi, S. J., Dittmore, A., & Saleh, O. A. (2013). A high-speed magnetic tweezer beyond 10,000 frames per second. *Review of Scientific Instruments*, *84*, 044301.

Lauritzen, I., Chemin, J., Honoré, E., Jodar, M., Guy, N., Lazdunski, M., et al. (2005). Cross-talk between the mechano-gated K2P channel TREK-1 and the actin cytoskeleton. *EMBO Reports*, *6*, 642–648.

Lee, C.-H., & Gouaux, E. (2011). Amino terminal domains of the NMDA receptor are organized as local heterodimers. *PLoS One*, *6*, e19180.

Lee, C., Kim, K. H., & Cohen, P. (2016). MOTS-c: A novel mitochondrial-derived peptide regulating muscle and fat metabolism. *Free Radical Biology and Medicine*, *100*, 182–187.

Lee, J. S. H., Panorchan, P., Hale, C. M., Khatau, S. B., Kole, T. P., Tseng, Y., et al. (2006). Ballistic intracellular nanorheology reveals ROCK-hard cytoplasmic stiffening response to fluid flow. *Journal of Cell Science*, *119*, 1760–1768.

Legant, W. R., Miller, J. S., Blakely, B. L., Cohen, D. M., Genin, G. M., & Chen, C. S. (2010). Measurement of mechanical tractions exerted by cells in three-dimensional matrices. *Nature Methods*, 7, 969.

Levina, N., Tötemeyer, S., Stokes, N. R., Louis, P., Jones, M. A., & Booth, I. R. (1999). Protection of Escherichia coli cells against extreme turgor by activation of MscS and MscL mechanosensitive channels: Identification of genes required for MscS activity. *The EMBO Journal*, *18*, 1730–1737.

Lewis, A. H., & Grandl, J. (2015). Mechanical sensitivity of Piezo1 ion channels can be tuned by cellular membrane tension. *eLife*, *4*, e12088.

Li, J., Guo, J., Xiaomin, O., Zhang, M., Li, Y., & Liu, Z. (2015). Mechanical coupling of the multiple structural elements of the large-conductance mechanosensitive channel during expansion. *Proceedings of the National Academy of Sciences*, *112*, 10726–10731.

Lieber, A. D., Schweitzer, Y., Kozlov, M. M., & Keren, K. (2015). Front-to-rear membrane tension gradient in rapidly moving cells. *Biophysical Journal*, *108*, 1599–1603.

Lieber, A. D., Yehudai-Resheff, S., Barnhart, E. L., Theriot, J. A., & Keren, K. (2013). Membrane tension in rapidly moving cells is determined by cytoskeletal forces. *Current Biology*, *23*, 1409–1417.

Lin, S.-H., Cheng, Y.-R., Banks, R. W., Min, M.-Y., Bewick, G. S., & Chen, C.-C. (2016). Evidence for the involvement of ASIC3 in sensory mechanotransduction in proprioceptors. *Nature Communications*, 7, 1–15.

Lin, S.-Y., & Corey, D. P. (2005). TRP channels in mechanosensation. *Current Opinion in Neurobiology*, *15*, 350–357.

Lipfert, J., Lee, M., Ordu, O., Kerssemakers, J. W. J., & Dekker, N. H. (2014). Magnetic tweezers for the measurement of twist and torque. *Journal of Visualized Experiments*, *87*. https://doi.org/10.3791/51503.

Litvinov, R. I., Bennett, J. S., Weisel, J. W., & Shuman, H. (2005). Multi-step fibrinogen binding to the integrin αIIbβ3 detected using force spectroscopy. *Biophysical Journal*, *89*, 2824–2834.

Liu, H., Wen, J., Xiao, Y., Liu, J., Hopyan, S., Radisic, M., et al. (2014). In situ mechanical characterization of the cell nucleus by atomic force microscopy. *ACS Nano*, *8*, 3821–3828.

Lombardi, M. L., Knecht, D. A., Dembo, M., & Lee, J. (2007). Traction force microscopy in dictyostelium reveals distinct roles for myosin II motor and actin-crosslinking activity in polarized cell movement. *Journal of Cell Science*, *120*, 1624–1634.

López-Lara, I. M., & Geiger, O. (2017). Bacterial lipid diversity. *Biochimica et Biophysica Acta (BBA)-Molecular and Cell Biology of Lipids*, *1862*, 1287–1299.

Loukin, S., Zhou, X., Zhenwei, S., Saimi, Y., & Kung, C. (2010). Wild-type and brachyolmia-causing mutant TRPV4 channels respond directly to stretch force. *Journal of Biological Chemistry*, *285*, 27176–27181.

Lu, H., Koo, L. Y., Wang, W. M., Lauffenburger, D. A., Griffith, L. G., & Jensen, K. F. (2004). Microfluidic shear devices for quantitative analysis of cell adhesion. *Analytical Chemistry*, *76*, 5257–5264.

Maity, K., Heumann, J. M., McGrath, A. P., Kopcho, N. J., Hsu, P.-K., Lee, C.-W., et al. (2019). Cryo-EM structure of OSCA1. 2 from Oryza sativa elucidates the mechanical basis of potential membrane hyperosmolality gating. *Proceedings of the National Academy of Sciences*, *116*, 14309–14318.

Malcolm, H. R., Blount, P., & Maurer, J. A. (2015). The mechanosensitive channel of small conductance (MscS) functions as a Jack-in-the box. *Biochimica et Biophysica Acta (BBA)-Biomembranes*, *1848*, 159–166.

Maneshi, M. M., Maki, B., Gnanasambandam, R., Belin, S., Popescu, G. K., Sachs, F., et al. (2017). Mechanical stress activates NMDA receptors in the absence of agonists. *Scientific Reports*, *7*, 39610.

Mannix, R. J., Kumar, S., Cassiola, F., Montoya-Zavala, M., Feinstein, E., Prentiss, M., et al. (2008). Nanomagnetic actuation of receptor-mediated signal transduction. *Nature Nanotechnology*, *3*, 36.

Marganski, W. A., De Biase, V. M., Burgess, M. L., & Dembo, M. (2003). Demonstration of altered fibroblast contractile activity in hypertensive heart disease. *Cardiovascular Research*, *60*, 547–556.

Marsh, D. (2007). Lateral pressure profile, spontaneous curvature frustration, and the incorporation and conformation of proteins in membranes. *Biophysical Journal*, *93*, 3884–3899.

Martinac, B. (2001). Mechanosensitive channels in prokaryotes. *Cellular Physiology and Biochemistry*, *11*, 61–76.

Martinac, B. (2004). Mechanosensitive ion channels: Molecules of mechanotransduction. *Journal of Cell Science*, *117*, 2449–2460.

Martinac, B. (2014). The ion channels to cytoskeleton connection as potential mechanism of mechanosensitivity. *Biochimica et Biophysica Acta (BBA)-Biomembranes*, *1838*, 682–691.

Martinac, B., Adler, J., & Kung, C. (1990). Mechanosensitive ion channels of E. coli activated by amphipaths. *Nature*, *348*, 261–263.

Martinac, A. D., Bavi, N., Bavi, O., & Martinac, B. (2017). Pulling MscL open via N-terminal and TM1 helices: A computational study towards engineering an MscL nanovalve. *PLoS One*, *12*, e0183822.

Martinac, B., Bavi, N., Ridone, P., Nikolaev, Y. A., Martinac, A. D., Nakayama, Y., et al. (2018). Tuning ion channel mechanosensitivity by asymmetry of the transbilayer pressure profile. *Biophysical Reviews*, *10*, 1377–1384.

Martinac, B., Buechner, M., Delcour, A. H., Adler, J., & Kung, C. (1987). Pressure-sensitive ion channel in Escherichia coli. *Proceedings of the National Academy of Sciences*, *84*, 2297–2301.

Martinac, B., & Cox, C. D. (2017). Mechanosensory transduction: Focus on ion channels. In H.-X. Zhou (Ed.), *Reference module in life sciences: Comprehensive biophysics* (pp. 1–47). Oxford, UK: Elsevier Inc.

Martinac, B., & Kloda, A. (2003). Evolutionary origins of mechanosensitive ion channels. *Progress in Biophysics and Molecular Biology*, *82*, 11–24.

Martinac, B., Nomura, T., Chi, G., Petrov, E., Rohde, P. R., Battle, A. R., et al. (2014). Bacterial mechanosensitive channels: Models for studying mechanosensory transduction. *Antioxidants & Redox Signaling*, *20*, 952–969.

Mason, T. G. (2000). Estimating the viscoelastic moduli of complex fluids using the generalized Stokes–Einstein equation. *Rheologica Acta*, *39*, 371–378.

Massiera, G., Van Citters, K. M., Biancaniello, P. L., & Crocker, J. C. (2007). Mechanics of single cells: Rheology, time dependence, and fluctuations. *Biophysical Journal*, *93*, 3703–3713.

Mata, A., Fleischman, A. J., & Roy, S. (2005). Characterization of polydimethylsiloxane (PDMS) properties for biomedical micro/nanosystems. *Biomedical Microdevices*, *7*, 281–293.

Matthews, B. D., Overby, D. R., Alenghat, F. J., Karavitis, J., Numaguchi, Y., Allen, P. G., et al. (2004). Mechanical properties of individual focal adhesions probed with a magnetic microneedle. *Biochemical and Biophysical Research Communications*, *313*, 758–764.

Meng, F., & Sachs, F. (2012). Orientation-based FRET sensor for real-time imaging of cellular forces. *Journal of Cell Science*, *125*, 743–750.

Minke, B., & Cook, B. (2002). TRP channel proteins and signal transduction. *Physiological Reviews*, *82*, 429–472.

Mio, C., Gong, T., Terray, A., & Marr, D. W. M. (2000). Design of a scanning laser optical trap for multiparticle manipulation. *Review of Scientific Instruments*, *71*, 2196–2200.

Mitchison, J. M., & Swann, M. M. (1954). The mechanical properties of the cell surface. *The Journal of Experimental Biology*, *31*, 443–460.

Moe, P. C., Blount, P., & Kung, C. (1998). Functional and structural conservation in the mechanosensitive channel MscL implicates elements crucial for mechanosensation. *Molecular Microbiology*, *28*, 583–592.

Moeendarbary, E., & Harris, A. R. (2014). Cell mechanics: Principles, practices, and prospects. *Wiley Interdisciplinary Reviews: Systems Biology and Medicine*, *6*, 371–388.

Mouritsen, O. G., & Zuckermann, M. J. (2004). What's so special about cholesterol? *Lipids*, *39*, 1101–1113.

Müller, D. J., & Dufrene, Y. F. (2010). Atomic force microscopy as a multifunctional molecular toolbox in nanobiotechnology. In *Nanoscience and technology: A collection of reviews from nature journals* (pp. 269–277). World Scientific.

Murase, K., Fujiwara, T., Umemura, Y., Suzuki, K., Iino, R., Yamashita, H., et al. (2004). Ultrafine membrane compartments for molecular diffusion as revealed by single molecule techniques. *Biophysical Journal*, *86*, 4075–4093.

Murthy, S. E., Dubin, A. E., & Patapoutian, A. (2017). Piezos thrive under pressure: Mechanically activated ion channels in health and disease. *Nature Reviews Molecular Cell Biology*, *18*, 771.

Murthy, S. E., Dubin, A. E., Whitwam, T., Jojoa-Cruz, S., Cahalan, S. M., Mousavi, S. A. R., et al. (2018). OSCA/TMEM63 are an evolutionarily conserved family of mechanically activated ion channels. *eLife*, 7, e41844.

Na, S., Collin, O., Chowdhury, F., Tay, B., Ouyang, M., Wang, Y., et al. (2008). Rapid signal transduction in living cells is a unique feature of mechanotransduction. *Proceedings of the National Academy of Sciences*, *105*, 6626–6631.

Nakayama, Y., Komazawa, K., Bavi, N., Hashimoto, K.-i., Kawasaki, H., & Martinac, B. (2018). Evolutionary specialization of MscCG, an MscS-like mechanosensitive channel, in amino acid transport in Corynebacterium glutamicum. *Scientific Reports*, *8*, 1–13.

Nakayama, Y., Mustapić, M., Ebrahimian, H., Wagner, P., Kim, J. H., Hossain, M. S., et al. (2015). Magnetic nanoparticles for "smart liposomes". *European Biophysics Journal*, *44*, 647–654.

Nakayama, Y., Yoshimura, K., & Iida, H. (2012). Organellar mechanosensitive channels in fission yeast regulate the hypo-osmotic shock response. *Nature Communications*, *3*, 1–10.

Neumeister, J. M., & Ducker, W. A. (1994). Lateral, normal, and longitudinal spring constants of atomic force microscopy cantilevers. *Review of Scientific Instruments*, *65*, 2527–2531.

Nicolson, G. L. (2014). The fluid-mosaic model of membrane structure: Still relevant to understanding the structure, function and dynamics of biological membranes after more than 40 years. *Biochimica et Biophysica Acta (BBA)-Biomembranes*, *1838*, 1451–1466.

Nikolaev, Y. A., Cox, C. D., Ridone, P., Rohde, P. R., Cordero-Morales, J. F., Vásquez, V., et al. (2019). Mammalian TRP ion channels are insensitive to membrane stretch. *Journal of Cell Science*, *132*. https://doi.org/10.1242/jcs.23836.

Nomura, T., Cranfield, C. G., Deplazes, E., Owen, D. M., Macmillan, A., Battle, A. R., et al. (2012). Differential effects of lipids and lyso-lipids on the mechanosensitivity of the mechanosensitive channels MscL and MscS. *Proceedings of the National Academy of Sciences*, *109*, 8770–8775.

Noreng, S., Bharadwaj, A., Posert, R., Yoshioka, C., & Baconguis, I. (2018). Structure of the human epithelial sodium channel by cryo-electron microscopy. *eLife*, 7, e39340.

Ollila, O. H. S., Róg, T., Karttunen, M., & Vattulainen, I. (2007). Role of sterol type on lateral pressure profiles of lipid membranes affecting membrane protein functionality:

Comparison between cholesterol, desmosterol, 7-dehydrocholesterol and ketosterol. *Journal of Structural Biology*, *159*, 311–323.

Pace, N. R. (1997). A molecular view of microbial diversity and the biosphere. *Science*, *276*, 734–740.

Pan, B., Akyuz, N., Liu, X.-P., Asai, Y., Nist-Lund, C., Kurima, K., et al. (2018). TMC1 forms the pore of mechanosensory transduction channels in vertebrate inner ear hair cells. *Neuron*, *99*, 736–753.e6.

Paoletti, P., & Ascher, P. (1994). Mechanosensitivity of NMDA receptors in cultured mouse central neurons. *Neuron*, *13*, 645–655.

Park, K.-H., Berrier, C., Martinac, B., & Ghazi, A. (2004). Purification and functional reconstitution of N- and C-halves of the MscL channel. *Biophysical Journal*, *86*, 2129–2136.

Park, S., Koch, D., Richard, C., Käs, J., & Shih, C.-K. (2005). Cell motility and local viscoelasticity of fibroblasts. *Biophysical Journal*, *89*, 4330–4342.

Patel, A. J., & Honoré, E. (2001). Properties and modulation of mammalian 2P domain K+ channels. *Trends in Neurosciences*, *24*, 339–346.

Patel, A. J., Honoré, E., Maingret, F., Lesage, F., Fink, M., Duprat, F., et al. (1998). A mammalian two pore domain mechano-gated S-like K+ channel. *The EMBO Journal*, *17*, 4283–4290.

Patel, A., Sharif-Naeini, R., Folgering, J. R. H., Bichet, D., Duprat, F., & Honoré, E. (2010). Canonical TRP channels and mechanotransduction: from physiology to disease states. *Pflügers Archiv/European Journal of Physiology*, *460*, 571–581.

Patkunarajah, A., Stear, J. H., Moroni, M., Schroeter, L., Blaszkiewicz, J., Tearle, J. L. E., et al. (2020). TMEM87a/Elkin1, a component of a novel mechanoelectrical transduction pathway, modulates melanoma adhesion and migration. *eLife*, *9*, e53308.

Perozo, E., Kloda, A., Cortes, D. M., & Martinac, B. (2002). Physical principles underlying the transduction of bilayer deformation forces during mechanosensitive channel gating. *Nature Structural Biology*, *9*, 696–703.

Peruzzi, G., Sinibaldi, G., Silvani, G., Ruocco, G., & Casciola, C. M. (2018). Perspectives on cavitation enhanced endothelial layer permeability. *Colloids and Surfaces B: Biointerfaces*, *168*, 83–93.

Petrov, A. G. (2007). Flexoelectricity and mechanotransduction. *Current Topics in Membranes*, *58*, 121–150.

Peyronnet, R., Tran, D., Girault, T., & Frachisse, J.-M. (2014). Mechanosensitive channels: Feeling tension in a world under pressure. *Frontiers in Plant Science*, *5*, 558.

Phan-Thien, N., & Mai-Duy, N. (2017). *Understanding viscoelasticity: An introduction to rheology*. Springer.

Picard, C., & Donald, A. (2009). The impact of environmental changes upon the microrheological response of adherent cells. *The European Physical Journal E*, *30*, 127.

Picas, L., Rico, F., & Scheuring, S. (2012). Direct measurement of the mechanical properties of lipid phases in supported bilayers. *Biophysical Journal*, *102*, L01–L03.

Pleshakova, T. O., Bukharina, N. S., Archakov, A. I., & Ivanov, Y. D. (2018). Atomic force microscopy for protein detection and their physicochemical characterization. *International Journal of Molecular Sciences*, *19*, 1142.

Pliotas, C., Caroline, A., Dahl, E., Rasmussen, T., Mahendran, K. R., Smith, T. K., et al. (2015). The role of lipids in mechanosensation. *Nature Structural & Molecular Biology*, *22*, 991–998.

Plotnikov, S. V., Sabass, B., Schwarz, U. S., & Waterman, C. M. (2014). High-resolution traction force microscopy. In J. C. Waters, & T. Wittman (Eds.), *Methods in cell biology: 123* (pp. 367–394). Elsevier.

Pogoda, K., Jaczewska, J., Wiltowska-Zuber, J., Klymenko, O., Zuber, K., Fornal, M., et al. (2012). Depth-sensing analysis of cytoskeleton organization based on AFM data. *European Biophysics Journal*, *41*, 79–87.

Pontes, B., Ayala, Y., Fonseca, A. C. C., Romão, L. F., Amaral, R. F., Salgado, L. T., et al. (2013). Membrane elastic properties and cell function. *PLoS One*, *8*(7). https://doi.org/10.1371/journal.pone.0067708.

Pontes, B., Monzo, P., & Gauthier, N. C. (2017). Membrane tension: A challenging but universal physical parameter in cell biology. In R. Zaidel-Bar (Ed.), *Seminars in cell & developmental biology* (pp. 30–41). Elsevier.

Poole, K., Herget, R., Lapatsina, L., Ngo, H.-D., & Lewin, G. R. (2014). Tuning Piezo ion channels to detect molecular-scale movements relevant for fine touch. *Nature Communications*, *5*, 1–14.

Qi, Y., Andolfi, L., Frattini, F., Mayer, F., Lazzarino, M., & Jing, H. (2015). Membrane stiffening by STOML3 facilitates mechanosensation in sensory neurons. *Nature Communications*, *6*, 1–13.

Rahimzadeh, J., Meng, F., Sachs, F., Wang, J., Verma, D., & Hua, S. Z. (2011). Real-time observation of flow-induced cytoskeletal stress in living cells. *American Journal of Physiology-Cell Physiology*, *301*, C646–C652.

Ranade, S. S., Syeda, R., & Patapoutian, A. (2015). Mechanically activated ion channels. *Neuron*, *87*, 1162–1179.

Rand, R. P., & Burton, A. C. (1964). Mechanical properties of the red cell membrane: I. Membrane stiffness and intracellular pressure. *Biophysical Journal*, *4*, 115–135.

Reddy, B., Lu, A., Bavi, N., Hsu, A., Borgnia, M., & Perozo, E. (2019). High-resolution structures of MscS in a lipid bilayer: Reinterpreting "force from lipids" activation in mechanosensitive channels. *Biophysical Journal*, *116*, 459a.

Reitsma, S., Slaaf, D. W., Vink, H., Van Zandvoort, M. A. M. J., & Egbrink, M. G. A. O. (2007). The endothelial glycocalyx: Composition, functions, and visualization. *Pflügers Archiv/European Journal of Physiology*, *454*, 345–359.

Retailleau, K., Duprat, F., Arhatte, M., Ranade, S. S., Peyronnet, R., Martins, J. R., et al. (2015). Piezo1 in smooth muscle cells is involved in hypertension-dependent arterial remodeling. *Cell Reports*, *13*, 1161–1171.

Rettler, E., Hoeppener, S., Sigusch, B. W., & Schubert, U. S. (2013). Mapping the mechanical properties of biomaterials on different length scales: Depth-sensing indentation and AFM based nanoindentation. *Journal of Materials Chemistry B*, *1*, 2789–2806.

Ridone, P., Grage, S. L., Patkunarajah, A., Battle, A. R., Ulrich, A. S., & Martinac, B. (2018). "Force-from-lipids" gating of mechanosensitive channels modulated by PUFAs. *Journal of the Mechanical Behavior of Biomedical Materials*, *79*, 158–167.

Ridone, P., Vassalli, M., & Martinac, B. (2019). Piezo1 mechanosensitive channels: What are they and why are they important. *Biophysical Reviews*, *11*, 795–805.

Rief, M., Oesterhelt, F., Heymann, B., & Gaub, H. E. (1997). Single molecule force spectroscopy on polysaccharides by atomic force microscopy. *Science*, *275*, 1295–1297.

Rodriguez, M. L., McGarry, P. J., & Sniadecki, N. J. (2013). Review on cell mechanics: Experimental and modeling approaches. *Applied Mechanics Reviews*, *65*(6), 060801.

Rösel, D., Brábek, J., Tolde, O., Mierke, C. T., Zitterbart, D. P., Raupach, C., et al. (2008). Up-regulation of Rho/ROCK signaling in sarcoma cells drives invasion and increased generation of protrusive forces. *Molecular Cancer Research*, *6*, 1410–1420.

Rosenbluth, M. J., Lam, W. A., & Fletcher, D. A. (2006). Force microscopy of nonadherent cells: A comparison of leukemia cell deformability. *Biophysical Journal*, *90*, 2994–3003.

Rosenmund, C., & Westbrook, G. L. (1993). Calcium-induced actin depolymerization reduces NMDA channel activity. *Neuron*, *10*, 805–814.

Van Rosmalen, M. G. M., Kamsma, D., Biebricher, A. S., Li, C., Zlotnick, A., Roos, W. H., et al. (2020). Revealing in real-time a multistep assembly mechanism for SV40 virus-like particles. *Science Advances*, *6*, eaaz1639.

Le Roux, A.-L., Quiroga, X., Walani, N., Arroyo, M., & Roca-Cusachs, P. (2019). The plasma membrane as a mechanochemical transducer. *Philosophical Transactions of the Royal Society B*, *374*, 20180221.

Sabass, B., Gardel, M. L., Waterman, C. M., & Schwarz, U. S. (2008). High resolution traction force microscopy based on experimental and computational advances. *Biophysical Journal*, *94*, 207–220.

Sachs, F. (2010). Stretch-activated ion channels: What are they? *Physiology*, *25*, 50–56.

Saldaña, C., Naranjo, D., Coria, R., Peña, A., & Vaca, L. (2002). Splitting the two pore domains from TOK1 results in two cationic channels with novel functional properties. *Journal of Biological Chemistry*, *277*, 4797–4805.

Saotome, K., Murthy, S. E., Kefauver, J. M., Whitwam, T., Patapoutian, A., & Ward, A. B. (2018). Structure of the mechanically activated ion channel Piezo1. *Nature*, *554*, 481–486.

Sarkar, A., Robertson, R. B., & Fernandez, J. M. (2004). Simultaneous atomic force microscope and fluorescence measurements of protein unfolding using a calibrated evanescent wave. *Proceedings of the National Academy of Sciences*, *101*, 12882–12886.

Schitter, G., Allgöwer, F., & Stemmer, A. (2003). A new control strategy for high-speed atomic force microscopy. *Nanotechnology*, *15*, 108.

Schoen, I., Hu, W., Klotzsch, E., & Vogel, V. (2010). Probing cellular traction forces by micropillar arrays: Contribution of substrate warping to pillar deflection. *Nano Letters*, *10*, 1823–1830.

Servin-Vences, M. R., Moroni, M., Lewin, G. R., & Poole, K. (2017). Direct measurement of TRPV4 and PIEZO1 activity reveals multiple mechanotransduction pathways in chondrocytes. *eLife*, *6*, e21074.

Shi, Z., Graber, Z. T., Baumgart, T., Stone, H. A., & Cohen, A. E. (2018). Cell membranes resist flow. *Cell*, *175*, 1769–1779.e13.

Silvani, G., Scognamiglio, C., Caprini, D., Marino, L., Chinappi, M., Sinibaldi, G., et al. (2019). Reversible cavitation-induced junctional opening in an artificial endothelial layer. *Small*, *15*, 1905375.

Singer, S. J., & Nicolson, G. L. (1972). The fluid mosaic model of the structure of cell membranes. *Science*, *175*, 720–731.

Sinha, B., Köster, D., Ruez, R., Gonnord, P., Bastiani, M., Abankwa, D., et al. (2011). Cells respond to mechanical stress by rapid disassembly of caveolae. *Cell*, *144*, 402–413.

Sirghi, L., Ponti, J., Broggi, F., & Rossi, F. (2008). Probing elasticity and adhesion of live cells by atomic force microscopy indentation. *European Biophysics Journal*, *37*, 935–945.

Sitters, G., Kamsma, D., Thalhammer, G., Ritsch-Marte, M., Peterman, E. J. G., & Wuite, G. J. L. (2015). Acoustic force spectroscopy. *Nature Methods*, *12*, 47–50.

Sorkin, R., Bergamaschi, G., Kamsma, D., Brand, G., Dekel, E., Ofir-Birin, Y., et al. (2018). Probing cellular mechanics with acoustic force spectroscopy. *Molecular Biology of the Cell*, *29*, 2005–2011.

Spector, A. A., & Yorek, M. A. (1985). Membrane lipid composition and cellular function. *Journal of Lipid Research*, *26*, 1015–1035.

Steltenkamp, S., Rommel, C., Wegener, J., & Janshoff, A. (2006). Membrane stiffness of animal cells challenged by osmotic stress. *Small*, *2*, 1016–1020.

Stewart, M. P., Sharei, A., Ding, X., Sahay, G., Langer, R., & Jensen, K. F. (2016). In vitro and ex vivo strategies for intracellular delivery. *Nature*, *538*, 183–192.

Stricker, J., Sabass, B., Schwarz, U. S., & Gardel, M. L. (2010). Optimization of traction force microscopy for micron-sized focal adhesions. *Journal of Physics: Condensed Matter*, *22*, 194104.

Style, R. W., Boltyanskiy, R., German, G. K., Hyland, C., MacMinn, C. W., Mertz, A. F., et al. (2014). Traction force microscopy in physics and biology. *Soft Matter*, *10*, 4047–4055.

Sukharev, S. (2002). Purification of the small mechanosensitive channel of Escherichia coli (mscS): The subunit structure, conduction, and gating characteristics in liposomes. *Biophysical Journal*, *83*, 290–298.

Sukharev, S. I., Blount, P., Martinac, B., Blattner, F. R., & Kung, C. (1994). A large-conductance mechanosensitive channel in E. coli encoded by mscL alone. *Nature*, *368*, 265–268.

Sukharev, S. I., Blount, P., Martinac, B., Kung, & Ching. (1997). Mechanosensitive channels of Escherichia coli: The MscL gene, protein, and activities. *Annual Review of Physiology*, *59*, 633–657.

Syeda, R., Florendo, M. N., Cox, C. D., Kefauver, J. M., Santos, J. S., Martinac, B., et al. (2016). Piezo1 channels are inherently mechanosensitive. *Cell Reports*, *17*, 1739–1746.

Tanase, M., Biais, N., & Sheetz, M. (2007). Magnetic tweezers in cell biology. *Methods in Cell Biology*, *83*, 473–493.

Teng, J., Loukin, S., Anishkin, A., & Kung, C. (2015). The force-from-lipid (FFL) principle of mechanosensitivity, at large and in elements. *Pflügers Archiv/European Journal of Physiology*, *467*, 27–37.

Teng, J., Loukin, S., Zhou, X., & Kung, C. (2013). Yeast luminometric and Xenopus oocyte electrophysiological examinations of the molecular mechanosensitivity of TRPV4. *Journal of Visualized Experiments*, *82*, e50816.

Titushkin, I., & Cho, M. (2006). Distinct membrane mechanical properties of human mesenchymal stem cells determined using laser optical tweezers. *Biophysical Journal*, *90*, 2582–2591.

Tokarev, A., Aprelev, A., Zakharov, M. N., Korneva, G., Gogotsi, Y., & Kornev, K. G. (2012). Multifunctional magnetic rotator for micro and nanorheological studies. *Review of Scientific Instruments*, *83*, 065110.

Tolić-Nørrelykke, I. M., & Wang, N. (2005). Traction in smooth muscle cells varies with cell spreading. *Journal of Biomechanics*, *38*, 1405–1412.

Trepat, X., Deng, L., An, S. S., Navajas, D., Tschumperlin, D. J., Gerthoffer, W. T., et al. (2007). Universal physical responses to stretch in the living cell. *Nature*, *447*, 592–595.

Tseng, Y., Kole, T. P., & Wirtz, D. (2002). Micromechanical mapping of live cells by multiple-particle-tracking microrheology. *Biophysical Journal*, *83*, 3162–3176.

Tuson, H. H., Auer, G. K., Renner, L. D., Hasebe, M., Tropini, C., Salick, M., et al. (2012). Measuring the stiffness of bacterial cells from growth rates in hydrogels of tunable elasticity. *Molecular Microbiology*, *84*, 874–891.

Ushiyama, A., Kataoka, H., & Iijima, T. (2016). Glycocalyx and its involvement in clinical pathophysiologies. *Journal of Intensive Care*, *4*, 59.

Van Meer, G., Voelker, D. R., & Feigenson, G. W. (2008). Membrane lipids: Where they are and how they behave. *Nature Reviews Molecular Cell Biology*, *9*, 112–124.

Veatch, S. L., & Keller, S. L. (2002). Organization in lipid membranes containing cholesterol. *Physical Review Letters*, *89*, 268101.

Vilfan, I. D., Lipfert, J., Koster, D. A., Lemay, S. G., & Dekker, N. H. (2009). Magnetic tweezers for single-molecule experiments. In P. Hinterdorfer, & A. van Oijen (Eds.), *371. Handbook of single-molecule biophysics* (pp. 371–395). Springer.

Vlahakis, N. E., Schroeder, M. A., Pagano, R. E., & Hubmayr, R. D. (2002). Role of deformation-induced lipid trafficking in the prevention of plasma membrane stress failure. *American Journal of Respiratory and Critical Care Medicine*, *166*, 1282–1289.

Vriens, J., Nilius, B., & Voets, T. (2014). Peripheral thermosensation in mammals. *Nature Reviews Neuroscience*, *15*, 573–589.

Wakatsuki, T., Kolodney, M. S., Zahalak, G. I., & Elson, E. L. (2000). Cell mechanics studied by a reconstituted model tissue. *Biophysical Journal*, *79*, 2353–2368.

Walker, L. M., Holm, Å., Cooling, L., Maxwell, L., Öberg, Å., Sundqvist, T., et al. (1999). Mechanical manipulation of bone and cartilage cells with 'optical tweezers'. *FEBS Letters*, *459*, 39–42.

Wang, N., Butler, J. P., & Ingber, D. E. (1993). Mechanotransduction across the cell surface and through the cytoskeleton. *Science*, *260*, 1124–1127.

Wang, Y., & Dimitrakopoulos, P. (2006). Nature of the hemodynamic forces exerted on vascular endothelial cells or leukocytes adhering to the surface of blood vessels. *Physics of Fluids*, *18*, 087107.

Wang, L., Zhou, H., Zhang, M., Liu, W., Deng, T., Zhao, Q., et al. (2019). Structure and mechanogating of the mammalian tactile channel PIEZO2. *Nature*, *573*, 225–229.

Wang, Y., Liu, Y., DeBerg, H. A., Nomura, T., Hoffman, M. T., Rohde, P. R., … Selvin, P. R. (2014). Single molecule FRET reveals pore size and opening mechanism of a mechano-sensitive ion channel. *eLife*, *3*, e01834.

Williams, M. C. (2002). Optical tweezers: Measuring piconewton forces. In *Biophysics textbook online*. http://www.biophysics.org/btol.

Wirtz, D. (2009). Particle-tracking microrheology of living cells: Principles and applications. *Annual Review of Biophysics*, *38*, 301–326.

Woese, C. R. (1994). There must be a prokaryote somewhere: Microbiology's search for itself. *Microbiology and Molecular Biology Reviews*, *58*, 1–9.

Wojcikiewicz, E. P., Zhang, X., & Moy, V. T. (2004). Force and compliance measurements on living cells using atomic force microscopy (AFM). *Biological Procedures Online*, *6*, 1.

Wu, J., Lewis, A. H., & Grandl, J. (2017). Touch, tension, and transduction–The function and regulation of piezo ion channels. *Trends in Biochemical Sciences*, *42*, 57–71.

Xue, F., Cox, C. D., Bavi, N., Rohde, P. R., Nakayama, Y., & Martinac, B. (2020). Membrane stiffness is one of the key determinants of E. coli MscS channel mechanosensitivity. *Biochimica et Biophysica Acta (BBA)-Biomembranes*, *1862*, 183203.

Yamamoto, H., Teramoto, H., Uetani, K., Igawa, K., & Shimizu, E. (2002). Cyclic stretch upregulates interleukin-8 and transforming growth factor-β1 production through a protein kinase C-dependent pathway in alveolar epithelial cells. *Respirology*, 7, 103–109.

Yan, J., Skoko, D., & Marko, J. F. (2004). Near-field-magnetic-tweezer manipulation of single DNA molecules. *Physical Review E*, *70*, 011905.

Yan, Z., Zhang, W., He, Y., Gorczyca, D., Yang, X., Cheng, L. E., et al. (2013). Drosophila NOMPC is a mechanotransduction channel subunit for gentle-touch sensation. *Nature*, *493*, 221–225.

Yilmaz, O., Afsar, B., Ortiz, A., & Kanbay, M. (2019). The role of endothelial glycocalyx in health and disease. *Clinical Kidney Journal*, *12*, 611–619.

Yin, J., & Kuebler, W. M. (2010). Mechanotransduction by TRP channels: General concepts and specific role in the vasculature. *Cell Biochemistry and Biophysics*, *56*, 1.

Yost, M. J., Simpson, D., Wrona, K., Ridley, S., Ploehn, H. J., Borg, T. K., et al. (2000). Design and construction of a uniaxial cell stretcher. *American Journal of Physiology. Heart and Circulatory Physiology*, *279*, H3124–H3130.

Zhang, W., Cheng, L. E., Kittelmann, M., Li, J., Petkovic, M., Cheng, T., et al. (2015). Ankyrin repeats convey force to gate the NOMPC mechanotransduction channel. *Cell*, *162*, 1391–1403.

Zhong, D., & Blount, P. (2013). Phosphatidylinositol is crucial for the mechanosensitivity of Mycobacterium tuberculosis MscL. *Biochemistry*, *52*, 5415–5420.

Zhu, R., Avsievich, T., Popov, A., & Meglinski, I. (2020). Optical tweezers in studies of red blood cells. *Cell*, *9*, 545.

Zoratti, M., & Ghazi, A. (1993). Stretch-activated channels in prokaryotes. In E. P. Bakker (Ed.), *Alkali transport systems in prokaryotes* (pp. 349–358). Boca Raton: CRC Press.

CHAPTER FOUR

Mechanics of the cell: Interaction mechanisms and mechanobiological models

Caleb Liebman[a,†], Andrew McColloch[a,†], Manoochehr Rabiei[b,†], Alan Bowling[b,*], and Michael Cho[a,*]

[a]Department of Bioengineering, University of Texas at Arlington, Arlington, TX, United States
[b]Department of Mechanical and Aerospace Engineering, University of Texas at Arlington, Arlington, TX, United States
[*]Corresponding authors: e-mail address: bowling@uta.edu; michael.cho@uta.edu

Contents

[†] The authors made equal contribution and their names are listed in alphabetical order.

Current Topics in Membranes, Volume 86
ISSN 1063-5823
https://doi.org/10.1016/bs.ctm.2020.09.001

Abstract

The importance of cell mechanics has long been recognized for the cell development and function. Biomechanics plays an important role in cell metabolism, regulation of mechanotransduction pathways and also modulation of nuclear response. The mechanical properties of the cell are likely determined by, among many others, the cytoskeleton elasticity, membrane tension and cell-substrate adhesion. This coordinated but complex mechanical interplay is required however, for the cell to respond to and influence in a reciprocal manner the chemical and mechanical signals from the extracellular matrix (ECM). In an effort to better and more fully understand the cell mechanics, the role of nuclear mechanics has emerged as an important contributor to the overall cellular mechanics. It is not too difficult to appreciate the physical connection between the nucleus and the cytoskeleton network that may be connected to the ECM through the cell membrane. Transmission of forces from ECM through this connection is essential for a wide range of cellular behaviors and functions such as cytoskeletal reorganization, nuclear movement, cell migration and differentiation. Unlike the cellular mechanics that can be measured using a number of biophysical techniques that were developed in the past few decades, it still remains a daunting challenge to probe the nuclear mechanics directly. In this paper, we therefore aim to provide informative description of the cell membrane and cytoskeleton mechanics, followed by unique computational modeling efforts to elucidate the nucleus-cytoskeleton coupling. Advances in our knowledge of complete cellular biomechanics and mechanotransduction may lead to clinical relevance and applications in mechanodiseases such as atherosclerosis, stem cell-based therapies, and the development of tissue engineered products.

1. Introduction

Cell mechanics is focused on quantification of the mechanical properties and responses of living cells and determination of the coupling mechanisms that detect, mediate and impart external mechanical signals to regulate physiological cell function. The importance of maintaining homeostasis of the cell mechanics has been recognized in pathophysiology of, just to name a few, musculoskeletal disorders, acute traumas, diseased heart, and vascular diseases such as atherosclerosis (Cocciolone et al., 2018; Lyon, Zanella, Omens, & Sheikh, 2015). Because the cell can detect mechanical signals and respond to them by exerting mechanical forces to the extracellular matrix (ECM), the term reciprocal mechanical interplay was coined to acknowledge the complexity that is involved in the regulation of homeostasis of the cell mechanics. Although invented more than 100 years, the field of mechanobiology has enjoyed great advances only recently to elucidate the connection between the mechanical cues and mechanosensitive signaling pathways that stimulate and reorganize the cell membrane, cytoskeletal network, and more recently the nuclear remodeling. It should be noted here factors other than mechanical cues can also induce and modulate mechanical changes in the cell. For example, oxidation of the membrane cholesterol has been demonstrated the increase the cell membrane stiffness (Shentu et al., 2012). It is interesting to observe that non-mechanical factors such as cholesterol can change the biochemical composition of the cell membrane and modulate the cell membrane mechanics to initiate both the biochemical signaling and mechanical signaling pathways. Elucidation of cross-signaling mechanisms between these two sets of pathways has indeed been the subject of numerous studies (Saha, Nagy, & Weiner, 2018) and continues to be further investigated.

A better understanding of the cell membrane and cytoskeletal mechanics has been made possible through powerful biophysical techniques that allowed systematic studies of the major mechanical components of the cell, including microtubules, microfilaments, and intermediate filaments (see reviews from Jensen, Morris, & Weitz, 2015; Kazsa et al., 2007; Pegoraro, Janmey, & Weitz, 2017). Determination of the mechanical properties of cytoskeletal polymers has revealed that the cytoskeleton network is not an elastic entity but rather a highly complex nonlinear system. Mechanical nonlinear behavior is thought to provide the cell with the ability to undergo morphological deformation, cell migration and other important cellular functions (Fletcher & Mullins, 2010). Recent efforts are now aimed at the nuclear mechanics (Agrawal & Lele, 2019; Mao, Gavara, & Song, 2015; Maurer & Lammerding, 2019; Pfeifer, Irianto, & Discher, 2019). Mechanotransduction mediated by the cytoskeleton is presumed to modulate nuclear mechanics and, in turn, the cell can respond to its mechanical microenvironment by epigenetic changes and altering gene expression (Alisafaei, Jokhun, Shivashankar, & Shenoy, 2019). Not only the nucleus is the largest and a very stiff organelle in the cell, it is intrinsically integrated with the cytoskeleton through the linker structures (LINC complexes) that are thought to mediate the force transmission between them. However, the nuclear events are transmitted to the ECM not only through the cytoskeleton network but also via the cell membrane as well. Such an integrated mechanical communication is believed to be essential for critical cellular functions including cell migration and movement and positioning of the chromosomes (Aureille, Belaadi, & Guilluy, 2017). A complex picture is beginning to emerge in that the cell mechanics should be discussed and understood in a larger context of at least three separate mechanics at the cell membrane, cytoskeleton, and the nucleus. The aim of this paper is to provide a brief description of each of the three biomechanics and then present several attempts to establish interconnections by incorporating recent experimental and computational results. Interdisciplinary approaches involving biological and biomechanical techniques with numerical simulation models to encompass larger spatial scales (e.g., on the order of the cell size) and longer time scales (e.g., days and not subseconds) may pave the way to better understand the cell mechanics at the different levels and provide experimental and computational strategies to unravel the complexities of cell mechanics.

2. Interconnection between extracellular matrix and cytoskeleton through cell membrane

2.1 ECM composition and function

Typical cell mechanical interaction with the extracellular matrix (ECM) begins at the cell membrane. Critical to understanding this first interaction step includes determination of the mechanical properties and identification of mechanotransduction. ECM is a meshwork of various proteins and glycans enveloping the cells within a tissue. The composition of the ECM varies greatly between tissues types based on their functional role (Silver, DeVore, & Siperko, 2003). While the ECM is primarily composed of collagen and proteoglycans, the basement membranes in arterial walls incorporate elastin to rebound after deformation by blood pressure dynamics (D'Armiento, 2003; Gentili & Cancedda, 2009). ECM is both constructed and maintained by the cells inhabiting it, and therefore the term reciprocal interplay is appropriate. Based on the phenotype of the cell, it secretes ECM composition elements that generally act in biofeedback to preserve its functionality (Yue, 2014). However, many pathologies are induced by either external influences on ECM structure or by genetic mutations in ECM proteins (Bateman, Boot-Handford, & Lamandé, 2009). While some conditions such as fibrosis occur due to an overproduction of collagen providing stiffness to the ECM in the process of scar formation, others include the overproduction of metalloproteinases (MMPs) leading to the degradation and weakening of the ECM such as osteoarthritis (Iannone & Lapadula, 2003; Lan, Huang, & Tan, 2013). Thus, an effective and regulated communication between the ECM and the resident cells must take place for optimal homeostasis.

2.2 Cellular sensation of ECM

In order to establish communication between the cell and ECM, the cell must be able to bind to and sense the ECM composition along with the mechanical stresses and strains occurring within it. Ubiquitous in virtually all adherent cells, a family of adhesion proteins, referred to as integrins, is known to mediate such communication. Integrins are transmembrane heterodimers composed of various isomers of α and β subunits (Campbell & Humphries, 2011). Through these combinations, there are at least several dozens of the known subtypes of integrins that bind to particular ECM ligands thereby imparting information to the cell about the local ECM microenvironment (Prowse, Chong, Gray, & Munro, 2010). Integrin binding to the ECM ligands initiates recruitment of more than 50 intracellular molecules near the cell membrane to form μm-sized subcellular structures, referred to as focal adhesion (Zamir & Geiger, 2001). In addition to creating mechanical tension through binding to microfilament, the focal adhesion can activate cytoskeletal enzymes such as focal adhesion kinases (FAK), Rho, and Rho-associated protein kinases (ROCK) (Huveneers & Danen, 2009). In addition, integrins interact with receptor tyrosine kinases (RTKs) and G protein-coupled receptors (GPCRs), thus modulating the downstream biochemical signaling of various growth factors derived from integrin-ECM interactions (Short, Boyer, & Juliano, 2000). The complexity of integrin-mediated biomechanical and mechanical signaling is schematically described in Fig. 1. While this description is incomplete and further investigation is warranted, it nonetheless summarizes the current understanding of one of the widely accepted mechanotransduction pathways.

In addition to the ECM composition, the matrix stiffness plays significant roles in proliferation, differentiation, and pathophysiology. The interactions between the mechanosensing signaling molecules such as FAKs, p130Cas, and Rac have been identified to enable the cell to sense the stiffness of the ECM and alter its proliferative state (Bae et al., 2014). The effect of FAKs on MAPK/ERK and associated signaling mechanisms, for example, have been shown to increase the expression of Cyclin D1 (CD1), thus inducing cell phase progressions and proliferation (Schwartz & Assoian, 2001). In association with FAKs, Rho can maintain the activation of ERK signaling and promote CD1 expression via actin polymerization and intracellular cytoskeletal stress, thus also affecting cellular proliferation (Assoian & Klein, 2008). While these types of proliferative interactions

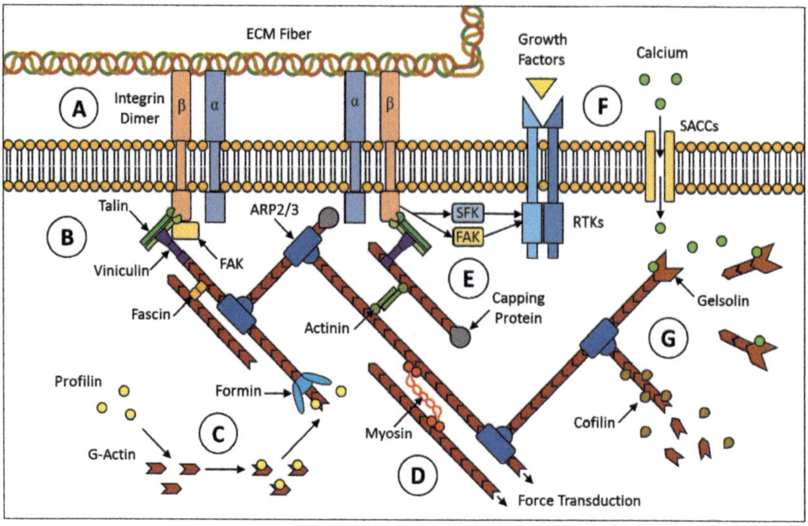

Fig. 1 (A) Integrins at the cell's surface bind to ECM components, linking the membrane to the ECM. (B) Focal adhesions form to physically and functionally bind integrins to the actin network below via protein complexes involving talin, viniculin, and FAKs. (C) Dynamics of F-actin formation involving profilin, G-actin, and formin. Additionally, crosslinking is performed by various proteins such as fascin and actinin, with branching by Arp-2/3. (D) Myosin interactions with actin filaments can be sensed downstream via direct mechanotransduction. (E) Interactions between integrins and RTKs (or GPCRs) via various kinases modulate downstream signaling events. (F) Membrane tension can open mechanically operated ion channels and allow ion influx (e.g., calcium). (G) Depolymerization of F-actin is commonly performed by cofilin binding or gelsolin cleavage spurred by an increased cytosolic calcium concentration.

may be beneficial, the ECM stiffness and cellular responses can promote pathophysiology of tumor progression and metastasis (Malik et al., 2019). Synthetic fibers were utilized in a study to investigate cellular responses to ECM alignment and mechanical stresses (Baker et al., 2015). This study demonstrated a strong connection between the ECM stiffness and cell migration and proliferation. Further, mathematical modeling studies have implicated a feedback loop between the migrating cells and their ECM, where the cells actively modulate the ECM stiffness and alignment which, in turn, affect their responses in a reciprocal manner (Ahmadzadeh et al., 2017). This line of investigations is important to understanding cellular responses to the ECM stiffness and composition and will likely continue to shed more insights into such a complex and dynamic reciprocity.

2.3 Cellular membrane

The molecules involved in ECM binding and transduction are incorporated by the cellular membrane, which itself can impart information about the mechanical environment to the nucleus (Poole, Moroni, & Lewin, 2014). The bulk structure of the cellular membrane is a lipid bilayer composed primarily of amphiphilic phospholipids. The strong hydrophobic interaction between these phospholipids accounts for a large portion of the elastic strength of the cellular surface (Frolov & Zimmerberg, 2010). Membrane-associated proteins are abundant and provide a vast variety of biological signaling functions. These proteins can greatly affect the mechanical behavior of the cellular membrane both directly and indirectly via downstream signaling (Los & Murata, 2004). The membrane-associated proteins commonly have an amphiphilic characteristics allowing them to form a transmembrane structure, or by hydrophobic regions partially embedding the protein into one side of the bilayer (Khandelia, Ipsen, & Mouritsen, 2008). The cellular membrane itself is under constant tension that is estimated between 10 and 300 nN/m, dependent on the type of cell and its environmental conditions (Kozlov & Chernomordik, 2015). This tension is brought about primarily by the membrane's connection between the ECM and its internal cytoskeletal networks (Diz-Muñoz, Fletcher, & Weiner, 2012). This tension can alter the functionality of membrane bound-proteins including the opening or closing of ion channels (Anishkin, Loukin, Teng, & Kung, 2014). Alongside integrins, the cellular membrane adheres to the cytoskeleton via linking proteins such as ezrin, radixin, and moesin, collectively known as the ERM proteins. These proteins form FERM domains which function to attach the membrane to the filamentous actin networks below the surface (Fehon, McClatchey, & Bretscher, 2010). Taken together, changes in the membrane tension can be sensed both directly and indirectly by ERM- and integrin-mediated mechanotransduction to the cytoskeleton and altered membrane protein function.

While membrane tension affects embedded proteins, the expression levels of these proteins in the membrane also affect the stiffness of the membrane. Simulations have demonstrated that the presence of transmembrane proteins has a great effect on the stiffness of the membrane (Fowler et al., 2016). More active variations in membrane structure and tension are brought about by highly conserved BAR domains. BAR protein dimers form various curvatures and contain amphipathic helices allowing them to adhere to the cellular membrane. These domains thus exert forces on the membrane and control the membrane curvature (Simunovic, Evergren, Callan-Jones, & Bassereau, 2019). A vital function of protein manipulation of the membrane are the processes of endo- and exocytosis. In exocytosis, T-Snare and V-SNARE proteins combine into the trans-SNARE complex caught between the vesicle and the membrane. The calcium-dependent torsion in the SNARE helices provides a zippering function allowing for vesicle fusion to the membrane (Lou & Shin, 2016).

In addition to phospholipids, cholesterol has long been studied as an important component in cellular membranes with the functionality of altering membrane fluidity at various temperatures (Crockett, 1998). Cholesterol composition has also been shown to affect the membrane stiffness and tension in association with other lipid species such as sphingomyelin and galactosylceramide in the myelin sheath composition (Saeedimasine, Montanino, Kleiven, & Villa, 2019). Domains containing high concentrations of cholesterol are lipid rafts. These rafts influence receptor activation and subsequent signaling pathways. Proteins requiring cross-talk with each other likely associate in the same or nearby rafts so as to increase the efficiency of their signaling mechanisms (Simons & Ehehalt, 2002). These lipid rafts also incorporate FERM and integrin proteins to help regulate the interactions between the cellular membrane and the cytoskeleton beneath (Chichili & Rodgers, 2009; Fig. 2).

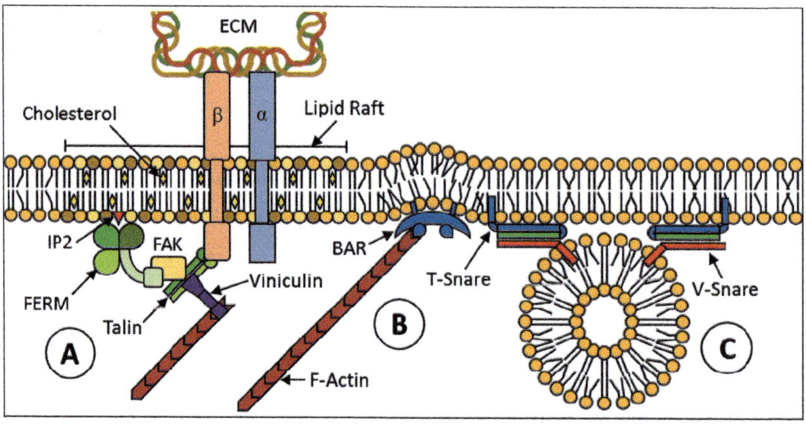

Fig. 2 (A) ERM proteins bind to IP_2 within the lipid membrane while also facilitates integration of actin with focal adhesions via FAK. (B) BAR proteins induce curvature in the membrane while also acting as actin nucleation sites. (C) Vesicles bind to the cellular membrane via the trans-SNARE complex created by an intertwining of T- and V-SNARE proteins.

3. Cytoskeletal structure and function

3.1 Anatomy of cytoskeleton

Underneath the cell membrane and throughout the cytosol, a network of polymerized structures is known as the cytoskeleton. It is the cell's internal structure by which the cell maintains its morphology and functionality. It is involved essentially in every aspect of the cell's physiology from proliferation and apoptosis to secretion and cell movement (Fletcher & Mullins, 2010). Unlike the rigid skeleton of vertebrates, the cytoskeleton is highly dynamic and can quickly adapt to either a rigid or an elastic structure (Gardel, Kasza, Brangwynne, Liu, & Weitz, 2008). The cytoskeleton is composed of three major categories of filamentous structures: microtubules, intermediate filaments, and actin networks (Fletcher & Mullins, 2010). Microtubules act as a more rigid support defining cellular morphology and acting as a guide for intracellular trafficking (Franker & Hoogenraad, 2013). While microtubules are approximately 25 nm in diameter, intermediate filaments are about 10 nm in width and play an intermediate role between microtubules and actin filaments (Godsel, Hobbs, & Green, 2008). These intermediate filaments also play an important role in nuclear structure in the form of lamin A/C filaments (Gruenbaum & Foisner, 2015). Finally, actins regulate a large array of cytoskeletal dynamics that, in turn, controls mechanotransduction (Harris, Jreij, & Fletcher, 2018). In fact, using atomic force microscopy (AFM), the microfilaments were observed to control the cell's bulk Young modulus, and a contribution to the cell's overall property by the microtubules was almost negligible in some types (Titushkin & Cho, 2007). Implication may be the dynamics of microfilament organization is sufficiently rapid to accommodate mechanical remodeling of the cell, while the microtubule organization facilitates the associated physiological changes (e.g., vesicle trafficking) to collectively respond to the external biochemical or biomechanical stimulation.

While molecular interactions with integrins inform the cell about the ECM composition, the binding between the integrin protein and the cytoskeleton can provide direct information about the mechanical stress in the ECM environment (Schwartz, 2010). Below the membrane surface lies the actin-myosin cortex, a dynamic network of actin and myosin that influences the behavior of the cell's membrane (Chugh & Paluch, 2018). Integrins bind to this actin network via proteins such as talin and viniculin in the focal adhesion which, in turn, link the focal adhesion to the cytoskeletal network (Vicente-Manzanares, Choi, & Horwitz, 2009). Not only this arrangement connects physically the cell membrane to the cytoskeleton, it is also found to be critical for cell migration, for example, by forming at the leading edge of lamellipodia during cellular locomotion and promoting survival by inhibiting apoptosis in anchorage dependent cells (Ishikawa, Ushida, Mori, & Shibanuma, 2015; Parsons, Horwitz, & Schwartz, 2010). The level of coordinated events that regulates the actin-myosin cortex is also interesting. As mentioned previously, the submembrane actin-myosin cortex has been identified to control cell migration and endo/exocytosis (Meunier & Gutiérrez, 2016). Actin filaments act as a scaffolding structure, while the myosin motor proteins act to pull actin filaments closer together (Carvalho et al., 2013). In migration, the actin cortex polymerizes at the lamellipodia's leading edge in tandem with myosin activation leading to a pulling force toward the leading edge. While this occurs, disassembly of FAs at the trailing end of the cell allows for forward movement in the direction of the lamellipodia formation (Frame, Fincham, Carragher, & Wyke, 2002). In endo/exocytosis, the actin-myosin cortex plays a role in receiving/docking vesicles and transporting them into or out of the cytosol (Bond, Brandstaetter, Sellers, Kendrick-Jones, & Buss, 2011).

3.2 Actin structure and activity

Actin proteins exist in either a globular (G-actin) or filamental (F-actin) state. In the globular state, actin remains as a polar monomer. Once ATP binds to G-actin, ATP bound G-actin will attempt to self-polymerize into shorter chains of F-actin (Dominguez & Holmes, 2011). These short oligomers are unstable and will spontaneously disassemble (Dominguez, 2009). For the cell to effectively form F-actin, other proteins are required such as formin (Courtemanche, 2018). Because actin is polar monomer, the polymerization of actin should progress faster on the "barbed end" of the chain than on the "pointed end." This property regulates and modulates the balance between polymerization and depolymerization of F-actin (Narita, Oda, & Maéda, 2011). It is not surprising that there is a large family of proteins that interact with actin to either promote or degrade F-actin formation (Lee & Dominguez, 2010). One such regulator of actin formation is profilin. By binding to the barbed end of G-actin, profilin is able to enhance F-actin formation by acting with formin. While promoting actin polymerization, it also serves to regulate actin branching by inhibiting Arp-2/3 (Suarez et al., 2014).

When actin polymerization should be ceased, protein complexes known as actin-capping proteins can prevent further polymerization of F-actin by binding to the barbed end of an actin filament, thereby stabilizing the filament (Edwards et al., 2014). When the disassembly of F-actin is preferred, proteins such as cofilin and gelsolin act to sever actin filaments (Pavlov, Muhlrad, Cooper, Wear, & Reisler, 2007). In particular, the gelsolin's activity is strongly activated by elevations in intracellular calcium (Choe et al., 2002). While intracellular calcium ions have a wide range of regulatory roles in the signal transduction pathways, a high concentration leads to an altered and disassembled F-actin organization throughout the cytoskeleton (Gremm & Wegner, 2000). In addition to the polymerization and degradation of filamentous actin, other proteins serve to induce actin branching and crosslinking, further altering the mechanical stiffness of those actin networks (Brill-Karniely, Ideses, Bernheim-Groswasser, & Ben-Shaul, 2009). For instance, the protein complex ARP-2/3 binds to existing actin filaments allowing further binding of ATP-actin. This results in branching filaments at various angles from the previously formed filaments (Smith, Daugherty-Clarke, Goode, & Gelles, 2013). Crosslinking proteins such as fascin, α-actinin, and filamin act to bind F-actin to one another (Lieleg, Claessens, & Bausch, 2010). The amount of crosslinking induced by these proteins further alters the stiffness of the actin networks and can increase myosin activity in cell motility (Laevsky & Knecht, 2003).

3.3 Actin network mechanical feedback

While the tensile forces exerted on these actin networks will directly be transferred to the nucleus and alter various types of protein expression, these tensile forces can induce feedback within the local protein environment (Stachowiak et al., 2014). When F-actin is under heavier loads, it can induce conformational changes in the actin filament and increase the binding affinity of various F-actin regulators (Gressin, Guillotin, Guérin, Blanchoin, & Michelot, 2015). For instance, single filaments experiencing tensile forces up to 30 pN did not affect cofilin binding, but torsional forces brought about by the initial binding of cofilin significantly enhanced further cofilin binding (Wioland, Jegou, & Romet-Lemonne, 2019). In addition, these increased forces on F-actin increased binding affinity to myosin II allowing for an increase in contractile force, alongside increased activation of Arp-2/3 complexes to allow for actin branching (Sonal et al., 2018). Thus the protein regulatory network for actin is itself able to sense and react to changes in cytoskeletal mechanical stress. The mechanical stresses at the membrane and the actin-myosin cortex can induce conformational changes in mechanically operated channels such as stretch-activated calcium channels (SACCs) (Sachs, 2010). When SACCs are activated, they allow calcium to influx into the cytosol. This can directly affect cytoskeletal dynamics through disassembly of F-actin by calcium-activated depolymerization of actin (Ito et al., 2009). Other ion channels can be activated by mechanical stress at the cellular membrane such as transient receptor potential ion channels (TRPs) (Clark, Middelbeek, & Leeuwen, 2008). The stretch-activated opening of SACCs and TRPs with subsequent calcium influx has been shown to be involved in neurite outgrowth and axon path finding (Jacques-Fricke, Seow, Gottlieb, Sachs, & Gomez, 2006).

To briefly summarize, the ability of cells to sense and respond to their surroundings is a key requirement for optimal functionality and their survival. They are able to sense the chemical environment typically by receptor binding and the mechanical environment through integrin binding. Both can generate mechanical stresses that are transduced by an interaction between the cell membrane and the actin-myosin cortex. From there, the stresses can be interpreted by the cell through either direct stresses transmitted through the cytoskeleton or by downstream secondary messengers. The regulatory protein network can also sense changes in cytoskeletal stress and respond quickly to these changes without nuclear responses or modulate the nuclear morphology and its mechanics in a prolonged stimulation. These feedback mechanisms are induced via, among other possibilities, conformational changes in the loaded F-actin and/or changes in local calcium dynamics. Through these and other mechanisms that are yet to be identified and validated, the cell is able to sense and respond to its environment via mechanotransduction in an intricate reciprocal interplay.

4. Nuclear membrane

The next level of mechanical interconnection that recently became a rapidly growing discipline is the nucleus. Indeed, the nuclear membrane is the hallmark differentiator between eukaryotic and prokaryotic cells. It is the boundary which separates a cell's genetic material from the rest of the cell, allowing a degree of regulation only found in eukaryotic organisms. The nuclear envelope provides a barrier for precise control over ingress and egress of molecules. In addition, the rigid lamin structure lining the inner nuclear membrane provides mechanical support for cells under physical stress. The nuclear membrane also interfaces with cytoskeletal elements and components of the nuclear interior, providing mechanism(s) to detect, interpret and respond to external mechanical stimuli delivered to the cell at the cell membrane level. The similar mechanism may also direct and control the regulation of genes. Not surprisingly, the importance of the nuclear envelope (its composition, organization, and degree of involvement in genetic activation) has been established in its contribution to many cellular processes and disease states (Dauer & Worman, 2009; Östlund & Worman, 2002; Worman & Courvalin, 2005).

Similar to the cell membrane, the main structure of the nuclear envelope is a phospholipid bilayer, generating an outer nuclear membrane (ONM), an inner nuclear membrane (INM), and subsequent intermembrane space. The ONM is contiguous with the endoplasmic reticulum and is thus marked with ribosomes as well as sites for cytoskeletal attachment (Dingwall & Laskey, 1992). The INM is home to a number of proteins which bind to internuclear structures to guide the nuclear shape and chromosome positioning (Dingwall & Laskey, 1992). Lining the nucleoplasmic INM sits the nuclear lamina, a meshwork of intermediate filaments (IFs) which provide the main structural component of the nuclear envelope (Gruenbaum & Foisner, 2015). Lamin, the proteins which make up the lamina, are type V IFs, and are specific to the nucleus (Leeuw, Gruenbaum, & Medalia, 2018). Generally, there exists two types of lamin proteins: A-type (lamins A and C, encoded from the *LMNA* gene) and B-type lamins (B1 and B2/3, encoded from *LMNB1* and *LMNB2*, respectively; Bridger, Foeger, Kill, & Herrmann, 2007). The ONM and INM are joined together in areas occupied by nuclear pore complexes (NPCs), which are responsible for directed transport of proteins and nucleic acids into and out of the internuclear space. NPCs are large (100 MDa) complexes comprised of a group of proteins demonstrating an eightfold rotational symmetry (Alber et al., 2007). The cytoplasmic surface of the NPCs possesses rod-like filaments which extend into the cytoplasmic space, which allows communication with the cytoplasmic elements of the cell. Passing through the nuclear lamina, INM, and ONM are an additional set of proteins which connect the cytoplasmic and nucleoplasmic elements of the cell. Together, these protein groups are known as linker of nucleoskeleton and cytoskeleton (LINC) complexes (Fig. 3). Comprising the LINC complexes are the SUN and

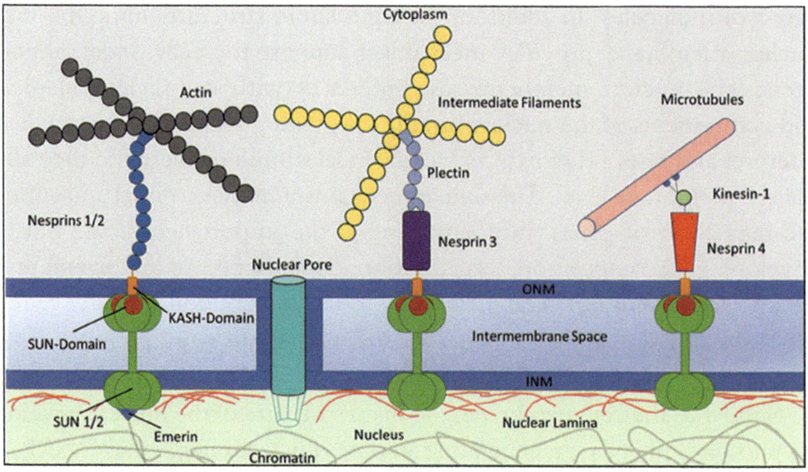

Fig. 3 Schematic diagram of SUN-KASH interactions bridging the nuclear envelope and cytoskeletal elements. SUN proteins (green) span the inner nuclear membrane and interface with internal nuclear components such as lamin (red), and chromosomes (gray line) through adaptors (blue diamond). In the intermembrane space, SUN domains (dark red) interact with KASH domains (orange) of KASH proteins in the outer nuclear membrane (ONM). KASH proteins span the ONM and protrude into the cytosol where they interface with various cytoskeletal elements. Giant KASH proteins Nesprin-1 and Nesprin-2 (blue circles) contain a multitude of spectrin repeats with an N-terminal actin binding domain to interface with cytoskeletal actin (gray circles). Nesprin-3 (purple) interacts with the cytoskeletal linker plectin (light blue circles) which further serves to link cytoskeletal elements such as intermediate filaments, microtubules, and actin (yellow circles). Nesprin-4 (red trapezoid) associates with the motor protein kinesin-1 (light green) which move along microfilaments and direct several cellular functions. Together, the bridge formed through the SUN-KASH interaction serves to convey mechanical signals from the cell's cytoskeleton to the nuclear interior.

KASH protein domains (Rothballer, Schwartz, & Kutay, 2013). SUN proteins are localized to the INM and connect to the nuclear lamina and chromosomes. KASH proteins span the ONM and connect SUN proteins to the cytoskeletal elements such as actin (Sosa, Kutay, & Schwartz, 2013). The LINC complexes appear to integrate with cytoskeletal elements and mediate mechanotransduction from the cell membrane to the nucleus and its internal components (Tapley & Starr, 2012).

4.1 SUN-KASH interactions

SUN and KASH interact with one another in the intermembrane space between the INM and ONM, forming a bridge across the nuclear envelope. SUN proteins form the INM half and recruit the KASH proteins of the

ONM half. It has been shown that KASH proteins require SUN interactions for localization to the ONM (Padmakumar et al., 2005). Along with interacting with each other, SUN and KASH proteins attach to several other proteins of the nuclear interior and cytoplasm. As integral members of the INM, SUN proteins interact with both KASH proteins in the intermembrane space, as well as several internal nuclear components. It has been shown that the N-terminal component of SUN proteins Sun1 and Sun2 in the nuclear interior form part of a multiple-interaction network with the nuclear lamina as well as chromatin. Associating with the lamin network provides an anchor for SUN in the nuclear envelope, as well as involvement in nuclear shape and chromosome positioning (Haque et al., 2009). Other than lamin, SUN proteins have additional modes of interaction with chromosomes. For example, SUN proteins have been found to interact with the INM protein emerin, a direct regulator of lamin and chromatin functionality (Berk, Tifft, & Wilson, 2013; Demmerle, Koch, & Holaska, 2013). In lower eukaryotes, SUN proteins have been found to associate with directly with DNA (Xiong et al., 2008).

Interactions between SUN and KASH proteins occur in the intermembrane space between the ONM and INM and are the bedrock of the LINC complex. As described, SUN proteins extend into the intermembrane space, where the SUN domains localize proximate to the ONM. Here, the SUN domains bind to C-terminal KASH domain of the various KASH proteins in the ONM. The SUN-KASH interaction is further stabilized by disulfide bonds, forming a complete bridge spanning the nuclear membrane (Jahed, Shams, & Mofrad, 2015). SUN domains assemble into multimers at the interior of the ONM, which in turn serve as a pocket for KASH domain binding (Cain et al., 2018). The crystal structure interaction between SUN and KASH has been demonstrated by Sosa et al., who describe the interface between Sun2 and Nesprin-1/Nesprin-2 (Sosa, Rothballer, Kutay, & Schwartz, 2012).

4.2 KASH interactions of cytoplasmic elements

Once localized to the ONM surface by interfacing with SUN proteins, KASH interacts with several cytoplasmic elements controlling a wide range of cellular activities. The largest isoforms of nesprin-1 and nesprin-2 are known as giant isoforms (800–1000 kDa) and feature an actin binding domain at the N-termini following the spectrin repeats (Liem, 2016). Nesprin-3 is similarly localized to the ONM in a SUN-dependent manner.

While much smaller than the giant isoforms of nesprin-1 and nesprin-2, nesprin-3 interfaces with plectin, forming a large structure which links the nuclear membrane to the main components of the cytoskeleton (intermediate filaments, actin, and microtubules; Wilhelmsen et al., 2005). An additional isoform of the KASH protein is nesprin-4, which interacts with kinesin-1 for directed motility of cellular microtubules (Roux et al., 2009). KASH protein interactions with all major components of the cytoskeleton are demonstrated in Fig. 4.

4.3 Cell-nucleus mechanotransduction

LINC complexes consisting of integral membrane proteins nesprin and SUN directly link the cytoskeletal architecture to that of the nucleus. Through this interaction, mechanical signals are transduced from extracellular stimuli to the nuclear interior. Several reports support this hypothesis, including disruption of KASH proteins resulting in a loss of cellular mechanical stiffness (Stewart-Hutchinson, Hale, Wirtz, & Hodzic, 2008). Additionally, the size and shape of the nucleus is dependent on the actin cytoskeleton, demonstrating deviations with cytoskeletal disruption (Antmen, Demirci, & Hasirci, 2019; Ross et al., 2019; Zhang et al., 2020).

Mechanotransduction begins with an outside force applied to focal adhesions at the cell surface through the ECM. Talin, the main connector between integrin and actin, transmits this force to actin fibers in the cellular interior (Jahed, Shams, Mehrbod, & Mofrad, 2014). Myosin II provides the actin fibers with tension, allowing stress propagation through the molecule (Wang, Tytell, & Ingber, 2009). Reducing actin tension by inhibiting myosin prevents cytoskeletal-to-nuclear mechanotransduction (Clark, Langeslag, Figdor, & Leeuwen, 2007; Wiggan, Schroder, Krapf, Bamburg, & DeLuca, 2017). Attached to actin through various means are other cytoskeletal elements such as intermediate filaments (attached through plectin) and microtubules (attached through microtubule-actin crosslinking factor-1, MACF1; Chen et al., 2006; García-Alvarez, Bobkov, Sonnenberg, & Pereda, 2003). The external force is then passed on to the initial element of the LINC complex, nesprin. As described, nesprins connect to all major cytoskeletal elements and contain a KASH domain for anchoring to the ONM. The KASH protein spans the ONM and interfaces with SUN proteins in the intermembrane space. The stability of the SUN-KASH interaction allows continuing propagation through the nuclear envelope and to the nuclear interior. SUN proteins on the

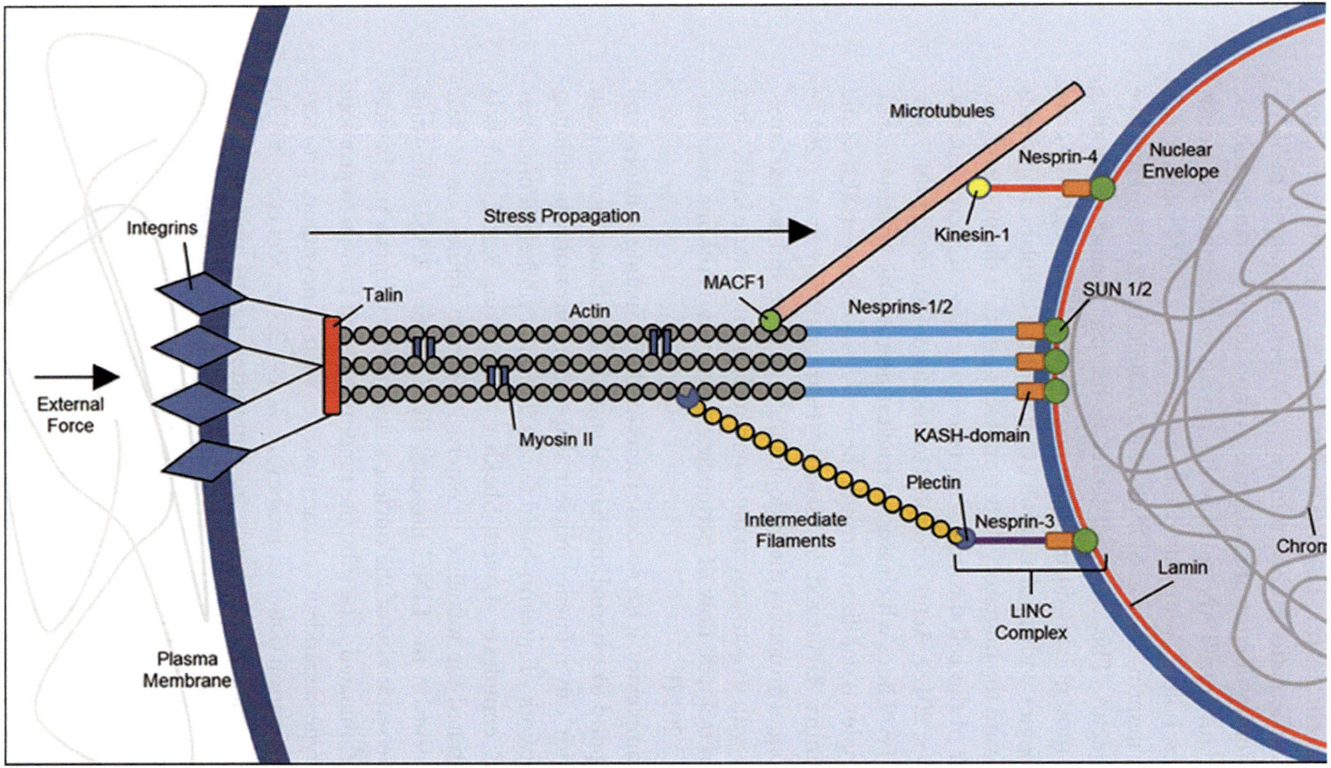

Fig. 4 Mechanotransduction from the extracellular matrix to the nuclear interior via cytoskeletal and LINC interactions. An external force is applied to integrins (blue diamonds) on the cell surface which is transmitted internally. Talin (red) connects surface integrins to the cytoskeletal network through actin (gray circles). Actin filaments are kept under tension by myosin II (blue =) and are connected to other major cytoskeletal elements intermediate filaments (yellow circles) and microtubules (pink cylinder) through the connectors plectin (blue partial circle) and MACF1 (light green circle), respectively. The cytoskeletal components propagate the force to the KASH proteins nesprin (red, light blue, purple lines) on the nuclear membrane surface. KASH and SUN (dark green circles) form a stable interaction in the intermembrane space between the ONM and INM, continuing the force transmission into the internuclear space. SUN proteins in the nuclear interior interface with the nuclear lamina (red line) and chromatin (gray line) to transmit the extracellular force to the nuclear interior.

nucleoplasmic face of the INM transmit the extracellular force to the nuclear lamina as well as chromatin. The mechanotransduction and stress propagation pathway is schematically shown in Fig. 4. A specific example of SUN-KASH mechanotransduction includes the composition of transmembrane actin-associated nuclear (TAN) lines, which have been described previously (Luxton, Gomes, Folker, Worman, & Gundersen, 2011). TAN lines include LINC proteins nesprin-2G and SUN2 which couple the nuclear envelope to cytoskeletal actin. This coupling serves to modify nuclear positioning within the cell, most notably during cell migration. For most migratory cells, cell migration begins with a polarization event wherein cellular components are oriented toward the migratory direction. Cells position the nuclear body toward the rear of the cell through actin retrograde flow (Calero-Cuenca, Janota, & Gomes, 2018). Actin is coupled to the aforementioned LINC proteins, forming TAN lines which assist in nuclear positioning within the cell. The LINC complex is stabilized through the interaction between SUN2 and the nuclear lamina, facilitated by Samp1 (Borrego-Pinto et al., 2012). TorsinA, an integral protein of the endoplasmic reticulum and nuclear envelope, integrates with its activator in the inner nuclear membrane LAP1, thus regulating the retrograde flow of perinuclear actin and TAN line assembly (Saunders et al., 2017).

In spite of complexity, it has become increasingly fascinating to understand how nuclei respond to external mechanical stimuli. Mechanical forces from the cell's exterior can influence nuclear shape and chromatin positioning through the network of integrin-based focal adhesion, cytoskeletal organization and LINC complexes. Lovett, Shekhar, Nickerson, Roux, and Lele (2013) demonstrated the influence of external substrate stiffness on nuclear shape is dependent on the LINC proteins. By disrupting KASH domains, for example, the sensitivity of the nuclear shape to the substrate mechanical properties was eliminated. The substrate rigidity has been known to promote a multitude of type-dependent cellular processes such as migration, proliferation, and differentiation (Discher, Janmey, & Wang, 2005; Wei et al., 2015). Ultimately, these mechanical signals are transmitted to the nucleus where genetic reorganization can occur. The control of mechano-induced gene transcription is regulated by LINC complexes. As described by Alam et al. (2016) mechanosensitive gene regulation is influenced by LINC disruption and myosin activity. The SUN proteins on the nuclear interior also interface with the nuclear lamina and other INM proteins which are implicated in chromatin organization and gene silencing (Mekhail & Moazed, 2010; Vosse, Wan, Wozniak, & Aitchison, 2011). Taken together, external mechanical forces propagate through a series of protein linkages and ultimately influence gene regulation. While one aims at elucidating this intricate network of interconnected and coordinated events, such efforts are hampered by a lack of experimental techniques that can probe directly the nuclear mechanics. Micropipette aspiration, laser optical tweezers and atomic microscopy all have been so instrumental in measuring the mechanical properties of the cell membrane and the cytoskeleton. These techniques were also utilized in measuring the cell's responses that were induced by biochemicals (e.g., stem cell differentiation by growth factors) or by mechanical cues (e.g., stretch mediated cell differentiation). However, there are no direct methods that can monitor mechanical changes in the nucleus. Although there are serious efforts underway to develop methods to elucidate the nuclear mechanics, one may turn to computational prowess to fill the gaps to better understand and appreciate the intricate mechanosensitive network(s) that ultimately transmit and transduce the external stimuli to the nucleus.

5. Computational modeling of intracellular biomechanics

In a biological system, many factors combine to influence the behavior of a cell. The cell membrane experiences external as well as internal forces. The external forces are transmitted to the cell membrane from the extracellular matrix (ECM), which can provide structural support for cells. ECMs of different tissues have different properties and topographies requiring different methods of study (Holle et al., 2018). Many computational models have been produced to study ECM mechanics, including the mechanical properties (Meyer & Lieber, 2011), behavior of ECM (Han et al., 2018), and its role in mechanotransduction (Weinberg, Mair, & Lemmon, 2017). This section focuses on the internal forces affecting the cell membrane including its interactions with the cytoskeleton and the nuclear membrane. However, the basic concepts in modeling and obtaining the solution for intracellular mechanics can be applied to modeling the cell's external environment.

The role of interconnected coupling mechanisms from the cell membrane to the nucleus is indeed complex and requires detailed studies. To better understand biomechanics of the cell as one contiguous biosystem, one must overcome several challenges. Unlike the cell membrane and cytoskeletal mechanics that may be probed using a variety of biophysical tools (e.g., micropipette aspiration, laser optical tweezers, atomic force microscopy and others), there are only a limited probing tools available to directly measure the nuclear mechanics (Muhamed, Chowdhury, & Maruthamuthu, 2017). In addition, the length and time scale of changes in biomechanics may not be similar. Instead, several laboratories have utilized mathematical modeling to help elucidate the mechanics of the cell. This approach involves defining a mathematical representation of the mechanics of biological phenomena. These models are composed of individual mathematical terms whose purpose are to capture an aspect of the mechanical behavior. The process of formulating these terms demands decomposing the mechanical behavior of the cellular processes into constituent parts. The ensemble of these terms combines to mathematically represent the mechanical behavior of the cell. A solution to this mathematical model can provide a prediction of the mechanical behavior of the cell. Comparing this prediction with experimentally reproducible observations facilitates a search for the parameters that define individual terms, which can lead to a better understanding of the mechanical behavior of the cell.

Physics-based models stem from Newton's Laws of Motion. The goal in physics-based modeling is to obtain a mathematical representation of a balance between applied forces and inertial forces. This theoretical basis for model construction allows for a wide range of applicability. However, it is difficult to model unknown physical behavior with this approach. One solution to this problem is to use mathematical models, which are constructed purely based on mathematical functions. The terms in such a model are chosen to correspond to experimental data. Yet the purely mathematical models have a narrow scope of applicability because they are based on specific experimental data. Physics-based models are often amended by purely mathematical terms to improve their predictive capability, especially when the physics driving cellular processes are not well known.

The complexity of cellular mechanics makes these mathematical models complicated and therefore difficult to solve. There are two broad categories of techniques used to address this issue: (1) model simplification/reduction and (2) computationally efficient techniques. Computational efficiency involves reducing the computational time required to evaluate the terms in the model. This can be done through different algorithms, such as parallel processing, or hardware that executes the computations faster, such as a supercomputer. Computational cost refers to the number of basic computations needed to evaluate the model, while computational time refers to the amount of time required to execute them.

A key technique in model reduction is coarse-graining, which involves reducing the resolution of details of cellular mechanics in the model. This typically involves grouping smaller bodies together and representing them by a larger body or representing discrete atomistic phenomena as a continuum such as viscosity. The second key model reduction technique involves efforts to minimize order, which means formulating a force balance while omitting inertial forces or other forces. This generally results in static, quasi-static, or dynamic models. Static models omit velocity- and acceleration-dependent forces, quasi-static models include velocity-dependent forces, and dynamic models include all forces. The computational models discussed herein are quasi-static or dynamic.

There are justifications for choosing different order reduction. It is easier to obtain static experimental data. This will allow a simpler comparison of the static model predictions with experimental results. However, transient behavior cannot be captured by a static model, but these transients may have a significant effect on the final static measurements. The quasi-static reduction can capture some of the transient behaviors, but the inertial forces are omitted. The justification for the quasi-static approach is a widely held notion that the inertial forces do not have appreciable effects at the cellular and subcellular scale (Lange & Grubmüller, 2006). However, recent studies suggest that this justification may not be appropriate, especially at the subcellular level (Haghshenas-Jaryani et al., 2014). These considerations led to the development of dynamic models, which incorporate inertial forces while capturing all effects included in the other models.

Coarse-graining and order reduction have direct effects on computational cost. The static order reduction has low computational costs, which increase for quasi-static models. Obtaining a solution for dynamic models requires the highest computational cost and time. Models with atomistic resolution, such as molecular dynamic (MD) simulations, may require supercomputers to obtain solutions. Furthermore, MD simulations typically can only predict the behavior of the system for a very short duration on the order of microseconds. If the process has a long duration, such as stem cell differentiation than can occur over a period of days or weeks, novel model reductions are needed.

Several physics-based mechanical models have been proposed to explain the mechanical behavior of cells. These models may have been amended by purely mathematical terms to allow for cellular processes that are not well understood. Some of the main models proposed are: (1) viscoelastic model, (2) tensegrity model, and (3) multibody dynamic model. More in-depth descriptions of each model are provided below. It should be noted that bioelectricity and biochemistry can affect the mechanical behavior of the cells (Dimitrov, 1984). However, pure bioelectrical processes, such as modeling of neural systems (Burkitt, 2006) and biochemical reactions, are not addressed in the following discussion.

5.1 Viscoelastic model

A widely used model for cell mechanics that relies on coarse-graining and order reduction techniques to simplify the complex cellular structures is referred to as the viscoelastic model. This approach focuses on the cell membrane. It treats the cell as a viscoelastic cell membrane encompassing incompressible cytoplasmic fluid in few main varieties referred to as: (1) dynamic systems, (2) finite element or (3) elastic modeling. In the dynamic systems approach, the model is coarse-grained by considering the bulk properties of one continuous viscoelastic membrane and incompressible cytoplasmic fluid inside; thus, the model simplifies down to one non-rigid body. In the finite element approach the cell is coarse-grained or meshed into numerous non-rigid bodies with various spatial resolutions. The mechanics of the membrane and the bulk characteristics of the cell are the focus of this model. In the elastic model, the cell membrane is coarse-grained into one or more linear or nonlinear mass-spring-damper elements. This model is mainly intended to assess the stiffness properties of the cell membrane.

The order of finite element models (FEM) is generally reduced to static by assuming certain velocity-dependent forces, such as the viscous behavior of the membrane, are constant for each solution of the model. Solving the model for different velocities gives something like a quasi-static solution. Therefore, FEM approaches are often referred to as quasi-static. However, solving the model in this way will not provide a prediction of the evolution of the mechanical state of the system over time, i.e., the time evolution/history.

The dynamic systems model is not actually a dynamic model of cellular processes. It is mostly a purely mathematical model of system properties added to the non-FEM viscoelastic model. Fig. 5 illustrates the dynamic systems model where the large circle represents the cell membrane. The annotations represent the processes that affect the bulk properties of the cell, for example *endo-* and *exo-*traffic of the cytoplasm, ion solute transport, internal and external pressure differences, etc. (Pawlowski, 2007). These processes are represented as purely mathematical functions based on experimental data

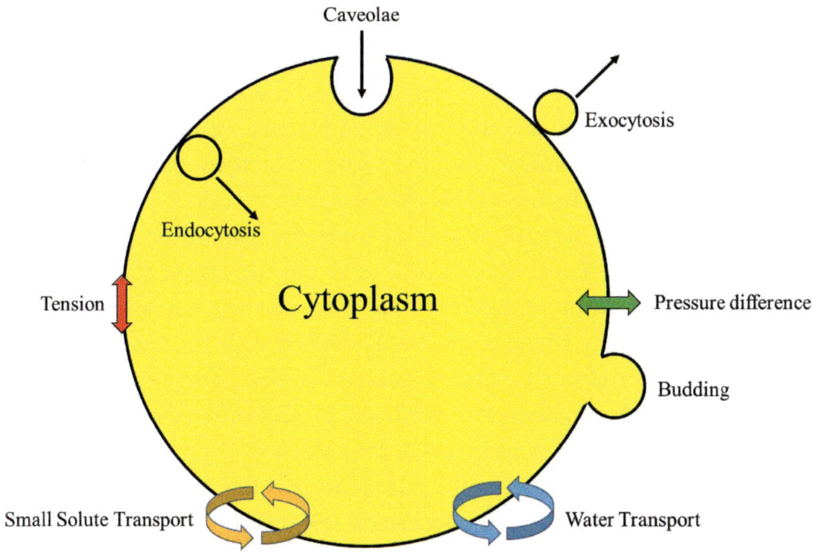

Fig. 5 Representative dynamic system to theoretical model the cell. Mathematical representation of the effects of various cellular processes and interactions can be included in the viscoelastic model. Shown in this schematic includes the examples of mechanical properties as well as plausible biochemical and bioelectrical processes.

analysis. The order of the model depends on the order of the physical and mathematical terms. Coarse-graining used in this approach results in fewer mathematical terms. The solution of this model is a prediction of the time evolution of the system to within the fidelity of terms included in the model (Lu et al., 2006). This approach can also be used to model the evolution of cell aggregates (Honda, Tanemura, & Nagai, 2004).

The computational cost of the FEM approaches directly corresponds to the level of coarse-graining, which maps to the resolution of the mesh. The addition of details such as heterogeneity between different parts of the mesh will increase the computational cost. FEM analysis is a well-developed field and computational solutions for these models are feasible, but these studies typically do not discuss computational time. The simplifications in the dynamics systems model generally reduce the computational time below that of the FEM approach.

Early elastic models were typically based on simple linear or nonlinear springs, such as the Hertz/Sneddon model which was first used to evaluate cell behavior (Kim, Kim, Shin, & Kim, 2011). This early approach assumed purely elastic behavior for the membrane. Advances in the use of atomic force microscopy (AFM) allow for precise application of force on the cell and measurements of its response (Guz, Dokukin, Kalaparthi, & Sokolov, 2014). Experimental results showed that the behavior of the cells diverges from the simple Hertz/Snodden model. Thus, the model was modified to close the gap between the theoretical and actual results (Guilak et al., 1999) by adding mechanical damping terms, which model viscous effects (Pawlowski, 2007). The use of this model allows for calculation of the bulk Young's modulus of the cell as well as some mechanical properties of the membrane (Darling, Topel, Zauscher, Vail, & Guilak, 2008). It also facilitates the precise identification of different cells based on their viscoelastic parameters (Rebelo, de Sousa, Mendes Filho, & Radmacher, 2013). Even viscoelastic properties of the nucleus can be examined (Guilak, Tedrow, & Burgkart, 2000). These results can be used to verify the predictions of the models as well as measurement of the key mechanical properties.

One of the main findings of the experiments is the modulation of membrane tension with time. That is, mechanical response is not only a function of the amount of force, but also its rate of application (Ekpenyong et al., 2012). This is the basis for modeling of the membrane lipid bilayer as a thin viscoelastic film (Darling, Zauscher, Block, & Guilak, 2007). Another modification of the Hertz/Snodden model is the addition of the effects of heterogeneity. This allows for more in-depth study of the different intracellular structures and their effects on cell behavior (Barreto, Clausen, Perrault, Fletcher, & Lacroix, 2013). Finite element methods are especially useful for modeling the response of a local area of the cell to stimuli, such as indentation (Karcher et al., 2003) and budding (Kosawada, Inoue, & Schmid-Schönbein, 2005) (Fig. 6). There are three main advantages to using the viscoelastic modeling approach. The first is the wide availability of experimental data on probing cells using AFM makes it an easy task to validate viscoelastic model predictions. Various experiments have been devised where mechanical, bioelectrical (Solovyova, Vikulova, Konovalov, Kohl, & Markhasin, 2004), or biochemical stimulus were applied to cells and the resulting mechanical behavior was observed (Marinkovic, Turner, Deng, Suresh, & Fredberg, 2007). Second, Both FEM and dynamic system modeling fields are well developed and multiple techniques for their formulation and solutions are standard. Third, the computational cost of solving these models can be very low. The main disadvantage of this approach is the difficulty in incorporating the intracellular structures and their mechanics in detail. The mechanical signal transduction and the biodynamic feedback systems cannot be fully studied. As with all quasi-static model reductions, the inertial forces are omitted, which prevents a true physics-based prediction of the time evolution of the system from being obtained.

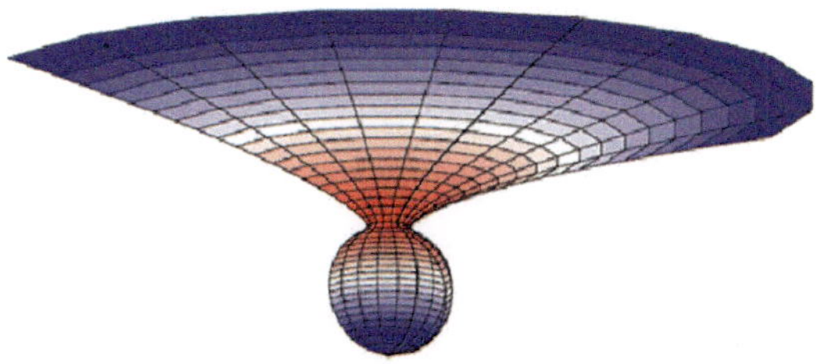

Fig. 6 Finite element model (FEM) of a region of the cell membrane during budding. The model demonstrates coarse-graining the cell membrane into non-rigid sections. This computational methodology allows to study the effects of forces in detail. The stress levels are represented as color spectrum, where darker red denotes a higher stress level to darker blue denotes a lower stress level.

5.2 Tensegrity model

This approach more closely examines the main structures of the cell, the cytoskeleton and membrane (Ingber, 2003). It coarse-grains the cytoskeleton into tension and compression members (Zhang et al., 2018). The choice of the number of compression and tension members, and membrane segments defines the level of coarse-graining. This choice is governed by the tradeoff between the resolution of the solution and the computational cost. These models may also include the cell nucleus with its membrane and internal structure in the same manner. The order of tensegrity models is generally reduced to quasi-static. The solution of the model may show changes in the structure, but the time dependency of these changes may not be well reproduced due to omission of inertial forces. The order of the model can be further reduced to static in the same manner as the viscoelastic models. A static structural analysis of the cell can be obtained in this manner.

The number of cytoskeletal elements and the order reduction method determines the computational cost of obtaining the solution. A static evaluation of the cellular structure with a limited number of members will have low computational cost. Reproducing the time evolution of the cell response to stimuli using a large number of bodies is computationally costly. The computational time of calculating the response of a large number of cytoskeletal elements can be reduced by employing the parallel processing techniques. However, studies of tensegrity do not typically discuss the computational time required to obtain the model's solution.

It is posited that the cytoskeletal structure behaves on the basis of tensegrity. That is, the discontinuous compression beams hold the structure so that the other actin filaments and the biomembrane are in tension (Fig. 7) (Ingber, 2008). This can be facilitated by the different polymerization of the

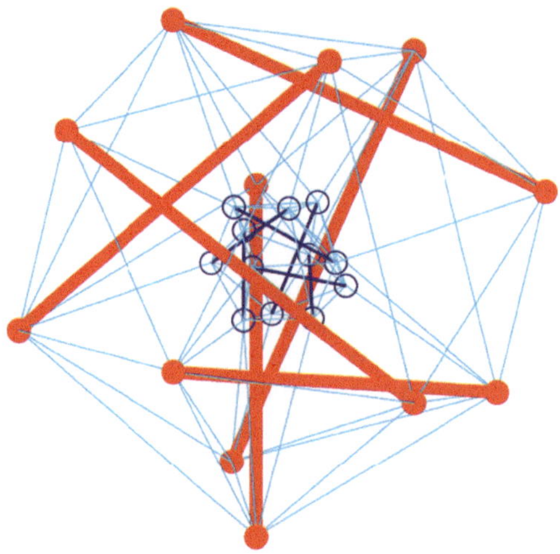

Fig. 7 Computer-generated schematic of a two-layer tensegrity hierarchy. The compression beams are depicted in red (outer layer), and blue (inner layer). The tension elements are depicted as cyan ropes. The end points of the outer layer components are depicted as filled spheres, and the end points of the inner layer are depicted as open spheres. The hierarchy level can be increased by repetition of the model into lower tiers. Macroscale stresses applied to the outer layer are presumed to be transmitted to the intracellular layers that induce structural rearrangement in the cell and subsequently initiate the mechanotransduction signaling pathways.

actin filaments in tension and compression. Evidence suggests that the actin filaments in compression are polymerized such that the bulk stiffness of the cell increases. The actin filaments do not show this behavior under tension (Ronan, Deshpande, McMeeking, & McGarry, 2012). The biological polymerization feedback process is a slow dynamic process. Fast application of force does not result in rapid reaction. This contrasts with presumed fast chemical process of polymerization itself.

The cytoskeleton is connected to the cell membrane and to the nuclear membrane. Mechanical signals from outside of the cell will change the stress and configuration of the cytoskeletal structure. This in turn changes the force applied to the nuclear membrane. The nuclear membrane is under tension. The chromatin structure and lamin A/C filaments act as the equivalent of the cytoskeletal structure inside the nucleus (Dai, Heo, & Mauck, 2020). Changes in applied force from the cytoskeletal structures will changes the stress and the configuration of intranuclear structures. This constitutes the transmission of mechanical signals to the nucleus (Ingber, 2008). The assumption is that this change in chromatin force fields will alter gene expression and guide mechanical feedback.

It is generally accepted that the cell membrane is under tension. The tensegrity model can explain how the membrane remains under tension in the presence of outside compression. The macro scale dynamics of cell membrane behavior suggest a slow tension regulation. When the membrane experiences steady applied force, it polymerizes or depolymerizes to maintain its tension within a threshold, much like a viscoelastic material. This allows the cell to conform to the shape and stiffness of extracellular matrices (Xu, Li, Feng, & Gao, 2016). The advantage of the tensegrity approach is that it facilitates a better understanding of mechanical signal transduction as well as cell to extracellular matrix conformity. The computational costs are generally higher than viscoelastic models because of more detailed inclusion of subcellular structures. Quasi-static and static model reductions of these models result in computationally feasible analysis. The tensegrity model has difficulties simulating viscoelastic bulk effects. The cytoplasm and its interactions with the cytoskeletal structure as well as *endo*-exocytosis do affect cell behavior. This causes difficulties in validating the model with experimental observations that show viscoelastic characteristics of the cell. The inherent problems of quasi-static order reduction in reproduction of the time dependency of dynamic cellular processes still remain.

5.3 Multibody dynamics model

This approach consists of modeling the cell as a collection of rigid bodies acted on by forces. The model can be constructed at the atomistic scale or coarse-grained to achieve larger scales (Karplus & Petsko, 1990). Most of the multibody dynamic models used to model cellular processes are coarse-grained at the molecular level (Fig. 8), which means individual atoms are grouped into molecule-sized bodies. The interactions between the rigid bodies, bioelectrical charges, and biochemical reaction effects, are modeled as forces and force fields. The physical model of the molecular interactions is well developed and the molecular construction of the intracellular structures and organelles are becoming more available (Bao et al., 2010). As such, this model of the cell can reproduce the elastic behavior of cytoskeleton, the viscoelasticity of the membrane, and the viscous effects of the cytoplasm in discrete fashion (Chang, Ayton, & Voth, 2005). This coarse-graining leads to the well-known molecular dynamics (MD) simulation techniques. The system can be further coarse-grained, increasing the scale above molecular level into subcellular structures by combining multiple molecules into larger structures (Nielson, Lopex, Srinivas, & Klein, 2004). This requires an

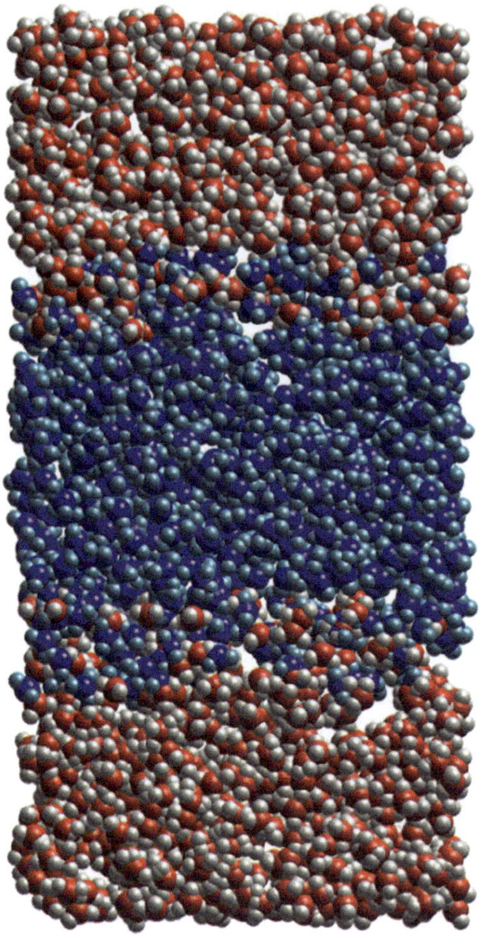

Fig. 8 An atomic scale schematic illustration of a region of the phospholipid bilayer (dimyristoylphosphatidylcholine). Red and white spheres depict the hydrated hydrophilic outer layer of the membrane, and the blue spheres depict the hydrophobic inner layer. The movement of each molecule can be calculated using, for example, molecular dynamics (MD) simulation, and produce a time evolution of the system for a very short time period (e.g., 10^{-3} s).

understanding of intracellular dynamics (Poursina, 2011). Some of these dynamics are unknown making this approach difficult to implement. This issue deters coarse-graining above the molecular level. However, a full cell model at the molecular scale is complex and computationally intractable.

The inclusion of viscous effects in these models produces further technical challenges in obtaining the solution. Current approaches reduce the order of the model to quasi-static by omitting the inertial forces

(Gillespie, 2008). However, this omission eliminates the capability of the model to reproduce some behaviors of the cell (Haghshenas-Jaryani et al., 2014). Only recently, a feasible method was developed to produce a dynamic model of a cell nucleus and cytoskeleton using a novel method of model reduction. This dynamic model retains the inertial forces and therefore can capture all possible dynamic behaviors of these structures. Obtaining the solution of molecular dynamic (MD) simulations are computationally costly, however (Huggins et al., 2019). The very fine resolution, the molecular lengths in angstrom and nanometers, causes very high computational times. This large computational time makes it impractical to solve the model for long time histories, e.g., greater than one millisecond (Lane, Shukla, Beauchamp, & Pande, 2013). The simulation of a folding protein, composed of thousands of molecules, requires the employment of massive parallelization techniques and supercomputer hardware to achieve a 1.112ms time history of behavior (Lindorff-Larsen, Piana, Dror, & Shaw, 2011). Therefore, simulation of full cell models is not attempted.

The main approach for addressing this computational issue has been to reduce the resolution of the model by coarse-graining. However, just using coarse-graining does not reduce computational time to the orders necessary to predict biologically significant longer periods, i.e., hours to days. Coarse-graining combined with a new order reduction approach, which consists of scaling the force balance between inertial and applied forces, has addressed this problem, allowing prediction of time histories on the order of days and weeks for the cytoskeleton and nucleus (McColloch, Rabiei, Rabbani, Bowling, & Cho, 2019). Cellular processes occur at deferent time scales. Biochemical reactions, for example, are extremely fast (Ischenko, Weber, & Miller, 2017) while cellular level processes, such as stem cell differentiation, are orders of magnitude slower. Cellular processes are therefore multiscale in nature. Methods used for capturing the dynamics of fast processes, such as MD simulations, will be impractical in capturing slow processes. The inclusion of inertial forces, which are orders of magnitude smaller than other forces, substantially increases the computational cost and time that cannot be remedied using conventional algorithms, model reductions, or use of the fastest computational hardware. A novel approach (Haghshenas-Jaryani & Bowling, 2015) based on method of multiple scales (Nayfeh, 1973) has been shown to significantly reduce computational time while capturing processes in multiple time scales simultaneously. This approach has allowed the multibody dynamic simulation of a coarse-grained cytoskeleton and nucleus for a longer period of time to be obtained (Fig. 9; McColloch et al., 2019; Rabiei et al., 2020). Current efforts are to expand

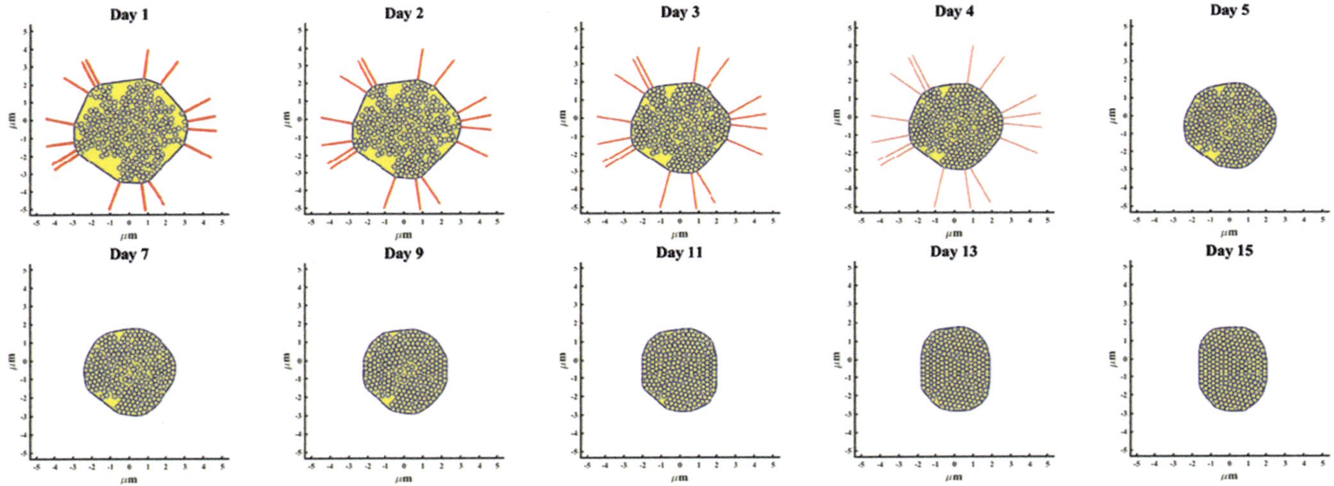

Fig. 9 Computer simulation of human mesenchymal stem cell nucleus during adipogenic differentiation using a multibody dynamics model. The nucleoplasm is shown in yellow. Blue circles represent the nuclear components. The nuclear membrane is depicted in blue line. The red lines represent the microfilament-mediated physical connection to the nucleus. As the stem cells undergo adipogenesis, the fading red lines represent actin depolymerization at different days of differentiation and, as a result, the nuclear morphology is remodeled (McColloch et al., 2019; Rabiei, McColloch, Rabbani, Cho, & Bowling, 2020). This multibody dynamic simulation was proven capable of describing and predicting the mechanical remodeling of the stem cell and its nuclear reorganization over more than 2 weeks. Such computational modeling greatly advances our understanding of the dynamical remodeling of the nucleus that is likely to determine the fate and lineage of stem cells.

this approach to the study of a combined cell membrane, cytoskeleton and nucleus model, to allow more detailed study of the long-term behavior of a cell.

One of the advantages of the multibody dynamic model is that it can produce accurate predictions of biological dynamic behavior. Typically, these models are developed at molecular scale and solved using MD simulation techniques which are well established. New techniques are allowing for accurate long-term predictions of cellular and subcellular level dynamic processes. The disadvantage of MD simulations is the short duration of time history that can be obtained. The computational costs can be prohibitively high. Even with the aid of parallel processing techniques and supercomputer hardware, obtaining evaluation of the system for long time periods are out of reach of MD simulations (Ma et al., 2019). Although new methods allow long-term simulations, they require knowledge of characteristics of intracellular structures and processes at higher than molecular level, which are currently unknown. More experiments are needed to inform the mechanics of intracellular structures and forces for long time periods.

5.4 Summary

Mechanical modeling approaches differ in many regards and therefore are used for different applications. The viscoelastic and tensegrity models have shown reliable affinity with experimental data generated by AFM and electrostatic force microscopy (EFM). Useful understanding of the physical and structural properties of cellular processes has been gained using these methods. The viscoelastic and tensegrity models are mainly constructed for structural analysis and evaluated in quasi-static manner. While the final state of the model can be estimated, these quasi-static approaches cannot capture all the possible dynamic behavior. Reproduction of the dynamic process is hindered by omission of inertial forces. This limits the usefulness of these methods' predictions of cellular behavior over the time period of biological processes. Imaging tools have difficulty capturing the dynamic changes of biological processes. Therefore physics-based dynamic models are needed to assist in providing insights into the dynamic behavior of biological systems.

MD simulations can produce very accurate physics-based results. Evaluation of the MD models is computationally costly. The computational time required for evaluating cellular substructures are high enough that more than obtaining milliseconds of computational solution of the model are impractical to achieve even with the aid of modern supercomputers. Methods and procedures have been developed to increase the time scale of the analysis but obtaining the solution of the model for slow biological phenomena is still out of reach. Therefore, a truly dynamic representation of cellular processes for large time periods is warranted.

References

Agrawal, A., & Lele, T. P. (2019). Mechanics of nuclear membrane. *Journal of Cell Science, 132*, 1–8. https://doi.org/10.1242/jcs.229245.

Ahmadzadeh, H., Webster, M., Behera, R., Jimenez Valencia, A., Wirtz, D., Weeraratna, A., et al. (2017). Modeling the two-way feedback between contractility and matrix realignment reveals a nonlinear mode of cancer cell invasion. *Proceedings of the National Academy of Sciences of the United States of America, 114*, E1617–E1626. https://doi.org/10.1073/pnas.1617037114.

Alam, S. G., Zhang, Q., Prasad, N., Li, Y., Chamala, S., Kuchibhotla, R., et al. (2016). The mammalian LINC complex regulates genome transcriptional responses to substrate rigidity. *Scientific Reports, 6*. https://doi.org/10.1038/srep38063.

Alber, F., Dokudovskaya, S., Veenhoff, L. M., Zhang, W., Kipper, J., Devos, D., et al. (2007). The molecular architecture of the nuclear pore complex. *Nature, 450*, 695–701. https://doi.org/10.1038/nature06405.

Alisafaei, F., Jokhun, D. S., Shivashankar, G. V., & Shenoy, V. B. (2019). Regulation of nuclear architecture, mechanics, and nucleocytoplasmic shuttling of epigenetic factors by cell geometric constraints. *Proceedings of the National Academy of Sciences of the United States of America, 116*, 13200–13209. https://doi.org/10.1073/pnas.1902035116.

Anishkin, A., Loukin, S. H., Teng, J., & Kung, C. (2014). Feeling the hidden mechanical forces in lipid bilayer is an original sense. *Proceedings of the National Academy of Sciences of the United States of America, 111*, 7898–7905. https://doi.org/10.1073/pnas.1313364111.

Antmen, E., Demirci, U., & Hasirci, V. (2019). Loss of nuclear deformability of breast cancer cells by the disruption of actin filaments. *bioRxiv*, 1–28. https://doi.org/10.1101/854562. 2.

Assoian, R., & Klein, E. (2008). Growth control by intracellular tension and extracellular stiffness. *Trends in Cell Biology, 18*, 347–352. https://doi.org/10.1016/j.tcb.2008.05.002.

Aureille, J., Belaadi, N., & Guilluy, C. (2017). Mechanotransduction via the nuclear envelope: A distant reflection of the cell surface. *Current Opinion in Cell Biology, 44*, 59–67. https://doi.org/10.1016/j.ceb.2016.10.003.

Bae, Y., Mui, K., Hsu, B., Liu, S., Cretu, A., Razinia, Z., et al. (2014). A FAK-Cas-Rac-lamellipodin signaling module transduces extracellular matrix stiffness into mechanosensitive cell cycling. *Science Signaling*, 7, ra57. https://doi.org/10.1126/scisignal.2004838.

Baker, B., Trappmann, B., Wang, W. Y., Sakar, M. S., Kim, I. L., Shenoy, V. B., et al. (2015). Cell-mediated fibre recruitment drives extracellular matrix mechanosensing in engineered fibrillar microenvironments. *Nature Materials, 14*, 1262–1268. https://doi.org/10.1038/nmat4444.

Bao, G., Kamm, R. D., Thomas, W., Hwang, W., Fletcher, D. A., Grodzinsky, A. J., et al. (2010). Molecular biomechanics: The molecular basis of how forces regulate cellular function. *Cellular and Molecular Bioengineering, 3*, 91–105.

Barreto, S., Clausen, C. H., Perrault, C. M., Fletcher, D. A., & Lacroix, D. (2013). A multistructural single cell model of force-induced interactions of cytoskeletal components. *Biomaterials, 34*, 6119–6126.

Bateman, J. F., Boot-Handford, R. P., & Lamandé, S. R. (2009). Genetic diseases of connective tissues: Cellular and extracellular effects of ECM mutations. *Nature Reviews Genetics*, *10*, 173–183. https://doi.org/10.1038/nrg2520.

Berk, J. M., Tifft, K. E., & Wilson, K. L. (2013). The nuclear envelope LEM-domain protein emerin. *Nucleus*, *4*, 298–314. https://doi.org/10.4161/nucl.25751.

Bond, L. M., Brandstaetter, H., Sellers, J. R., Kendrick-Jones, J., & Buss, F. (2011). Myosin motor proteins are involved in the final stages of the secretory pathways. *Biochemical Society Transactions*, *39*, 1115–1119. https://doi.org/10.1042/bst0391115.

Borrego-Pinto, J., Jegou, T., Osorio, D. S., Auradé, F., Gorjánácz, M., Koch, B., et al. (2012). Samp1 is a component of TAN lines and is required for nuclear movement. *Journal of Cell Science*, *125*, 1099–1105.

Bridger, J. M., Foeger, N., Kill, I. R., & Herrmann, H. (2007). The nuclear lamina: Both a structural framework and a platform for genome organization. *FEBS Journal*, *274*, 1354–1361. https://doi.org/10.1111/j.1742-4658.2007.05694.x.

Brill-Karniely, Y., Ideses, Y., Bernheim-Groswasser, A., & Ben-Shaul, A. (2009). From branched networks of actin filaments to bundles. *ChemPhysChem: European Journal of Chemical Physics and Physical Chemistry*, *10*, 2818–2827. https://doi.org/10.1002/cphc.200900615.

Burkitt, A. N. (2006). A review of the integrate-and-fire neuron model: I. Homogeneous synaptic input. *Biological Cybernetics*, *95*, 1–19.

Cain, N. E., Jahed, Z., Schoenhofen, A., Valdez, V. A., Elkin, B., Hao, H., et al. (2018). Conserved SUN-KASH interfaces mediate LINC complex-dependent nuclear movement and positioning. *Current Biology*, *28*, 3086–3097. https://doi.org/10.1016/j.cub.2018.08.001.

Calero-Cuenca, F. J., Janota, C. S., & Gomes, E. R. (2018). Dealing with the nucleus during cell migration. *Current Opinion in Cell Biology*, *50*, 35–41.

Campbell, I. D., & Humphries, M. J. (2011). Integrin structure, activation, and interactions. *Cold Spring Harbor Perspectives in Biology*, *3*. https://doi.org/10.1101/cshperspect.a004994.

Carvalho, K., Lemière, J., Faqir, F., Manzi, J., Blanchoin, L., Plastino, J., et al. (2013). Actin polymerization or myosin contraction: Two ways to build up cortical tension for symmetry breaking. *Philosophical Transactions of the Royal Society of London. Series B, Biological Sciences*, *368*. https://doi.org/10.1098/rstb.2013.0005.

Chang, R., Ayton, G. S., & Voth, G. A. (2005). Multiscale coupling of mesoscopic- and atomistic-level lipid bilayer simulations. *Journal of Chemical Physics*, *122*, 244716.

Chen, H.-J., Lin, C.-M., Lin, C.-S., Perez-Olle, R., Leung, C. L., & Liem, R. K. (2006). The role of microtubule actin cross-linking factor 1 (MACF1) in the Wnt signaling pathway. *Genes & Development*, *20*, 1933–1945. https://doi.org/10.1101/gad.1411206.

Chichili, G. R., & Rodgers, W. (2009). Cytoskeleton-membrane interactions in membrane raft structure. *Cellular and Molecular Life Sciences*, *66*, 2319–2328. https://doi.org/10.1007/s00018-009-0022-6.

Choe, H., Burtnick, L. D., Mejillano, M., Yin, H. L., Robinson, R. C., & Choe, S. (2002). The calcium activation of gelsolin: Insights from the 3Å structure of the G4–G6/actin complex. *Journal of Molecular Biology*, *324*, 691–702. https://doi.org/10.1016/s0022-2836(02)01131-2.

Chugh, P., & Paluch, E. K. (2018). The actin cortex at a glance. *Journal of Cell Science*, *131*. https://doi.org/10.1242/jcs.186254.

Clark, K., Langeslag, M., Figdor, C. G., & Leeuwen, F. N. (2007). Myosin II and mechanotransduction: A balancing act. *Trends in Cell Biology*, *17*, 178–186. https://doi.org/10.1016/j.tcb.2007.02.002.

Clark, K., Middelbeek, J., & Leeuwen, F. N. (2008). Interplay between TRP channels and the cytoskeleton in health and disease. *European Journal of Cell Biology*, *87*, 631–640. https://doi.org/10.1016/j.ejcb.2008.01.009.

Cocciolone, A., Hawes, J. Z., Staiculescu, M. C., Johnson, E. O., Murshed, M., & Wagenseil, J. E. (2018). Elastin, arterial mechanics, and cardiovascular disease. *American Journal of Physiology: Heart and Circulatory Physiology*, *315*, H189–H205. https://doi.org/10.1152/ajpheart.00087.2018.

Courtemanche, N. (2018). Mechanisms of formin-mediated actin assembly and dynamics. *Biophysical Reviews*, *10*, 1553–1569. https://doi.org/10.1007/s12551-018-0468-6.

Crockett, E. L. (1998). Cholesterol function in plasma membranes from ectotherms: Membrane-specific roles in adaptation to temperature. *American Zoologist*, *38*, 291–304. https://doi.org/10.1093/icb/38.2.291.

Dai, E. N., Heo, S.-J., & Mauck, R. L. (2020). "Looping in" mechanics: Mechanobiologic regulation of the nucleus and the epigenome. *Advanced Healthcare Materials*, *9*, 2000030.

Darling, E. M., Topel, M., Zauscher, S., Vail, T. P., & Guilak, F. (2008). Viscoelastic properties of human mesenchymally-derived stem cells and primary osteoblasts, chondrocytes, and adipocytes. *Journal of Biomechanics*, *41*, 454–464.

Darling, E. M., Zauscher, S., Block, J. A., & Guilak, F. (2007). A thin-layer model for viscoelastic, stress-relaxation testing of cells using atomic force microscopy: Do cell properties reflect metastatic potential? *Biophysical Journal*, *92*, 1784–1791.

D'Armiento, J. (2003). Decreased elastin in vessel walls puts the pressure on. *Journal of Clinical Investigation*, *112*, 1308–1310. https://doi.org/10.1172/jci20226.

Dauer, W. T., & Worman, H. J. (2009). The nuclear envelope as a signaling node in development and disease. *Developmental Cell*, *17*, 626–638. https://doi.org/10.1016/j.devcel.2009.10.016.

Demmerle, J., Koch, A. J., & Holaska, J. M. (2013). Emerin and histone deacetylase 3 (HDAC3) cooperatively regulate expression and nuclear positions of MyoD, Myf5, and Pax7 genes during myogenesis. *Chromosome Research; International Journal on the Molecular, Supramolecular and Evolutionary Aspects of Chromosome Biology*, *21*, 765–779. https://doi.org/10.1007/s10577-013-9381-9.

Dimitrov, D. S. (1984). Electric field-induced breakdown of lipid bilayers and cell membranes: A thin viscoelastic film model. *Journal of Membrane Biology*, *78*, 53–60.

Dingwall, C., & Laskey, R. (1992). The nuclear membrane. *Science*, *258*, 942–947. https://doi.org/10.1126/science.1439805.

Discher, D. E., Janmey, P., & Wang, Y.l. (2005). Tissue cells feel and respond to the stiffness of their substrate. *Science*, *310*, 1139–1143. https://doi.org/10.1126/science.1116995.

Diz-Muñoz, A., Fletcher, D. A., & Weiner, O. D. (2012). Use the force: Membrane tension as an organizer of cell shape and motility. *Trends in Cell Biology*, *23*, 47–53. https://doi.org/10.1016/j.tcb.2012.09.006.

Dominguez, R. (2009). Actin filament nucleation and elongation factors—structure-function relationships. *Critical Reviews in Biochemistry and Molecular Biology*, *44*, 351–366. https://doi.org/10.3109/10409230903277340.

Dominguez, R., & Holmes, K. C. (2011). Actin structure and function. *Annual Review of Biophysics*, *40*, 169–186. https://doi.org/10.1146/annurev-biophys-042910-155359.

Edwards, M., Zwolak, A., Schafer, D. A., Sept, D., Dominguez, R., & Cooper, J. A. (2014). Capping protein regulators fine-tune actin assembly dynamics. *Nature Reviews. Molecular Cell Biology*, *15*, 677–689. https://doi.org/10.1038/nrm3869.

Ekpenyong, A. E., Whyte, G., Chalut, K., Pagliara, S., Lautenschläger, F., Fiddler, C., et al. (2012). Viscoelastic properties of differentiating blood cells are fate- and function-dependent. *PLoS One*, 7. https://doi.org/10.1371/journal.pone.0045237.

Fehon, R. G., McClatchey, A. I., & Bretscher, A. (2010). Organizing the cell cortex: The role of ERM proteins. *Nature Reviews Molecular Cell Biology*, *11*, 276–287. https://doi.org/10.1038/nrm2866.

Fletcher, D. A., & Mullins, R. D. (2010). Cell mechanics and the cytoskeleton. *Nature*, *463*, 485–492. https://doi.org/10.1038/nature08908.

Fowler, P. W., Hélie, J., Duncan, A., Chavent, M., Koldsø, H., & Sansom, M. S. (2016). Membrane stiffness is modified by integral membrane proteins. *Soft Matter, 12*, 7792–7803. https://doi.org/10.1039/c6sm01186a.

Frame, M. C., Fincham, V. J., Carragher, N. O., & Wyke, J. A. (2002). v-SRC'S hold over actin and cell adhesions. *Nature Reviews Molecular Cell Biology, 3*, 233–245. https://doi.org/10.1038/nrm779.

Franker, M. A., & Hoogenraad, C. C. (2013). Microtubule-based transport—Basic mechanisms, traffic rules and role in neurological pathogenesis. *Journal of Cell Science, 126*, 2319–2329. https://doi.org/10.1242/jcs.115030.

Frolov, V. A., & Zimmerberg, J. (2010). Cooperative elastic stresses, the hydrophobic effect, and lipid tilt in membrane remodeling. *FEBS Letters, 584*, 1824–1829. https://doi.org/10.1016/j.febslet.2010.01.039.

Garcı´a-Alvarez, B., Bobkov, A., Sonnenberg, A., & Pereda, J. M. (2003). Structural and functional analysis of the actin binding domain of plectin suggests alternative mechanisms for binding to F-actin and integrin β4. *Structure, 11*, 615–625. https://doi.org/10.1016/s0969-2126(03)00090-x.

Gardel, M. L., Kasza, K. E., Brangwynne, C. P., Liu, J., & Weitz, D. A. (2008). Mechanical response of cytoskeletal networks. *Methods in Cell Biology, 89*, 487–519. https://doi.org/10.1016/s0091-679x(08)00619-5.

Gentili, C., & Cancedda, R. (2009). Cartilage and bone extracellular matrix. *Current Pharmaceutical Design, 15*, 1334–1348. https://doi.org/10.2174/138161209787846739.

Gillespie, D. T. (2008). Simulation methods in systems biology. In *International school on formal methods for the design of computer, communication and software systems* (pp. 125–167). Springer.

Godsel, L. M., Hobbs, R. P., & Green, K. J. (2008). Intermediate filament assembly: Dynamics to disease. *Trends in Cell Biology, 18*, 28–37. https://doi.org/10.1016/j.tcb.2007.11.004.

Gremm, D., & Wegner, A. (2000). Gelsolin as a calcium-regulated actin filament-capping protein: Gelsolin as a calcium-regulated actin filament-capping protein. *European Journal of Biochemistry, 267*, 4339–4345. https://doi.org/10.1046/j.1432-1327.2000.01463.x.

Gressin, L., Guillotin, A., Guérin, C., Blanchoin, L., & Michelot, A. (2015). Architecture dependence of actin filament network disassembly. *Current Biology, 25*, 1437–1447. https://doi.org/10.1016/j.cub.2015.04.011.

Gruenbaum, Y., & Foisner, R. (2015). Lamins: Nuclear intermediate filament proteins with fundamental functions in nuclear mechanics and genome regulation. *Annual Review of Biochemistry, 84*, 131–164. https://doi.org/10.1146/annurev-biochem-060614-034115.

Guilak, F., Tedrow, J. R., & Burgkart, R. (2000). Viscoelastic properties of the cell nucleus. *Biochemical and Biophysical Research Communications, 269*, 781–786.

Guilak, F., Ting-Beall, H. P., Baer, A. E., Trickey, W. R., Erickson, G. R., & Setton, L. A. (1999). Viscoelastic properties of intervertebral disc cells: Identification of two biomechanically distinct cell populations. *Spine, 24*, 2475.

Guz, N., Dokukin, M., Kalaparthi, V., & Sokolov, I. (2014). If cell mechanics can be described by elastic modulus: Study of different models and probes used in indentation experiments. *Biophysical Journal, 107*, 564–575.

Haghshenas-Jaryani, M., Black, B., Ghaffari, S., Drake, J., Bowling, A., & Mohanty, S. (2014). Dynamics of microscopic objects in optical tweezers: Experimental determination of underdamped regime and numerical simulation using multiscale analysis. *Nonlinear Dynamics, 76*, 1013–1030.

Haghshenas-Jaryani, M., & Bowling, A. (2015). Modeling flexibility in myosin V using a multiscale articulated multi-rigid body approach. *ASME Journal of Computational and Nonlinear Dynamics, 10*, 011015 (11 pages).

Han, Y. L., Ronceray, P., Xu, G., Malandrino, A., Kamm, R. D., Lenz, M., et al. (2018). Cell contraction induces long-ranged stress stiffening in the extracellular matrix. *Proceedings of the National Academy of Sciences of the United States of America, 115*, 4075–4080.

Haque, F., Mazzeo, D., Patel, J. T., Smallwood, D. T., Ellis, J. A., Shanahan, C. M., et al. (2009). Mammalian SUN protein interaction networks at the inner nuclear membrane and their role in laminopathy disease processes. *Journal of Biological Chemistry, 285*, 3487–3498. https://doi.org/10.1074/jbc.m109.071910.

Harris, A. R., Jreij, P., & Fletcher, D. A. (2018). Mechanotransduction by the actin cytoskeleton: Converting mechanical stimuli into biochemical signals. *Annual Review of Biophysics, 47*, 617–631. https://doi.org/10.1146/annurev-biophys-070816-033547.

Holle, A. W., Young, J. L., Van Vilet, K. J., Kamm, R. D., Discher, D., Janmey, P., et al. (2018). Cell-extracellular matrix mechanobiology: Forceful tools and emerging needs for basic and translational research. *Nano Letters, 18*, 1–8.

Honda, H., Tanemura, M., & Nagai, T. (2004). A three-dimensional vertex dynamics cell model of space-filling polyhedra simulating cell behavior in a cell aggregate. *Journal of Theoretical Biology, 226*, 439–453.

Huggins, D. J., Biggin, P. C., Dämgen, M. A., Essex, J. W., Harris, S. A., Henchman, R. H., et al. (2019). Biomolecular simulations: From dynamics and mechanisms to computational assays of biological activity. *Computational Molecular Science, 9*, e1393.

Huveneers, S., & Danen, E. H. (2009). Adhesion signaling—Crosstalk between integrins, Src and Rho. *Journal of Cell Science, 122*, 1059–1069. https://doi.org/10.1242/jcs.039446.

Iannone, F., & Lapadula, G. (2003). The pathophysiology of osteoarthritis. *Aging Clinical and Experimental Research, 15*, 364–372. https://doi.org/10.1007/bf03327357.

Ingber, D. E. (2003). Tensegrity I. Cell structure and hierarchical systems biology. *Journal of Cell Science, 116*, 1157–1173.

Ingber, D. E. (2008). Tensegrity-based mechanosensing from macro to micro. *Progress in Biophysics and Molecular Miology, 97*, 163–179.

Ischenko, A. A., Weber, P. M., & Miller, R. D. (2017). Capturing chemistry in action with electrons: Realization of atomically resolved reaction dynamics. *Chemical Reviews, 117*, 11066–11124.

Ishikawa, F., Ushida, K., Mori, K., & Shibanuma, M. (2015). Loss of anchorage primarily induces non-apoptotic cell death in a human mammary epithelial cell line under atypical focal adhesion kinase signaling. *Cell Death and Disease, 6*. https://doi.org/10.1038/cddis.2014.583.

Ito, S., Suki, B., Kume, H., Numaguchi, Y., Ishii, M., Iwaki, M., et al. (2009). Actin cytoskeleton regulates stretch-activated Ca2+ influx in human pulmonary microvascular endothelial cells. *American Journal of Respiratory Cell and Molecular Biology, 43*, 26–34. https://doi.org/10.1165/rcmb.2009-0073oc.

Jacques-Fricke, B. T., Seow, Y., Gottlieb, P. A., Sachs, F., & Gomez, T. M. (2006). Ca2+ Influx through mechanosensitive channels inhibits neurite outgrowth in opposition to other influx pathways and release from intracellular stores. *Journal of Neuroscience, 26*, 5656–5664. https://doi.org/10.1523/jneurosci.0675-06.2006.

Jahed, Z., Shams, H., Mehrbod, M., & Mofrad, M. R. (2014). Mechanotransduction pathways linking the extracellular matrix to the nucleus. *International Review of Cell and Molecular Biology, 310*, 171–220. https://doi.org/10.1016/b978-0-12-800180-6.00005-0.

Jahed, Z., Shams, H., & Mofrad, M. R. (2015). A disulfide bond is required for the transmission of forces through SUN-KASH complexes. *Biophysical Journal, 109*, 501–509. https://doi.org/10.1016/j.bpj.2015.06.057.

Jensen, M. H., Morris, E. J., & Weitz, D. A. (2015). Mechanics and dynamics of reconstituted cytoskeletal systems. *Biochimica et Biophysics Acta, 1853*, 3038–3042. https://doi.org/10.1016/j.bbamcr.2015.06.013.

Karcher, H., Lammerding, J., Huang, H., Lee, R. T., Kamm, R. D., & Kaazempur-Mofrad, M. R. (2003). A three-dimensional viscoelastic model for cell deformation with experimental verification. *Biophysical Journal*, *85*, 3336–3349.

Karplus, M., & Petsko, G. A. (1990). Molecular dynamics simulations in biology. *Nature*, *347*, 631–639. Retrieved from https://doi.org/10.1038/347631a0.

Kazsa, K. E., Rowart, A. C., Liu, J., Angelini, T. E., Brangwynne, C. P., Koenderink, G. H., et al. (2007). The cells as a material. *Current Opinion in Cell Biology*, *19*, 101–107. https://doi.org/10.1016/j.ceb.2006.12.002.

Khandelia, H., Ipsen, J. H., & Mouritsen, O. G. (2008). The impact of peptides on lipid membranes. *Biochimica et Biophysica Acta*, *1778*, 1528–1536. https://doi.org/10.1016/j.bbamem.2008.02.009.

Kim, Y., Kim, M., Shin, J. H., & Kim, J. (2011). Characterization of cellular elastic modulus using structure based double layer model. *Medical & Biological Engineering & Computing*, *49*, 453–462.

Kosawada, T., Inoue, K., & Schmid-Schönbein, G. W. (2005). Mechanics of curved plasma membrane vesicles: Resting shapes, membrane curvature, and in-plane shear elasticity. *Journal of Biomechanical Engineering*, *127*, 229–236. https://doi.org/10.1115/1.1865197.

Kozlov, M. M., & Chernomordik, L. V. (2015). Membrane tension and membrane fusion. *Current Opinion in Structural Biology*, *33*, 61–67. https://doi.org/10.1016/j.sbi.2015.07.010.

Laevsky, G., & Knecht, D. A. (2003). Cross-linking of actin filaments by myosin II is a major contributor to cortical integrity and cell motility in restrictive environments. *Journal of Cell Science*, *116*, 3761–3770. https://doi.org/10.1242/jcs.00684.

Lan, T.-H., Huang, X.-Q., & Tan, H.-M. (2013). Vascular fibrosis in atherosclerosis. *Cardiovascular Pathology*, *22*, 401–407. https://doi.org/10.1016/j.carpath.2013.01.003.

Lane, T. J., Shukla, D., Beauchamp, K. A., & Pande, V. S. (2013). To milliseconds and beyond: Challenges in the simulation of protein folding. *Current Opinion in Structural Biology*, *23*, 58–65.

Lange, O. F., & Grubmüller, H. (2006). Collective Langevin dynamics of conformational motions in proteins. *Journal of Chemical Physics*, *124*, 214903.

Lee, S. H., & Dominguez, R. (2010). Regulation of actin cytoskeleton dynamics in cells. *Molecules and Cells*, *29*, 311–325. https://doi.org/10.1007/s10059-010-0053-8.

Leeuw, R. D., Gruenbaum, Y., & Medalia, O. (2018). Nuclear lamins: Thin filaments with major functions. *Trends in Cell Biology*, *28*, 34–45. https://doi.org/10.1016/j.tcb.2017.08.004.

Lieleg, O., Claessens, M. M., & Bausch, A. R. (2010). Structure and dynamics of cross-linked actin networks. *Soft Matter*, *6*, 218–225. https://doi.org/10.1039/b912163n.

Liem, R. K. (2016). Cytoskeletal integrators: The spectrin superfamily. *Cold Spring Harbor Perspectives in Biology*, *8*. https://doi.org/10.1101/cshperspect.a018259.

Lindorff-Larsen, K., Piana, S., Dror, R. O., & Shaw, D. E. (2011). How fast-folding proteins fold. *Science*, *334*, 517–520.

Los, D. A., & Murata, N. (2004). Membrane fluidity and its roles in the perception of environmental signals. *Biochimica et Biophysica Acta—Biomembranes*, *1666*, 142–157. https://doi.org/10.1016/j.bbamem.2004.08.002.

Lou, X., & Shin, Y.-K. (2016). SNARE zippering. *Bioscience Reports*, *36*. https://doi.org/10.1042/bsr20160004.

Lovett, D. B., Shekhar, N., Nickerson, J. A., Roux, K. J., & Lele, T. P. (2013). Modulation of nuclear shape by substrate rigidity. *Cellular and Molecular Bioengineering*, *6*, 230–238. https://doi.org/10.1007/s12195-013-0270-2.

Lu, Y.-B., Franze, K., Seifert, G., Steinhäuser, C., Kirchhoff, F., Wolburg, H., et al. (2006). Viscoelastic properties of individual glial cells and neurons in the CNS. *Proceedings of the National Academy of Sciences of the United States of America*, *103*, 17759–17764.

Luxton, G. G., Gomes, E. R., Folker, E. S., Worman, H., & Gundersen, G. G. (2011). TAN lines: A novel nuclear envelope structure involved in nuclear positioning. *Nucleus, 2*, 956–959.

Lyon, R. C., Zanella, F., Omens, J. H., & Sheikh, F. (2015). Mechanotransduction in cardiac hypertrophy and failure. *Circulation Research, 116*, 1462–1476. https://doi.org/10.1161/CIRCRESAHA.116.304937.

Ma, H., Bhowmik, D., Lee, H., Turilli, M., Young, M. T., Jha, S., et al. (2019). Deep generative model driven protein folding simulation. *arXiv*. preprint arXiv:1908.00496.

Malik, R., Luong, T., Cao, X., Han, B., Shah, N., Franco-Barraza, J., et al. (2019). Rigidity controls human desmoplastic matrix anisotropy to enable pancreatic cancer cell spread via extracellular signal-regulated kinase 2. *Matrix Biology, 81*, 50–69. https://doi.org/10.1016/j.matbio.2018.11.001.

Mao, X., Gavara, N., & Song, G. (2015). Nucluar mechanics and stem cell differentiation. *Stem Cell Research and Reports, 6*, 804–812. https://doi.org/10.1007/s12015-015-9610-z.

Marinkovic, M., Turner, K. T., Deng, L., Suresh, S., & Fredberg, J. J. (2007). Viscoelasticity of the human red blood cell. *American Journal of Physiology: Cell Physiology, 293*, C597–C605. https://doi.org/10.1152/ajpcell.00562.2006.

Maurer, M., & Lammerding, J. (2019). The driving force: Nuclear mechanotransduction in cellular function, fate, and disease. *Annual Review of Biomedical Engineering, 21*, 443–468. https://doi.org/10.1146/annurev-bioeng-060418-052139.

McColloch, A., Rabiei, M., Rabbani, P., Bowling, A., & Cho, M. (2019). Correlation between nuclear morphology and adipogenic differentiation: Application of a combined experimental and computational modeling approach. *Scientific Reports, 9*, 1–13.

Mekhail, K., & Moazed, D. (2010). The nuclear envelope in genome organization, expression and stability. *Nature Reviews Molecular Cell Biology, 11*, 317–328. https://doi.org/10.1038/nrm2894.

Meunier, F. A., & Gutiérrez, L. M. (2016). Captivating new roles of F-actin cortex in exocytosis and bulk endocytosis in neurosecretory cells. *Trends in Neurosciences, 39*, 605–613. https://doi.org/10.1016/j.tins.2016.07.003.

Meyer, G. A., & Lieber, R. L. (2011). Elucidation of extracellular matrix mechanics from muscle fibers and fiber bundles. *Journal of Biomechanics, 44*, 771–773.

Muhamed, I., Chowdhury, F., & Maruthamuthu, V. (2017). Biophysical tools to study cellular mechanotransduction. *Bioengineering, 4*, 12. https://doi.org/10.3390/bioengineering4010012.

Narita, A., Oda, T., & Maéda, Y. (2011). Structural basis for the slow dynamics of the actin filament pointed end. *The EMBO Journal, 30*, 1230–1237. https://doi.org/10.1038/emboj.2011.48.

Nayfeh, A. H. (1973). *Perturbation methods*. Wiley.

Nielson, S. O., Lopex, C. F., Srinivas, G., & Klein, M. L. (2004). Coarse grain models and the computer simulation of soft materials. *Journal of Physics Condensed Matter, 16*, R481.

Östlund, C., & Worman, H. J. (2002). Nuclear envelope proteins and neuromuscular diseases: Nuclear envelope and disease. *Muscle & Nerve, 27*, 393–406. https://doi.org/10.1002/mus.10302.

Padmakumar, V. C., Libotte, T., Lu, W., Zaim, H., Abraham, S., Noegel, A. A., et al. (2005). The inner nuclear membrane protein Sun1 mediates the anchorage of Nesprin-2 to the nuclear envelope. *Journal of Cell Science, 118*, 3419–3430. https://doi.org/10.1242/jcs.02471.

Parsons, J. T., Horwitz, A. R., & Schwartz, M. A. (2010). Cell adhesion: Integrating cytoskeletal dynamics and cellular tension. *Nature Reviews Molecular Cell Biology, 11*, 633–643. https://doi.org/10.1038/nrm2957.

Pavlov, D., Muhlrad, A., Cooper, J., Wear, M., & Reisler, E. (2007). Actin filament severing by cofilin. *Journal of Molecular Biology, 365*, 1350–1358. https://doi.org/10.1016/j.jmb.2006.10.102.

Pawlowski, P. H. (2007). Mechanokinetic model of cell membrane: Theoretical analysis of plasmalemma homeostasis, growth and division. *Journal of Theoretical Biology*, *249*, 67–76.

Pegoraro, A. F., Janmey, P., & Weitz, D. A. (2017). Mechanical properties of the cytoskeleton and cells. *Cold Spring Harbor Perspectives in Biology*, *9*. https://doi.org/10.1101/cshperspect.a022038.

Pfeifer, C. R., Irianto, J., & Discher, D. E. (2019). Nuclear mechanics and cancer cell migration. *Advances in Experimental Medicine and Biology*, *1146*, 117–130. https://doi.org/10.1007/978-3-030-17593-1_8.

Poole, K., Moroni, M., & Lewin, G. R. (2014). Sensory mechanotransduction at membrane-matrix interfaces. *Pflügers Archiv: European Journal of Physiology*, *467*, 121–132. https://doi.org/10.1007/s00424-014-1563-6.

Poursina, M. (2011). *Robust framework for the adaptive multiscale modeling of biopolymers*. PhD dissertation Rensselaer Polytechnic Institute.

Prowse, A. B., Chong, F., Gray, P. P., & Munro, T. P. (2010). Stem cell integrins: Implications for ex-vivo culture and cellular therapies. *Stem Cell Research*, *6*, 1–12. https://doi.org/10.1016/j.scr.2010.09.005.

Rabiei, M., McColloch, A., Rabbani, P., Cho, M., & Bowling, A. (2020). Long term dynamic simulation of a stem cell nucleus. *Journal of Computational and Nonlinear Dynamics*. in press.

Rebelo, L. M., de Sousa, J. S., Mendes Filho, J., & Radmacher, M. (2013). Comparison of the viscoelastic properties of cells from different kidney cancer phenotypes measured with atomic force microscopy. *Nanotechnology*, *24*, 055102.

Ronan, W., Deshpande, V. S., McMeeking, R. M., & McGarry, J. P. (2012). Numerical investigation of the active role of the actin cytoskeleton in the compression resistance of cells. *Journal of the Mechanical Behavior of Biomedical Materials*, *14*, 143–157.

Ross, J. A., Levy, Y., Ripolone, M., Kolb, J. S., Turmaine, M., Holt, M., et al. (2019). Impairments in contractility and cytoskeletal organisation cause nuclear defects in nemaline myopathy. *Acta Neuropathologica*, *138*, 477–495. https://doi.org/10.1007/s00401-019-02034-8.

Rothballer, A., Schwartz, T. U., & Kutay, U. (2013). LINCing complex functions at the nuclear envelope: What the molecular architecture of the LINC complex can reveal about its function. *Nucleus*, *4*, 29–36. https://doi.org/10.4161/nucl.23387.

Roux, K. J., Crisp, M. L., Liu, Q., Kim, D., Kozlov, S., Stewart, C. L., et al. (2009). Nesprin 4 is an outer nuclear membrane protein that can induce kinesin-mediated cell polarization. *Proceedings of the National Academy of Sciences of the United States of America*, *106*, 2194–2199. https://doi.org/10.1073/pnas.0808602106.

Sachs, F. (2010). Stretch-activated ion channels: What are they? *Physiology*, *25*, 50–56. https://doi.org/10.1152/physiol.00042.2009.

Saeedimasine, M., Montanino, A., Kleiven, S., & Villa, A. (2019). Role of lipid composition on the structural and mechanical features of axonal membranes: A molecular simulation study. *Scientific Reports*, *9*, 8000. https://doi.org/10.1038/s41598-019-44318-9.

Saha, S., Nagy, T. L., & Weiner, O. D. (2018). Joining forces: Crosstalk between biochemical signalling and physical forces orchestrates cellular polarity and dynamics. *Philosophical Transactions of the Royal Society of London. Series B, Biological Sciences*, *373*. https://doi.org/10.1098/rstb.2017.0145.

Saunders, C. A., Harris, N. J., Willey, P. T., Woolums, B. M., Wang, Y., McQuown, A. J., et al. (2017). TorsinA controls TAN line assembly and the retrograde flow of dorsal perinuclear actin cables during rearward nuclear movement. *Journal of Cell Biology*, *216*, 657–674.

Schwartz, M. A. (2010). Integrins and extracellular matrix in mechanotransduction. *Cold Spring Harbor Perspectives in Biology*, *2*. https://doi.org/10.1101/cshperspect.a005066.

Schwartz, M. A., & Assoian, R. K. (2001). Integrins and cell proliferation: Regulation of cyclin-dependent kinases via cytoplasmic signaling pathways. *Journal of Cell Science*, *114*, 2553–2560.

Shentu, T. P., Singh, D. K., Oh, M.-J., Sun, S., Sadaat, L., Makino, A., et al. (2012). The role of oxysterols in control of endothelial stiffness. *Journal of Lipid Research, 53*, 1348–1358. https://doi.org/10.1194/jlr.m027102.

Short, S. M., Boyer, J. L., & Juliano, R. L. (2000). Integrins regulate the linkage between upstream and downstream events in G protein-coupled receptor signaling to mitogen-activated protein kinase. *Journal of Biological Chemistry, 275*, 12970–12977. https://doi.org/10.1074/jbc.275.17.12970.

Silver, F. H., DeVore, D., & Siperko, L. M. (2003). Invited review: Role of mechanophysiology in aging of ECM: Effects of changes in mechanochemical transduction. *Journal of Applied Physiology, 95*, 2134–2141. https://doi.org/10.1152/japplphysiol.00429.2003.

Simons, K., & Ehehalt, R. (2002). Cholesterol, lipid rafts, and disease. *Journal of Clinical Investigation, 110*, 597–603. https://doi.org/10.1172/jci16390.

Simunovic, M., Evergren, E., Callan-Jones, A., & Bassereau, P. (2019). Curving cells inside and out: Roles of BAR domain proteins in membrane shaping and its cellular implications. *Annual Review of Cell and Developmental Biology, 35*, 111–129. https://doi.org/10.1146/annurev-cellbio-100617-060558.

Smith, B. A., Daugherty-Clarke, K., Goode, B. L., & Gelles, J. (2013). Pathway of actin filament branch formation by Arp2/3 complex revealed by single-molecule imaging. *Proceedings of the National Academy of Sciences of the United States of America, 110*, 1285–1290. https://doi.org/10.1073/pnas.1211164110.

Solovyova, O. E., Vikulova, N. A., Konovalov, P. V., Kohl, P., & Markhasin, V. S. (2004). Mathematical modelling of mechano-electric feedback in cardiomyocytes. *Russian Journal of Numerical Analysis and Mathematical Modelling, 19*, 331–351.

Sonal, Ganzinger, K. A., Vogel, S. K., Mücksch, J., Blumhardt, P., & Schwille, P. (2018). Myosin-II activity generates a dynamic steady state with continuous actin turnover in a minimal actin cortex. *bioRxiv, 132*. https://doi.org/10.1101/312512.

Sosa, B. A., Kutay, U., & Schwartz, T. U. (2013). Structural insights into LINC complexes. *Current Opinion in Structural Biology, 23*, 285–291. https://doi.org/10.1016/j.sbi.2013.03.005.

Sosa, B. A., Rothballer, A., Kutay, U., & Schwartz, T. U. (2012). LINC complexes form by binding of three KASH peptides to domain interfaces of trimeric SUN proteins. *Cell, 149*, 1035–1047. https://doi.org/10.1016/j.cell.2012.03.046.

Stachowiak, M. R., Smith, M. A., Blankman, E., Chapin, L. M., Balcioglu, H. E., Wang, S., et al. (2014). A mechanical-biochemical feedback loop regulates remodeling in the actin cytoskeleton. *Proceedings of the National Academy of Sciences of the United States of America, 111*, 17528–17533. https://doi.org/10.1073/pnas.1417686111.

Stewart-Hutchinson, P. J., Hale, C. M., Wirtz, D., & Hodzic, D. (2008). Structural requirements for the assembly of LINC complexes and their function in cellular mechanical stiffness. *Experimental Cell Research, 314*, 1892–1905. https://doi.org/10.1016/j.yexcr.2008.02.022.

Suarez, C., Carroll, R. T., Burke, T. A., Christensen, J. R., Bestul, A. J., Sees, J. A., et al. (2014). Profilin regulates F-actin network homeostasis by favoring formin over Arp2/3 complex. *Developmental Cell, 32*, 43–53. https://doi.org/10.1016/j.devcel.2014.10.027.

Tapley, E. C., & Starr, D. A. (2012). Connecting the nucleus to the cytoskeleton by SUN-KASH bridges across the nuclear envelope. *Current Opinion in Cell Biology, 25*, 57–62. https://doi.org/10.1016/j.ceb.2012.10.014.

Titushkin, I., & Cho, M. (2007). Modulation of cellular mechanics during osteogenic differentiation of human mesenchymal stem cells. *Biophysical Journal, 93*, 3693–3702. https://doi.org/10.1529/biophysj.107.107797.

Vicente-Manzanares, M., Choi, C. K., & Horwitz, A. R. (2009). Integrins in cell migration—The actin connection. *Journal of Cell Science, 122*, 1473. https://doi.org/10.1242/jcs.052894.

Vosse, D. W., Wan, Y., Wozniak, R. W., & Aitchison, J. D. (2011). Role of the nuclear envelope in genome organization and gene expression. *Interdisciplinary Reviews Systems Biology and Medicine*, *3*, 147–166. https://doi.org/10.1002/wsbm.101.

Wang, N., Tytell, J. D., & Ingber, D. E. (2009). Mechanotransduction at a distance: Mechanically coupling the extracellular matrix with the nucleus. *Nature Reviews Molecular Cell Biology*, *10*, 75–82. https://doi.org/10.1038/nrm2594.

Wei, S. C., Fattet, L., Tsai, J. H., Guo, Y., Pai, V. H., Majeski, H. E., et al. (2015). Matrix stiffness drives epithelial-mesenchymal transition and tumour metastasis through a TWIST1-G3BP2 mechanotransduction pathway. *Nature Cell Biology*, *17*, 678–688. https://doi.org/10.1038/ncb3157.

Weinberg, S. H., Mair, D. B., & Lemmon, C. A. (2017). Mechanotransduction dynamics at the cell-matrix interface. *Biophysical Journal*, *112*, 1962–1974.

Wiggan, O., Schroder, B., Krapf, D., Bamburg, J. R., & DeLuca, J. G. (2017). Cofilin regulates nuclear architecture through a Myosin-II dependent mechanotransduction module. *Scientific Reports*, 7, 40953. https://doi.org/10.1038/srep40953.

Wilhelmsen, K., Litjens, S. H., Kuikman, I., Tshimbalanga, N., Janssen, H., Bout, I. V., et al. (2005). Nesprin-3, a novel outer nuclear membrane protein, associates with the cytoskeletal linker protein plectin. *Journal of Cell Biology*, *171*, 799–810. https://doi.org/10.1083/jcb.200506083.

Wioland, H., Jegou, A., & Romet-Lemonne, G. (2019). Torsional stress generated by ADF/cofilin on cross-linked actin filaments boosts their severing. *Proceedings of the National Academy of Sciences of the United States of America*, *116*, 2595–2602. https://doi.org/10.1073/pnas.1812053116.

Worman, H. J., & Courvalin, J.-C. (2005). Nuclear envelope, nuclear lamina, and inherited disease. *International Review of Cytology*, *246*, 231–279. https://doi.org/10.1016/s0074-7696(05)46006-4.

Xiong, H., Rivero, F., Euteneuer, U., Mondal, S., Mana-Capelli, S., Larochelle, D., et al. (2008). Dictyostelium Sun-1 connects the centrosome to chromatin and ensures genome stability. *Traffic*, *9*, 708–724. https://doi.org/10.1111/j.1600-0854.2008.00721.x.

Xu, G.-K., Li, B., Feng, X.-Q., & Gao, H. (2016). A tensegrity model of cell reorientation on cyclically stretched substrates. *Biophysical Journal*, *111*, 1478–1486.

Yue, B. (2014). Biology of the extracellular matrix: An overview. *Journal of Glaucoma*, *23*, S20–S23. https://doi.org/10.1097/ijg.0000000000000108.

Zamir, E., & Geiger, B. (2001). Molecular complexity and dynamics of cell-matrix adhesions. *Journal of Cell Science*, *114*, 3583–3590.

Zhang, J., Alisafaei, F., Nikolić, M., Nou, X. A., Kim, H., Shenoy, V. B., et al. (2020). Nuclear mechanics within intact cells is regulated by cytoskeletal network and internal nanostructures. *Small*, *16*. https://doi.org/10.1002/smll.201907688.

Zhang, Q., Zhang, D., Dobah, Y., Scarpa, F., Fraternali, F., & Skelton, R. E. (2018). Tensegrity cell mechanical metamaterial with metal rubber. *Applied Physics Letters*, *113*, 031906.

Further reading

Discher, D. E., Mooney, D. J., & Zandstra, P. W. (2009). Growth factors, matrices, and forces combine and control stem cells. *Science*, *324*, 1673–1677.

Dreger, M., Bengtsson, L., Schoneberg, T., Otto, H., & Hucho, F. (2001). Nuclear envelope proteomics: Novel integral membrane proteins of the inner nuclear membrane. *Proceedings of the National Academy of Sciences of the United States of America*, *98*, 11943–11948. https://doi.org/10.1073/pnas.211201898.

Ingber, D. E. (1997). Integrins, tensegrity, and mechanotransduction. *Gravitational and Space Research*, *10*, 49–55. http://gravitationalandspaceresearch.org/index.php/journal/article/view/179.

Lippincott-Schwartz, J., & Fambrough, D. M. (1987). Cycling of the integral membrane glycoprotein, LEP100, between plasma membrane and lysosomes: Kinetic and morphological analysis. *Cell*, *49*, 669–677.

Tito, P., Azzolin, L., Cordenonsi, M., & Piccolo, S. (2017). Mechanobiology of YAP and R+TAZ in physiology and disease. *Nature Review Molecular Cell Biology*, *18*, 758–770. https://doi.org/10.1038/nrm.2017.87.

CHAPTER FIVE

Mechanisms of endothelial stiffening in dyslipidemia and aging: Oxidized lipids and shear stress

Elizabeth Le Master, Sang Joon Ahn, and Irena Levitan*
Department of Medicine, Division of Pulmonary, Critical Care and Sleep Medicine, University of Illinois at Chicago, Chicago, IL, United States
*Corresponding author: e-mail address: levitan@uic.edu

Contents

Abstract

Vascular stiffening of the arterial walls is well-known as a key factor in aging and the development of cardiovascular disease; however, the role of endothelial stiffness in vascular dysfunction is still an emerging topic. In this review, the authors discuss the impact of dyslipidemia, oxidized lipids, substrate stiffness, age and pro-atherogenic disturbed flow have on endothelial stiffness. Furthermore, we investigate several mechanistic pathways that are key contributors in endothelial stiffness and discuss their physiological effects in the onset of atherogenesis in the disturbed flow regions of the aortic vasculature. The findings in this chapter describe a novel paradigm of synergistic interaction of plasma dyslipidemia/oxidized lipids and pro-atherogenic disturbed shear stress, as well as aging has on endothelial stiffness and vascular dysfunction.

Current Topics in Membranes, Volume 86
ISSN 1063-5823
https://doi.org/10.1016/bs.ctm.2020.08.006

1. Introduction

The vascular endothelium is a single cellular layer of the blood vessel which separates the lumen from the vascular wall (Chistiakov, Orekhov, & Bobryshev, 2015). It has several functions in vascular homeostasis including the regulation of vascular tone, the control of permeability, anti-thrombotic activity, the control of vascular inflammatory mechanism, and the regulation of vascular integrity (Claesson-Welsh, 2015; Murakami & Simons, 2009). The vascular endothelium is also unique in being constantly exposed to the hemodynamic forces of fluid shear stress generated by blood flow. In addition to the frictional forces of shear stress, pulsatile blood pressure also exposes the endothelial monolayer to the stretch and compressive forces on the vasculatures' circumference. Numerous studies explored the roles of the hemodynamic forces in endothelial function and it is well-established that these forces play key roles in multiple functions, including endothelium-dependent vasodilation, barrier function and inflammation, as discussed in multiple excellent reviews (Collins & Tzima, 2011; Gimbrone & Garcia-Cardena, 2013; Kutys & Chen, 2016). Our studies focus on endothelial biomechanical properties, such as endothelial stiffness or elastic modulus and contractility, which clearly may significantly affect the ability of endothelial cells to respond to external mechanical forces, as well as affect other endothelial functions, such as permeability barrier or cell-cell interactions. More specifically, we focus on the impact of dyslipidemia and oxidized lipids on endothelial stiffness and briefly, we demonstrated that exposure to several types of oxidized lipids in vitro or diet-induced plasma dyslipidemia in vivo result in significant endothelial stiffening (Le Master, Huang, et al., 2018). Significant changes in endothelial stiffness were also identified by several groups including ours in response to shear stress, physical properties of the underlying substrate and aging. The goal of this chapter is to provide a comprehensive review of what is currently known about endothelial stiffness under physiological and pathophysiological conditions, such as dyslipidemia and aging, and its role in endothelial function.

2. Discriminating between the stiffness of the arterial wall and the stiffness of the endothelium

Stiffening of the arterial walls is a well-known characteristic of cardiovascular disease (CVD), a class of diseases including hypertension, heart attack, stroke, and aortic aneurisms. Studies show that increased arterial stiffness resulting from changes in vessel compliance can increase blood pulse wave velocity, fluid shear stress, aortic pulse pressure, and contributes to changes in vascular geometry (Kohn, Lampi, & Reinhart-King, 2015). These changes to the geometry of the vasculature, including stenosis, further alter regional blood hemodynamics to create distinct areas of disturbed flow (DF) patterns which have been shown to increase inflammatory oxidative stress and adhesion molecule expression, as well as decrease antioxidant genes expression, nitric oxide production and reduce endothelial regeneration (Chiu & Chien, 2011). It is important to note, however, the difference between whole artery stiffness and the stiffness of the endothelial cell monolayer itself, which on its own can be a major factor in the functioning of the endothelium. Research by Martinelli et al. (2014) and Schaefer et al. (2014) has shown that endothelial stiffness on its own has a role in increased adhesion, spreading and transendothelial migration of proinflammatory leukocytes which have been shown to be a key initiating factor in the onset of atherosclerosis (Martinelli et al., 2014; Schaefer et al., 2014).

Several studies have investigated not only the biomechanical properties of whole human aortas but also have focused on the compressive mechanical properties of the heterogeneous atherosclerotic plaque tissues located at these regions of DF. Plaques reside below the endothelial membrane and are categorized according to their cellular components. In general, the plaque's lipid core and areas of hematoma are soft as compared to the calcified regions and the smooth muscle cell hyperproliferative regions (Chai, Speelman, Oomens, & Baaijens, 2014). The elastic properties of different plaque components are emerging as a quantifiable property. As determined by Tracqui et al. (2011), the plaques found in ApoE KO (apolipoproteinE knockout) mice, a hyperlipidemic atherosclerotic mouse model fed a high fat, high cholesterol diet, contained stiffer components including the hypo-cellular fibrosis area [23 ± 4 kPa] with increased scarring of connective tissues, increased calcification and decreased cellular composition and hyper-cellular fibrosis regions [12 ± 2 kPa] with smooth muscle cell hyperproliferation, whereas the lipid rich foam cell regions produced in part by macrophages was relatively compliant [3 ± 3 kPa] (Tracqui et al., 2011). Knowledge of plaque biomechanics is emerging as an indicator of plaque stability (Chai et al., 2014). Furthermore, since the developing plaques reside below the endothelial monolayer, the stiffness of the endothelial cells may play a significant role in plaque stability as well.

3. Endothelial stiffening in atheroprone regions: Impact of dyslipidemia

Early studies initially compared the biomechanical properties of EC from atheroprone versus athero-resistant regions in the vasculature and have found increased EC stiffness in the atheroprone regions. Miyazaki and Hayashi (1999) showed altered EC stiffening in intact rabbit aortas resulted from changes in cytoskeletal structures such as stress fiber formation and F-actin filament distribution (Miyazaki & Hayashi, 1999). There was increased stiffening in the ECs from the inner (medial) bifurcation of the common iliac artery (CIA) shown to have disturbed flow patterns, as compared to ECs from the outer (lateral) curvature of the CIA characterized by higher shear rates and more uniform/unidirectional blood flow patterns. A similar effect was observed by Collins et al. (2014) in murine models. Collins' research attributed the regionally specific biomechanical changes in intact ECs due to the differential activation of the mechanosensor PECAM1 (platelet endothelial cell adhesion molecule 1) in vitro in the PKA, protein kinase A/RhoA signaling pathway. This pathway is altered due to differential focal adhesion dynamics observed in the regionally specific sub-EC matrix beds of the aortic arch compared to that of the descending aorta. Namely, that the large depositions of collagen found in the athero-resistant descending aorta hinder integrin growth and binding, $G\alpha_s$-dependent PKA activation and RhoA inhibition which facilities EC compliance. In contrast, they show increased EC stiffness (via increased relative displacement of magnetic bead displacement) in the arch resulting from increased focal adhesions, PKA inhibition/RhoA activation in part due to differential responsiveness of ECs on the increased fibronectin depositions in the arch's extracellular matrix. Collins showed differential mechanical responsiveness to applied tension on PECAM1 in intact ECs using permanent ceramic magnet or pulsatile magnetic tweezers, whereas Miyazaki and Hayashi (1999), Le Master, Fancher, Lee, and Levitan (2018) and Le Master, Huang, et al. (2018) quantified biomechanical changes in intact ECs using atomic force microscopy (AFM).

Our studies focus on the effects of dyslipidemia on endothelial stiffness. In our early study, we investigated the effect of high fat/high cholesterol diet on endothelial stiffness in a porcine model of atherosclerosis (Byfield, Tikku, Rothblat, Gooch, & Levitan, 2006). Pigs were fed a high fat/high cholesterol (10% lard/0.5% cholesterol) diet for 3–6 months, which increased cholesterol level in the plasma to dyslipidemic conditions, greater than

300 mg/dL total cholesterol, whereas control pigs that were fed normal laboratory diet had a total cholesterol level of <100 mg/dL. Endothelial stiffness was assessed using micropipette aspiration, a technique that measures the deformation of the plasma membrane under well-defined negative pressure into a glass pipette (Fig. 1A, top; Byfield, Aranda-Espinoza, Romanenko,

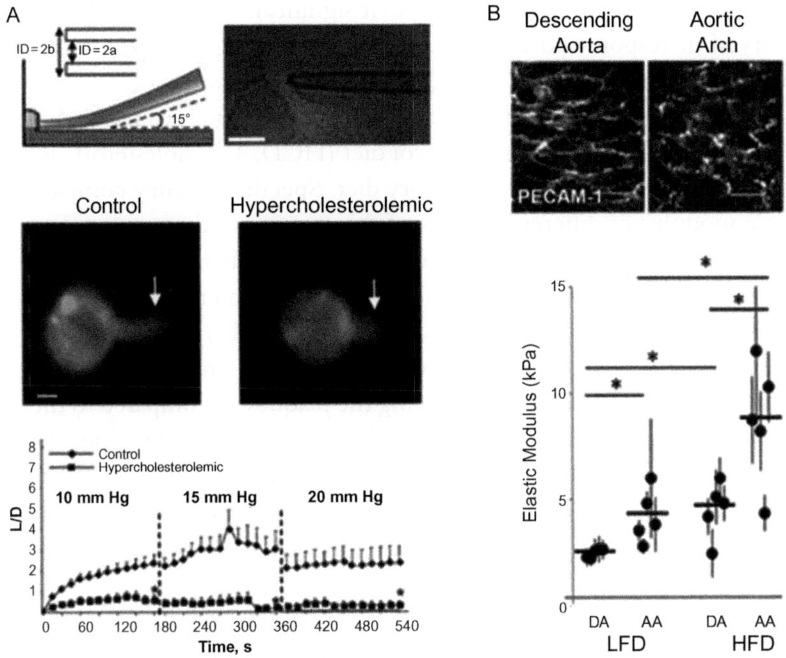

Fig. 1 Diet-induced dyslipidemia increases the elastic modulus of aortic endothelial cells in pigs (A) and mice (B). (A, top) Micropipette aspiration technique shown. (Middle) Fluorescently labeled ECs showing membrane deformation. Bar = 40 μm. (Bottom) Average time course of membrane deformation (length/diameter) between control and hypercholesterolemic pig aortic ECs. (B, top) *En face* images of freshly isolated aorta showing PECAM1 staining. Scale bar = 20 μm. (Bottom) Average elastic modulus values for the descending aorta (DA) and aortic arch (AA) from low fat and high fat diet-fed WT mice. *$P < 0.05$. *Images reprinted with permission from Byfield FJ, Aranda-Espinoza H, Romanenko VG, Rothblat GH & Levitan I. (2004). Cholesterol depletion increases membrane stiffness of aortic endothelial cells. Biophysical Journal 87, 3336–3343., Byfield FJ, Tikku S, Rothblat GH, Gooch KJ & Levitan I. (2006). OxLDL increases endothelial stiffness, force generation, and network formation. Journal of Lipid Research 47, 715–723.; Le Master E, Huang RT, Zhang C, Bogachkov Y, Coles C, Shentu TP, Sheng Y, Fancher IS, Ng C, Christoforidis T, Subbaiah PV, Berdyshev E, Qain Z, Eddington DT, Lee J, Cho M, Fang Y, Minshall RD & Levitan I. (2018). Proatherogenic flow increases endothelial stiffness via enhanced CD36-mediated uptake of oxidized low-density lipoproteins. Arteriosclerosis, Thrombosis, and Vascular Biology 38, 64–75..*

Rothblat, & Levitan, 2004). The experiments were performed on freshly isolated aortic endothelial cells. We found that endothelial cells isolated from the aorta of high fat/high cholesterol fed pigs had significantly less membrane deformation at several levels of negative pressure, which indicated that the cells were stiffer than cells isolated from the aortas of pigs fed a normal chow diet (Fig. 1A, middle, bottom; Byfield et al., 2006). This was the first study that discovered endothelial stiffening under dyslipidemic conditions. In this study, we also demonstrated that significant endothelial stiffening is observed in response to oxidized modifications of LDL (oxLDL) and the mechanistic basis of this effect is described later in the chapter.

In contrast, Hayashi and Higaki (2017) did not observe endothelial stiffening in rabbits fed a high cholesterol diet (HCD; 1% cholesterol), as compared to animals on normal laboratory diet. Specifically, they compared the elastic modulus of different regions of the aorta, the atheroprone disturbed flow region of the media CIA as compared to the athero-protective laminar flow region of the outer CIA and abdominal aorta in rabbits fed a high cholesterol diet versus controls (Hayashi & Higaki, 2017). In the HCD group they separated each regionally specific sample into an 'off-plaque' region, the EC monolayer directly surrounding the plaques, as compared to the stiffness of the ECs directly covering the plaques ('on-plaque' region). Within each diet group and consistent with their previous study (Miyazaki & Hayashi, 1999), ECs from intact rabbit vessels were stiffer in the disturbed flow, pro-atherogenic regions of the medial CIA as compared to the higher flow, athero-protective regions of the lateral CIA. However, they did not observe any diet-induced changes in EC stiffness (off-plaque) in any of the three regional groups. Interestingly, the ECs from the medial CIA from the HCD condition were softer 'on' the plaque as compared to its surrounding area.

Our lab, however, discovered significant endothelial stiffening in a mouse model of high-fat diet-induced dyslipidemia and obesity (Le Master, Huang, et al., 2018) as well as increased EC stiffening with aging (Le Master, Fancher, et al., 2018). Similar to the findings of Miyazaki and Hayashi (1999) and Collins et al. (2014), our data shows that the endothelial surface of the intact mouse aortas (Fig. 1B, top) from pro-atherogenic regions (inner curvature of the aortic arch) are stiffer than those from athero-protective regions (descending aorta) in mice fed a regular laboratory diet (Fig. 1B, bottom left). Notably, however, we found that this differential EC stiffness of the aortic arch as compared to the descending aorta is abrogated in mice deficient in scavenger receptor CD36 (cluster of differentiation 36) (Le Master, Huang, et al., 2018), known to mediate the uptake of oxidized modifications of low-density lipoproteins (oxLDL) and fatty acids (Adachi & Tsujimoto, 2006; Endemann et al., 1993). Much research has shown that mice globally deficient in CD36 confers protection against the formation of atherosclerotic lesions (Brown, Kennedy, Morton, & Febbraio, 2015; Febbraio et al., 2000; Guy, Kuchibhotla, Silverstein, & Febbraio, 2007) though one study showed increased lesion area in their aortic sinus (Moore et al., 2005). The critical dependence of the endothelial stiffening of the arch on the presence of the CD36 receptor suggests that lipid uptake plays a major role in this process. Indeed, we also found increased CD36 expression and increased lipid deposition in the aortic arch of wild type mice even when fed a normal laboratory diet (Le Master, Huang, et al., 2018). Increased lipid depositions in the arch were absent in the aortas of CD36-deficient mice (Le Master, Huang, et al., 2018). Notably, the elastic modulus we measure reflects the deformability (stiffness) in the endothelium without external intervention, whereas Collins et al. measured a stiffening response to the local application of force via permanent ceramic magnet and magnetic microbeads, significantly different measurement parameters. These measurement parameters may have significantly different physiological consequences or may reflect different portions of the same signaling cascade, for example, the PKA/RhoA signaling observed in Collins et al. (2014) being downstream of oxLDL/CD36 effects our studies observed.

Moreover, we also found that a short-term 1 month high fat/high cholesterol diet (21% total fat/0.2% total cholesterol) further exacerbated endothelial stiffening in the vasculature, with a synergetic effect observed in the aortic arch (Fig. 1B, bottom right) due in part to the increased levels of oxidized LDL detected in the plasma as well as regionally disturbed flow patterns. This high fat, atherogenic diet has been shown to induce EC dysfunction in wild type mice and is translational to the type of EC dysfunction, atherosclerosis and progression of obesity to what is found in humans (Collins, Martin, Surwit, & Robidoux, 2004; Paigen, 1995; Park, Booth, Lee, Laye, & Zhang, 2012). Since central adiposity is markedly increased in wild type mice on a HF/HC diet and accompanied by gradual EC dysfunction (Park et al., 2012; Rebuffe-Scrive, Surwit, Feinglos, Kuhn, & Rodin, 1993), we used this translational HF/HC, atherogenic diet in mice to determine the effects of short-term diet-induced dyslipidemia on pro-atherogenic endothelial dysfunction and stiffening.

Moreover, as show in Fig. 2A we showed that diet-induced endothelial stiffening is also abrogated in CD36-deficient mice (Le Master, Huang, et al.,

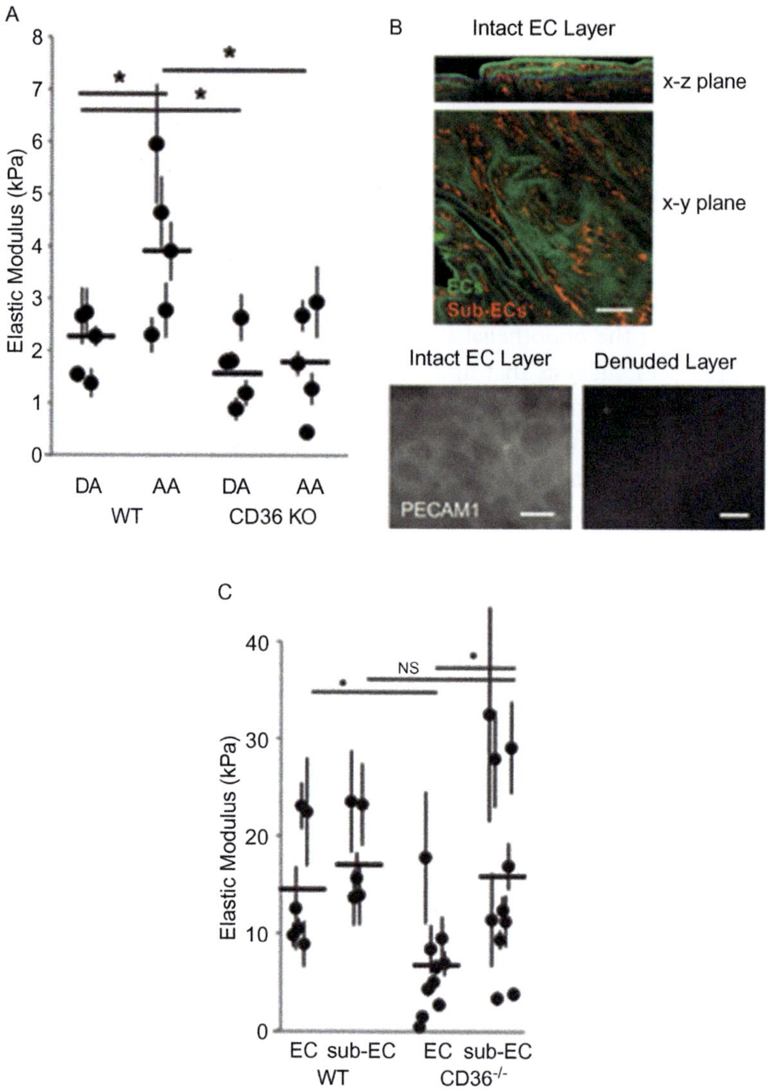

Fig. 2 Aortic endothelial stiffness depends on CD36 and aging. (A) Elastic moduli of aortic endothelium of WT and $CD36^{-/-}$ mice at 4–6 months in descending aorta (DA) and aortic arch (AA); (B, top) Confocal z-stacked images of intact EC layer (green) and the sub-EC layer (red) in the aortic arch of mice. (Bottom) Fluorescent staining of PECAM1 of intact EC monolayer (left) and denuded (right) layer from mouse aortas. Scale bars = 20 μm. (C) Elastic moduli of aortic vascular endothelium and of the sub-endothelial layer in aged WT and $CD36^{-/-}$ mice (22–24 months old). $^*P < 0.05$. NS = not significant. *Images reprinted with permission from Le Master E, Fancher IS, Lee J & Levitan I. (2018). Comparative analysis of endothelial cell and sub-endothelial cell elastic moduli in young and aged mice: Role of CD36. Journal of Biomechanics 76, 263–268. and Le Master E, Huang RT, Zhang C, Bogachkov Y, Coles C, Shentu TP, Sheng Y, Fancher IS, Ng C, Christoforidis T, Subbaiah PV, Berdyshev E, Qain Z, Eddington DT, Lee J, Cho M, Fang Y, Minshall RD & Levitan I. (2018). Proatherogenic flow increases endothelial stiffness via enhanced CD36-mediated uptake of oxidized low-density lipoproteins. Arteriosclerosis, Thrombosis, and Vascular Biology 38, 64–75..*

2018). It is also important to note that in this study, and in contrast to the previous studies in pigs, endothelial stiffness was measured on the intact aortas *en face*. Additional studies (Le Master, Fancher, et al., 2018) show that enhanced EC stiffness in intact mouse aortas depend on CD36 through myosin phosphorylation and that blocking CD36-mediated oxLDL uptake by a known pharmacological inhibitor of CD36, sulfosuccinimidyl oleate (SSO), abrogated the stiffening effect. Extensive in vitro studies (Byfield et al., 2006; Oh et al., 2016; Shentu et al., 2010) support these results by showing that oxLDL on its own increase endothelial stiffness. We show, however, that disturbed flow on its own does not increase endothelial stiffness, but rather facilitates oxLDL uptake leading to increased EC stiffness. Furthermore, consistent with our findings that there is a lack of endothelial stiffening in murine models of genetic deficiency of CD36, DF-induced enhanced oxLDL-mediated EC stiffening in vitro was also found to be mediated by the scavenger receptor CD36, but not by another well-known scavenger receptor LOX-1 (lectin-like oxidized LDL receptor-1) (Le Master, Huang, et al., 2018). Taken together, these results suggest that enhanced stiffness in the endothelial monolayer in vivo results from a synergetic effect of DF patterns and increased plasma oxLDL levels which are mediated by CD36.

Based on the observations in all three models, there is a connection between pro-atherogenic conditions and increased endothelial cell stiffness, either in response to high fat/high cholesterol diet (pigs, mice; Byfield et al., 2006; Le Master, Huang, et al., 2018), or in response to pro-atherogenic flow conditions (rabbits, mice; Hayashi & Higaki, 2017; Le Master, Huang, et al., 2018). Pigs, mice, and rabbits are well-known animal models for atherosclerosis studies including non-human primates (Getz & Reardon, 2012). Each of the species has different advantages and disadvantages. It is not exactly clear why the rabbit model did not increase endothelial stiffness in high cholesterol diet-fed animals. One possible explanation is the difference in the diet. In contrast to high fat diets used in both of our studies in pigs and mice, Hayashi et al. used high cholesterol rather than a high fat regimen. These different diets can lead to different metabolic profiles and physiological outcomes. It is also important to note that pigs are known for their highest similarity to human vascular system and lipid profiles whereas rabbits are very sensitive to cholesterol with their lipid metabolism being significantly different from humans (Fan et al., 2015; Getz & Reardon, 2012).

4. Endothelial stiffening in aging

The aging process is well-known to be associated with increased cardiovascular events, arterial stiffness, atherosclerotic lesion development and endothelial dysfunction. Several studies have shown that contributing factors of age-associated endothelial dysfunction include cell apoptosis, necrosis, and increased oxidative stress and inflammatory cell activation, yet the role of aging in endothelial stiffening per se is still being uncovered (Daiber et al., 2017).

A few studies have explored the effect of aging on the stiffness of the sub-endothelial matrix of the aorta, as assessed by denuding the endothelial monolayer. One study by Reinhart-King's group has shown in murine models that the sub-endothelial matrix (SEM) stiffens with age (21–25 months old versus 10 weeks) accompanied with an increase in the cell-cell junctional width between the endothelial cells and enhanced barrier permeability (Huynh et al., 2011). This was observed in the descending part of the aorta, which is relatively athero-protective and further studies are needed to compare the athero-protective and athero-susceptible regions of the aorta. They also found that inhibition of the Rho pathway by Y-27632, an inhibitor of Rho Kinase (ROCK), abrogated the effect of the sub-endothelial stiffening on the endothelial barrier: the junctional width tightened to the same extent as was seen in younger mice. This treatment did not affect the sub-endothelial stiffness indicating that the Rho pathway is responsible for the disruption of barrier function in response to the increased stiffness of the SEM. Age-induced stiffness in the SEM has been shown to be reversible by exercise (Kohn et al., 2016).

In an effort to understand the effect of aging on endothelial stiffening, we built upon our initial in vivo studies to include the effect that aging and the CD36 scavenger receptor have on EC and sub-EC stiffening in intact mouse aortas (Fig. 2B; Le Master, Fancher, et al., 2018). Consistent with previous studies (Huynh et al., 2011), our results show increased stiffness in the sub-endothelial layer in 22–25 month old mice, a commonly accepted age for advanced aged mice (Kohn et al., 2015) versus younger mice (4–6 months old) (17 ± 2 versus 7 ± 1 kPa in aged versus young mice). Our recent data show that aging also has a dramatic effect on stiffness of the endothelial layer of intact mouse aortas (15 ± 2 versus 4 ± 1 kPa in aged versus young mice) (Le Master, Fancher, et al., 2018).

Furthermore, as shown in Fig. 2C we found that the loss of CD36 resulted in decreased stiffening of the endothelial monolayer in aged mice, as compared to WT aged controls (6 ± 2 kPa in CD36 KOs versus 15 ± 2 kPa in WTs). In contrast, we showed that CD36 expression has no impact on the elastic modulus of the sub-endothelial layer of the aortic vasculature in both young and aged mice. These findings indicate that the stiffening of the endothelium in aged mice primarily depends upon the presence of CD36, but is independent of the stiffness from its sub-endothelial layer. We suggest, therefore, endothelial stiffening in aging is mediated by oxidized lipids.

5. Physiological factors that cause EC stiffening and mechanistic insights

Several factors, such as exposure to oxidized lipids, altered shear patterns and substrate stiffness activate several signaling cascades resulting in endothelial stiffening which in turn are implicated in inflammation and atherogenesis.

5.1 Oxidized lipoproteins

An oxidative environment facilitates inflammation and likewise the oxidatively modified form of low-density lipoprotein (oxLDL) and its components are well-known to create an inflammatory environment in vitro and in vivo. Elevated plasma levels of oxLDL are implicated as a major factor in the onset and progression of atherosclerosis (Itabe, 2012) with oxLDL as an established component in plaques from both human patients and in animal models of atherosclerosis (Ehara et al., 2001; Yla-Herttuala et al., 1989). The native, non-modified version, LDL, can be taken into cells, at which point it can be oxidized or cellular uptake of oxLDL can occur directly from the bloodstream via endothelial scavenger receptors. Both LDL and oxLDL particles bind to and become engulfed by macrophages in the sub-endothelial arterial space through species specific receptors. In this process, macrophages transform into lipid-laden foam cells which combine with hyperproliferized smooth muscle cells and calcified cells to form atherosclerotic plaques (Libby, Ridker, & Hansson, 2011).

Our lab has shown that endothelial exposure to oxidized LDL and specific lipid components of oxLDL, such as the oxysterols, 7-ketocholesterol (7KC) and 7-alpha-hydroxycholesterol, and oxidized phospholipids (oxPCs), such as POVPC (1-palmitoyl-2-(5-oxovaleroyl)-sn-glycero-3-phosphocholine) and PGPC (1-palmitoyl-2-glutaroyl-sn-glycero-3-phosphocholine), significantly increase the stiffness of endothelial cells (Fig. 3; Byfield et al., 2006; Oh et al., 2016; Shentu et al., 2012). All of these oxidized lipids, both

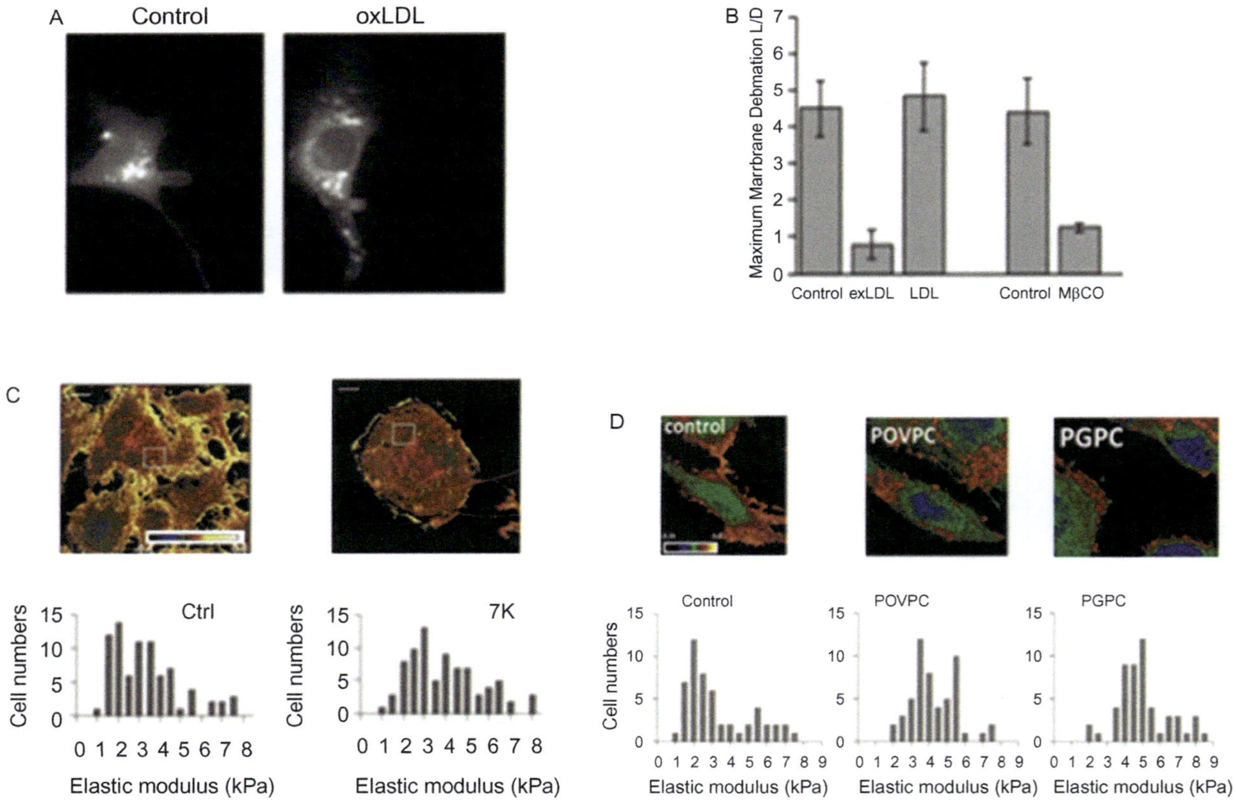

Fig. 3 See legend on next page.

oxysterols and oxPCs, occur naturally in the vascular system (Levitan, Volkov, & Subbaiah, 2010). As described below, we also elucidated the major mechanistic steps of this process: binding and internalization of oxLDL by its scavenger receptors (Le Master, Huang, et al., 2018); incorporations of the specific lipid moieties of oxLDL into the plasma membrane bilayer resulting in the disruption of lipid ordering of the membrane (Shentu et al., 2010); and activation of the RhoA/ROCK/MLCP pathway to induce the contractile response which underlies the stiffening (Oh et al., 2016).

Our early studies showed that cholesterol depletion (Fig. 3B, right), rather than enrichment, results in EC stiffening (Byfield et al., 2004). This was initially very surprising because it is well-known that an increase in the cholesterol content of the bilayer results in increased membrane bilayer stiffness/decrease in membrane deformability, as was demonstrated in lipid vesicles (Needham & Nunn, 1990). It is important to discriminate here, however, between the stiffness of the membrane bilayer of the plasma membrane and the stiffness of the cellular envelope that consists of both the membrane and the sub-membrane cytoskeleton and mechanically is dominated by the latter (Byfield et al., 2004; Sato, Theret, Wheeler,

---

Fig. 3 Increased endothelial stiffness results from exposure to oxidized LDL, cholesterol depletion, the oxysterol 7-ketocholesterol (7 K) and the oxPCs, POVPC and PGPC. (A) Fluorescent images of bovine aortic ECs showing micropipette aspiration with (left) more compliant control cell versus (right) decreased membrane deformation in oxLDL exposed cell and resulting (B) quantification showing decreased membrane deformation (increased EC stiffness) in oxLDL condition versus control and LDL exposed ECs (left) and decreased membrane deformation in cholesterol depleted ECs (MBCD) versus control ECs; (C, top) Multi-photon general polarization (GP) images of control (left) and 7 K-treated ECs (right). Scale bar = 11.2 μm. (Bottom) 7 K increases EC elastic modulus values (stiffness) as indicated by the rightward shift in histogram (right); (D, top) GP images showing POVPC- and PGPC-treated ECs with (bottom) their average elastic modulus histograms. *Images reprinted with permission from Levitan, I., & Gooch, K. J. (2007). Lipid rafts in membrane-cytoskeleton interactions and control of cellular biomechanics: Actions of oxLDL.* Antioxidants & Redox Signaling, *9(9), 1519–1534; Shentu, T.P., Titushkin, I., Singh, D.K., Gooch, K.J., Subbaiah, P.V., Cho, M., & Levitan, I. (2010). oxLDL-induced decrease in lipid order of membrane domains is inversely correlated with endothelial stiffness and network formation.* American Journal of Physiology. Cell Physiology, 299, *C218–C229; Shentu, T.P., Singh, D.K., Oh, M.J., Sun, S., Sadaat, L., Makino, A., Mazzone, T., Subbaiah, P.V., Cho, M., & Levitan, I. (2012). The role of oxysterols in control of endothelial stiffness.* Journal of Lipid Research, 53, *1348–1358; Ayee, M.A.A., LeMaster, E., Shentu, T.P., Singh, D.K., Barbera, N., Soni, D., Tiruppathi, C., Subbaiah, P.V., Berdyshev, E., Bronova, I., Cho, M., Akpa, B.S., & Levitan, I. (2017). Molecular-scale biophysical modulation of an endothelial membrane by oxidized phospholipids.* Biophysical Journal, 112, *325–338.*

Ohshima, & Nerem, 1990). Indeed, we found that EC stiffening mediated by cholesterol depletion depends on maintaining the structural integrity of F-actin even though we saw no increase in F-actin stress fibers formation after cholesterol depletion (Byfield et al., 2004). In a later study, however, we demonstrated that cholesterol depletion strengthens the adhesion force between the membrane and the underlying cytoskeleton (Sun et al., 2007). Additionally, Fessler et al. (2004) showed that cholesterol depletion induces actin polymerization in the cortical rim in human polymorphonuclear leukocytes.

Our next surprise was to discover that exposing endothelial cells to oxLDL results in the same effect as cholesterol depletion, endothelial stiffening (Byfield et al., 2006). The dominant belief at the time was that exposure to oxLDL should enrich endothelial cells with cholesterol rather than deplete cholesterol. Upon examination of this issue, however, we found that exposure to oxLDL does not have a significant effect on the level of membrane cholesterol of endothelial cells (Byfield et al., 2006) but instead results in the disruption of the lipid order of the membrane, presumably because of the incorporation of the oxidized lipids (Shentu et al., 2010). Both oxLDL-induced disruption of the lipid packing and EC stiffening are prevented by pre-loading the cells with cholesterol (Shentu et al., 2010).

It is important to note that oxLDL is a highly heterogenous particle that exists in multiple forms differing in the degree of oxidation from minimally/mildly to strongly oxidized, which also differ significantly in their lipid composition (Levitan et al., 2010). Minimally-oxidized LDL (m-oxLDL) is enriched predominantly in oxidized phosphatidylcholines (oxPCs) and can be recognized by both the cellular LDL receptor (Goldstein & Brown, 2009) and the scavenger receptor CD36 (Levitan & Shentu, 2011). M-oxLDL typically has relatively low levels of oxysterols, whereas as the degree of oxidation increases, the oxysterol content goes up (Levitan & Shentu, 2011). We compared the stiffening effects of mildly versus strongly oxidized LDL and found that the stiffening effect is observed for both mildly and strongly oxidized LDL and increases with the degree of oxidation (Shentu et al., 2012). Furthermore, we also identified the specific components of oxLDL that induce endothelial stiffening by comparing the five lipid fractions: cholesterol/cholesterol ester (Chol/CE), phosphatidylethanolamine (PE), phosphatidylcholine (PC), sphingomyelin (SM) and lysophosphatidylcholine (LPC). We found that two of these components, oxidized Chol/CE and oxPC, significantly increased endothelial stiffening whereas the other three components of oxLDL did not (Shentu et al., 2012).

Furthermore, we tested the impact of oxysterols on endothelial biomechanical properties. Oxysterols are bioactive components of oxLDL which are known to impair endothelial nitric oxide synthase (eNOS) activation, inhibit the synthesis of sterols and control sterol efflux and macrophage activation, thus playing an important role in atherosclerotic plaque development (Levitan & Shentu, 2011). We found that endothelial stiffening increases linearly as a function of oxysterol content in oxLDL (Shentu et al., 2012). Moreover, two major oxysterols components in oxLDL, 7-KC and 7-alpha-hydroxycholesterol, have a dominant stiffening effect on ECs whereas two other oxysterols, 5-alpha, 6-alpha epoxide and 5-beta, 6-beta epoxide isomers, showed no detectable role (Fig. 3C; Shentu et al., 2012). Further studies showed that 27-hydroxycholesterol, an oxysterol particularly abundant in human atherosclerotic plaques and lipid-laden foam cells, increased EC stiffening, whereas 25-hydroxycholesterol did not (Shentu et al., 2012).

We extended our studies of oxysterols to also include their potential incorporation into aortic tissues of young dyslipidemic mice, as compared to age and gender matched wild type controls. Earlier studies have shown that a known dyslipidemic atherogenic murine model, mice globally deficient in ApoE, can be used as a model for early stage atherogenesis at younger ages (<4 months old) (Nakashima, Plump, Raines, Breslow, & Ross, 1994). Interestingly, our results showed that two of the oxysterols identified to induce EC stiffening in vitro, 7-ketocholesterol and 7-alpha-hydroxycholesterol, were significantly increased in the aortic tissues of male ApoE KO mice as compared to their WT controls (Shentu et al., 2012), highlighting the role of oxysterols in plaque formation and control of cellular biomechanical properties.

Later studies from our lab (Ayee et al., 2017) investigated the role of other bioactive components of oxLDL: oxidized phospholipids, which include two short chain PAPC products, POVPC and PGPC, both of which are known to increase monocyte adhesion, induce the inflammatory mediator IL-8 (interleukin-8) and activate sterol regulatory element binding protein CD36 ligands (Levitan & Shentu, 2011). Ayee et al. (2017) showed that exposure to POVPC and PGPC increased the elastic modulus values and decreased the lipid membrane order in endothelial cells versus untreated ECs (Fig. 3D; Ayee et al., 2017). These results showed that a disruption in the membrane lipid ordering via introduction of these two bioactive oxidized phospholipids is accompanied by and correlated with increased endothelial cell stiffening. Furthermore, oxPAPC has been implicated as critical in oxLDL's recognition by scavenger receptor CD36 and to the internalization of caveolae, the macromolecular transport machinery comprised of specialized cholesterol-rich lipid raft formed into membrane invaginations (Levitan & Shentu, 2011).

OxLDL is known to activate small GTPases, such as RhoA and Rac 1 (Oh et al., 2016; Sugimoto et al., 2009). Small GTPases modulate cytoskeletal structural organization and regulate endothelial permeability (Birukov, 2009). Rac and another member of the Rho small GTPase family, CDC42, are involved in cell migration and restoration of the vascular integrity (Birukov et al., 2004), whereas RhoA induces contractile forces on intracellular stress fibers to disrupt the vascular barrier function (Shen, Wu, & Yuan, 2009; Fig. 4A). In our efforts to determine which signals of oxLDL activation affect endothelial stiffening, our lab showed that introducing dominant negative RhoA (DnRhoA^{T19N}) constructs abolished the oxLDL-induced stiffening of endothelial cells, while the overexpression of dominant negative Rac1 (DnRac1^{T17N}) had no effect (Oh et al., 2016). Additionally, the application of RhoA kinase (ROCK) inhibitor, Y27632, fully inhibited oxLDL-induced endothelial stiffening (Oh et al., 2016), which implies that it is mediated by the RhoA/ROCK pathway (Fig. 4). Furthermore, inhibiting myosin2 phosphorylation with blebbistatin also abrogated oxLDL-induced stiffening indicating that it was a result of RhoA/ROCK/myosin2 contractile response (Fig. 4E). We also showed that oxLDL-induced activation of the RhoA/ROCK/myosin2 pathway is dependent on the CD36 scavenger receptor (Oh et al., 2016). OxLDL and 7-ketocholesterol induced activation of RhoA/ROCK pathways was observed in both macro (human aortic ECs) and microvascular cells (HMVECs) (Oh et al., 2016).

The mechanism of RhoA activation by oxLDL, however, is not clear yet. In general, RhoA activation is mainly governed by three factors: guanine nucleotide exchange factors (GEFs), GTPase activating proteins (GAPs), and guanine nucleotide dissociation inhibitor protein (GDI) (Strassheim et al., 2019). Our lab showed that the activation of RhoA by oxLDL was disrupted by overexpression of GDI-1 (Zhang et al., 2017), which is known to bind to RhoA, sequestering its lipid anchor to the plasma membrane and thus preventing RhoA activation and GTP binding (Cherfils & Zeghouf, 2013). We also showed that oxLDL-induced activation of RhoA was not affected by p115RGS (Zhang et al., 2017), a known inhibitor of p115RhoGEF, which transfer GTP on RhoA to be activated in endothelial cells (Holinstat, Mehta, Kozasa, Minshall, & Malik, 2003). Our observation suggests that oxLDL may affect the interaction between GDI and RhoA, and regulate the activation. This observation alone is not

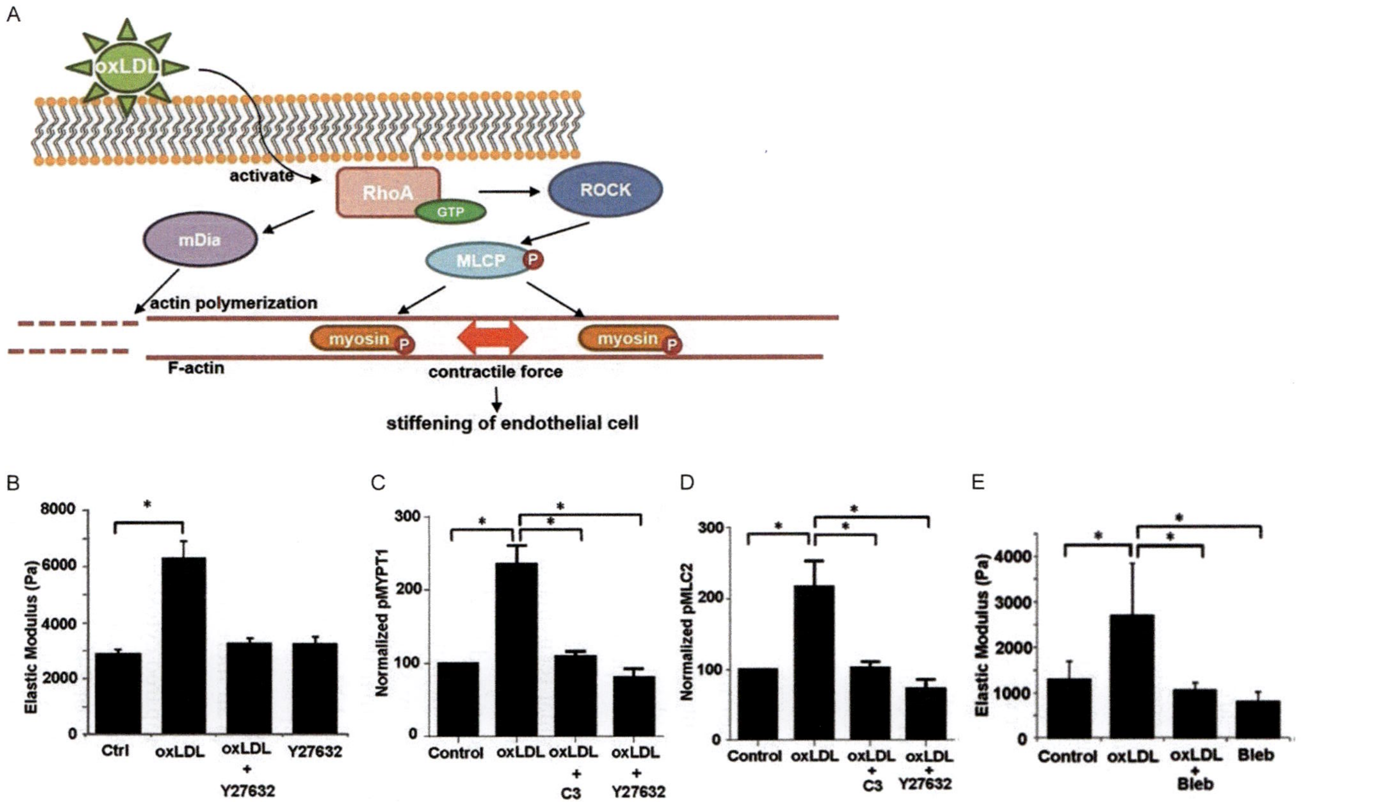

Fig. 4 Role of RhoA/ROCK/MLC pathway in controlling EC stiffness. (A) Schematic pathway of oxLDL-induced RhoA/ROCK/MLC. (B) oxLDL-induced EC stiffening is aborted by ROCK inhibitor, Y27632. (C and D) oxLDL-induced phosphorylations of MYPT1 (myosin phosphatase) and myosin light chain 2 (MLC2) are significantly decreased by RhoA inhibitor (C3) and ROCK inhibitor. (E) oxLDL-induced EC stiffening is abrogated by myosin inhibitor (blebbistatin). *Panel (A) modified from Huveneers, S., & Danen E.H. (2009). Adhesion signaling—Crosstalk between integrins, Src and Rho.* Journal of Cell Science, 122, *1059–1069.. Panels (B–E) reprinted with permission from Oh, M.J., Zhang, C., LeMaster, E., Adamos, C., Berdyshev, E., Bogachkov, Y., Kohler, E.E., Baruah, J., Fang, Y., Schraufnagel, D.E., Wary, K.K. & Levitan, I. (2016). Oxidized LDL signals through Rho-GTPase to induce endothelial cell stiffening and promote capillary formation.* Journal of Lipid Research, 57, *791–808.*

sufficient however to indicate that GDI-1 is the main mechanism for oxLDL-induced RhoA activation because over-expressing GDI-1 may not mimic the conditions in vivo.

Notably, there are other signaling pathways, such as ROS production that are induced by oxidized lipids and are known to lead to endothelial injury and atherogenesis, which could also contribute to endothelial stiffening via RhoA activation. Indeed, multiple studies showed that oxLDL induces an increase in ROS production (Fleming et al., 2005) and it was also shown that ROS activate RhoA (Li, Liu, Chen, & Fan, 2018). It is not known yet whether this pathway contributes to oxLDL-induced endothelial stiffening. It is also possible that an increase in the intracellular Ca^{2+} that is known to initiate the contractile response is involved in the stiffening effect (Kugelmann et al., 2018). More studies are needed to address these possibilities.

5.2 Shear stress

Another main factor in vascular stiffness and atherogenesis is the hemodynamic environment the endothelial monolayer experiences. It is well-known that high shear stress unidirectional laminar blood flow (LF) is athero-protective and promotes vasodilation, wound repair and decreases adhesion molecule expression, platelet aggregation and ROS production, whereas bidirectional recirculating disturbed flow (DF) is atherogenic and increases inflammation, adhesion molecule expression and promotes an environment of sustained oxidative stress (Chiu & Chien, 2011).

With respect to the impact of hemodynamics on endothelial biomechanical properties, a few studies have explored the effects of biomechanical changes due to laminar flow as compared to a static culture environment. Sato, Levesque, and Nerem (1987) found that bovine aortic EC exposed to increasing amounts of shear stress (10, 30 or 85 dynes/cm^2) for 24 h showed increasing alterations in F-actin filaments in the cells' cytoskeletal structure as well as an increase in cell stiffness as measured by micropipette aspiration (Sato et al., 1987). Later studies from this group (Sato, Ohshima, & Nerem, 1996) compared cytoskeletal changes, such as elongation and elastic modulus values of porcine aortic ECs exposed to a relatively low shear value (2 dynes/cm^2) for 6 or 24 h using a parallel plate flow system, as compared to static control PAECs. Similar to their previous studies, sheared ECs were stiffer than static control cells. Of those exposed to 24 h of shear stress, approximately 50% of these cells had elongated and thicker F-actin filaments aligned to the direction of flow. Cells with these characteristics had elastic modulus values that were 5–10 times higher than sheared ECs without elongated filaments and the static ECs. These results highlight the importance of cytoskeletal changes in the modulation of endothelial biomechanics.

A few groups have looked into the mechanistic properties affecting shear-induced changes in endothelial stiffening. One in vitro study showed increased cell stiffness due to unidirectional LF as compared to static cells, which was due to the development of centrally located thick stress fibers and increased VE-cadherin junctional protein expression (Sato & Ohashi, 2005). Another group (Collins et al., 2014) showed that application of unidirectional laminar flow activated endothelial PECAM1 and the PKA/RhoA pathway leading to changes in endothelial compliance.

Since our previous studies showed that oxLDL induces EC stiffening (Byfield et al., 2006; Oh et al., 2016; Shentu et al., 2012) and since it was recently shown that oscillatory flow facilitated increased oxLDL uptake in coronary ECs (Martorell et al., 2014), we hypothesized that the increased endothelial stiffness observed in disturbed flow region of the aortic arch may be mediated by increased oxLDL uptake and oxLDL-induced endothelial stiffening. To test this hypothesis, we compared oxLDL uptake in HAECs exposed to unidirectional LF versus nonunidirectional recirculating DF using a microfluidic flow chamber with a step barrier, described previously (DePaola et al., 1999). OxLDL uptake was quantified in HAECs exposed to physiological shear stress values of 10–12 dynes/cm^2 of laminar or <3 dynes/cm^2 disturbed flow (Chiu & Chien, 2011; Ku, Giddens, Zarins, & Glagov, 1985) for 48 h with fluorescently labeled-oxLDL (1 μg/mL) in the cell medium. Our data shows that there is increased oxLDL uptake in the disturbed flow regions, as compared to the laminar regions. Importantly, this DF-induced increase in oxLDL uptake was concurrent with significantly increased endothelial elastic modulus values (stiffness), as compared to cells from the LF environment within the same flow chamber.

To investigate the mechanism of increased oxLDL uptake under differential shear conditions, we focused on two types of scavenger receptors that were identified earlier to be responsible for oxLDL recognition, binding and internalization in endothelial cells: LOX1 and CD36 (Febbraio, Hajjar, & Silverstein, 2001; Lubrano & Balzan, 2014). LOX1 is found in ECs, vascular SMC, platelets and has been shown to play a significant role in oxLDL-induced inflammatory processes including EC and vascular SMC hyperproliferation, decreased nitric oxide production, increased ROS formation,

EC injury, and apoptosis (Yoshimoto et al., 2011). CD36 is a found in ECs, macrophages, and monocytes (Park, 2014) and is a multiple ligand receptor which in addition to oxLDL, also binds to collagen and long chain fatty acids. Binding of oxLDL to CD36 induces inflammatory cytokine production (Park, 2014) and has anti-angiogenic effects (Dawson et al., 1997). Notably, a number of studies suggested that LOX1 plays a more significant role in macrovascular endothelial cells whereas CD36 is more highly expressed and plays a more significant role in microvascular endothelium. We tested, therefore, whether these two receptors were responsible for DF-induced increased oxLDL uptake and EC stiffening.

First we tested the impact of differential flow patterns on the mRNA and protein expression of CD36 and LOX1 in HAECs. The mRNA expression in both scavenger receptors increased following 48 h of DF as compared to LF using a cone and plate model, previously described (Wu et al., 2015), with a notable 70-fold increase in CD36 expression as compared to LOX1. Moreover, protein expression of CD36 was greater under DF versus LF, a difference which was not observed in LOX1 protein expression. Our further studies showed the downregulation of CD36 significantly decreased oxLDL uptake under both LF and DF and abrogated the increase in oxLDL uptake in response to disturbed flow. In contrast, downregulating LOX1 expression had no effect on oxLDL uptake, leaving the effect of disturbed flow fully preserved. Since we show that CD36 was enhanced by and required for DF-induced oxidized LDL uptake into endothelial cells, but LOX1 had no detectable role (Le Master, Huang, et al., 2018), we therefore proposed that an increase in CD36 expression in response to disturbed flow is a main underlying mechanism of increased oxLDL uptake and subsequent increase in endothelial stiffness. Indeed, downregulation of CD36 abrogated oxLDL-induced EC stiffening under both laminar and disturbed flow conditions, highlighting that the DF-induced increase in oxLDL uptake and enhanced endothelial stiffening is dependent on scavenger receptor CD36.

Furthermore, our lab extended these studies of endothelial biomechanical changes to include a combination between substrates and flow environment. Indeed, earlier studies showed that the nature of the substrate has a major impact on endothelial biomechanics. Specifically, Collins et al. demonstrated that cells seeded on collagen have enhanced activation of mechanosensitive signals compared to cells seeded on fibronectin, two major ECM proteins in the vascular wall (Collins et al., 2014). In contrast to the Collins study, we showed that endothelial cells grown on collagen are stiffer than cells grown on fibronectin, an effect that may be attributed to concentration differences in the two ECM proteins (Le Master, Fancher, et al., 2018). However, regardless of the substrate, we observed no difference between cells exposed to laminar versus disturbed flow in the absence of oxLDL. These results highlight the importance of the combined effects of dyslipidemia and low shear stress DF in increased endothelial stiffness.

5.3 Substrate stiffness

Growing evidence has shown that substrate stiffness affects endothelial morphology, cell-cell junctional integrity (Kohn et al., 2015), actin organization and elongation (Byfield, Reen, Shentu, Levitan, & Gooch, 2009), controls angiogenic response (Lee, Bai, Smith, Bayless, & Yeh, 2013) and regulates endothelial cell adherence to the substrate (Jalali, Tafazzoli-Shadpour, Haghighipour, Omidvar, & Safshekan, 2015). Several reviews discuss the impact of substrate stiffness on endothelial cell function (Gray & Stroka, 2017; Karki & Birukova, 2018; Stroka & Aranda-Espinoza, 2010), but the control of endothelial stiffening by substrate stiffness has not been a well discussed topic.

Our early studies showed that endothelial cells become stiffer when grown on substrates of increased stiffness, both in 2D and 3D cultures (Byfield et al., 2009). To evaluate the effect of substrate stiffness on endothelial deformability in 2D, bovine aortic endothelial cells (BAECs) were seeded on top of polyacrylamide gels coated with fibronectin, a well-established method to test the effects of substrate stiffness (Elosegui-Artola et al., 2016; Yeung et al., 2005). The elastic modulus of the gels ranged between 1.7 and 9 kPa, which was chosen because this range allows for significant changes in stress fibers formation, cell adhesion and spreading (Byfield et al., 2009). This is also within the physiological range of the vascular wall (Woodrum et al., 2006). Under these conditions, the degree of endothelial deformability decreased about twofold in cells grown on the stiffer substrate. In addition, Byfield et al. extended their studies to a 3D environment to explore the role of substrate stiffness on endothelial cells during angiogenic or vasculogenic process. This was done by embedding the cells into collagen gels with stiffnesses ranging from 0.2 to 0.5 kPa, a limitation of the ability of endothelial cells to populate collagen gels. Interestingly, in spite of the very low range of the gel stiffness in the 3D culture, endothelial cells become significantly stiffer and the effect was comparable to that observed in the 2D environment.

Consistent with our earlier studies, Yi, Shen, Tang, Wang, and Zhang (2020) also observed endothelial cell stiffening in cells grown on stiffer substrates, using a wider range of substrate stiffness: 0.82–54.78 MPa, as measured by tensile strength (Yi et al., 2020). Yi et al. pointed out that this range is in the physiological range of human vascular tissue stiffness, when measured by tensile strength testing (Henry et al., 2017; Monson, Goldsmith, Barbaro, & Manley, 2003, 2005). This approach is different from the more standard AFM measurements and the values are not directly comparable. The difference between the two approaches is that the tensile strength is measured by pulling the sample unidirectionally, which gives a stress-strain curve to get the elastic modulus of the sample, whereas the AFM approach measures the stiffness of the surface using a probe, which indent the surface and convert the bending force of the probe to the stiffness. Notably, Yi et al. demonstrated that increased endothelial stiffness was accompanied with exacerbated endothelial permeability, endothelial-to-mesenchymal transition, and inflammatory activation of endothelial cells (Yi et al., 2020).

Mechanistically, the cellular response to substrate stiffness was attributed by outside-in signal transduction via integrins (Discher, Janmey, & Wang, 2005). On a stiffer substrate, more integrins bind to its extracellular matrix (ECM) resulting in increased cell spreading and stabilizing focal adhesions with higher concentrations of focal adhesion kinase (FAKs) (Yeh, Ling, Chen, Lin, & Tang, 2017). FAK activate Src family kinases (SFKs), resulting in RhoA activation, which in turn activates two parallel pathways: (i) mammalian diaphanous (mDia), that induces actin polymerization and F-actin formation, and (ii) RhoA kinase (ROCK), which activates myosin and triggers contractile forces (Gavara & Chadwick, 2016; Huveneers & Danen, 2009). Potentially, both an increase in F-actin and increased contraction can underlie and contribute to the increase in cell stiffness. In contrast, in cells on softer substrate, there is a decrease in integrin binding to ECM, smaller and fewer focal adhesions, and lower activity of RhoA (Discher et al., 2005).

In vivo, substrate stiffness is determined by the composition of ECM proteins, such as type I, III and IV collagen, fibronectin, and laminin. An increase in collagen I and III and fibronectin results in stiffer ECM (Kothapalli et al., 2012), whereas an increase in collagen IV and laminin result in a softer ECM. Importantly, the stiffening of sub-endothelial ECM occurs in vasculature in vivo under pathophysiological condition and aging (VanderBurgh & Reinhart-King, 2018). In both, aging and diet-induced atherosclerosis, there is sub-endothelial stiffening which occurs due to the thickening of intima by the deposition of fibronectin, which is then replaced with type I and III collagen (Hassanisaber, Rouleau, & Faucheux, 2019; VanderBurgh & Reinhart-King, 2018). For example, Rohwedder et al. showed that Apoe$^{-/-}$ mice fed a high-fat diet had increase depositions of fibronectin and collagen I and III in the vascular wall of the aorta and that a loss of fibronectin prevented the sub-endothelial stiffening (Rohwedder et al., 2012). Increased depositions of fibronectin and collagen I and III were also identified in athero-susceptible regions of the aorta with and without the diet (Palotie, Tryggvason, Peltonen, & Seppa, 1983; Rohwedder et al., 2012). Based on the dependence of endothelial stiffness on the stiffness of the substrate described above, an increase in the sub-endothelial stiffness may induce the stiffening of the endothelium.

6. Functional outcomes of EC stiffening

An increasing amount of research suggests that EC stiffness per se has significant physiological effects, including the recruitment of monocytes to sites of inflammation on the endothelial monolayer (Schaefer et al., 2014). The initial stages of atherogenesis is characterized by increased expression of chemotactic and adhesion molecule expression, such as ICAM-1, which in turn promote the capture and uptake of proinflammatory monocytes into the intimal and medial vascular layers directly below the endothelial monolayer (Libby et al., 2011). Schaefer et al. (2014) showed that increased stiffness of the EC monolayer, as regulated by endothelial actin-binding protein alpha-actinin-4, enhanced ICAM-1-mediated neutrophil rolling, adhesion and transmigration. This effect was supported by increased alpha-actinin-4 expression in the EC lining of human and murine atherosclerotic plaques. A similar effect was shown by Martinelli et al. (2014) who found that increased EC stiffness, as mediated via increased substrate stiffness, increased total lymphocyte transmigration as compared to ECs grown on a softer substrate. Moreover, Martinelli et al. (2014) investigated further whether the impact of endothelial stiffening on increased transmigration is mediated either via paracellular or transcellular pathways, two major pathways for the transport macromolecules and immune cells from circulating blood to the outer vascular tissues (Claesson-Welsh, 2015). Briefly, paracellular transport is mediated through the gap between endothelial cells and regulated by two inter-cellular junctional complexes, tight junctions with primary constituents being claudins, occludins, and junctional adhesion molecules

(JAMs) and adherent junctions with primary constituents being vascular-endothelial cadherins (VE-Cads) (Chistiakov et al., 2015). The transcellular transport is defined as endocytosis and transcytosis for macromolecules or monocytes mediated by intracellular vesicles (Predescu, Predescu, & Malik, 2007). Martinelli et al. (2014) found that increased lymphocyte migration resulted from increased EC permeability through a paracellular, rather than transcellular, mechanism (Martinelli et al., 2014). Thus, since increased EC stiffness modulates the transmigration of monocytes through the endothelial monolayer, which would be expected to contribute to foam cell formation and eventual plaque development (Schaefer & Hordijk, 2015), it is critical to determine the mechanisms that govern endothelial stiffness.

It is also important to note that as described in detail later, one of the major mechanisms of endothelial stiffening is activation of RhoA, a small GTPase responsible for cytoskeleton organization via ROCK, and activation of the endothelial contractile response (Shimokawa, Sunamura, & Satoh, 2016), and that contractile forces are well-known to disrupt the integrity of the endothelial barrier (Shen et al., 2009). Specifically, the junction proteins are connected to the cytoskeleton intracellularly and under physiological conditions, normal tension is maintained throughout the endothelial cytoskeleton and vascular integrity is secured (Vandenbroucke, Mehta, Minshall, & Malik, 2008). Multiple stimuli, however, such as histamines, thrombin, TNF-alpha, and lipopolysaccharide, induce activation of the RhoA cascade and endothelial cytoskeletal remodeling, which generates an increase in the contractile force, which in turn induces disruption of adherent junctions, and vascular leakage occurs (Shen et al., 2009; Vandenbroucke et al., 2008).

OxLDL has also been implicated in promoting EC injury and barrier permeability in vitro (May & Qu, 2010) and in vivo (Rangaswamy, Penn, Saidel, & Chisolm, 1997). Rangaswamy et al. (1997) found that oxLDL, but not native LDL or control PBS, increased barrier permeability in the aortic vasculature of rats in vivo. Similarly, both POVPC and PGPC were implicated in enhanced EC barrier permeability (Birukova et al., 2013), increased monocyte-endothelial cell adhesion (Huber et al., 2002) and was significantly increased in plaques from rabbits fed a high fat rather than a regular chow diet (Watson et al., 1997). The disruption of the EC barrier has been attributed to increased actin stress fiber formation and cell contraction due to m-oxLDL, but not native oxLDL, exposure in HUVECs (Essler et al., 1999). OxLDL activated the RhoA/ROCK pathway,

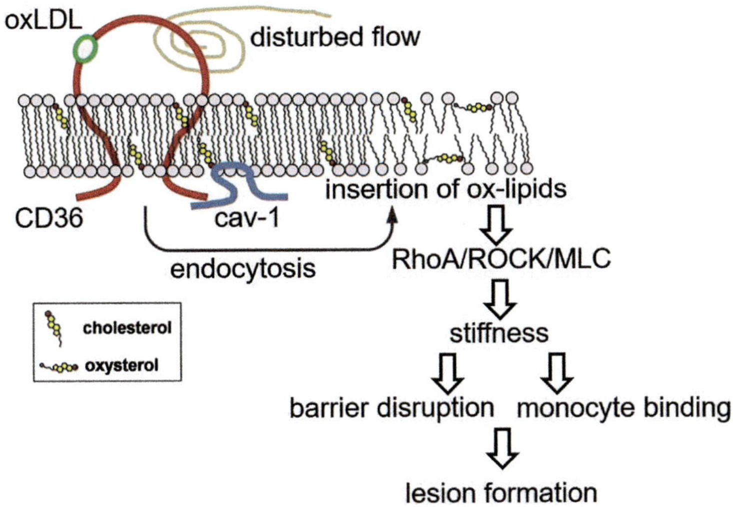

Fig. 5 Schematic pathway of oxLDL/CD36/RhoA/ROCK/MLC in EC stiffening and the functional outcomes.

inhibited MLC (myosin light chain) phosphatase which in turn promoted the phosphorylation of MLC (Essler et al., 1999), both processes which have been shown to be required for oxLDL-induced EC stiffening (Oh et al., 2016). Furthermore, activating the RhoA/ROCK/MLC pathway has been shown to enhance transendothelial migration of leukocytes (Schaefer & Hordijk, 2015) leading to lipid-laden foam cell formation, a beginning step in atherosclerotic plaque development (Fig. 5).

7. Conclusions

Overall, the investigation into endothelial stiffness of the aortic vasculature is an emerging topic shown to have significance in the onset and progression of atherosclerosis and cardiovascular events: from its role in the initial steps of monocyte adhesion and transcytosis to its potential role in modulating plaque stability. Here we discuss the impact of oxidized lipids, dyslipidemia, hemodynamics, substrate stiffness and aging have on the control of endothelial stiffness. Although a growing amount of evidence highlights the importance of these topics, as well as specific mechanistic pathways, as important contributors in endothelial stiffness, there are still some fundamental questions that remain unanswered. For example, at what levels of endothelial stiffness should be considered physiologically healthy as compared to pathological? Also, once a pathological condition such as sub-endothelial plaque formation has been created, would control over its endothelial layer help in preventing plaque rupture and eminent stroke? It is hoped that this chapter has provided a comprehensive review into the current knowledge of endothelial stiffness under physiological and pathological conditions which can begin to answer some of these fundamental questions.

References

Adachi, H., & Tsujimoto, M. (2006). Endothelial scavenger receptors. *Progress in Lipid Research*, *45*, 379–404.

Ayee, M. A. A., LeMaster, E., Shentu, T. P., Singh, D. K., Barbera, N., Soni, D., et al. (2017). Molecular-scale biophysical modulation of an endothelial membrane by oxidized phospholipids. *Biophysical Journal*, *112*, 325–338.

Birukov, K. G. (2009). Small GTPases in mechanosensitive regulation of endothelial barrier. *Microvascular Research*, *77*, 46–52.

Birukov, K. G., Bochkov, V. N., Birukova, A. A., Kawkitinarong, K., Rios, A., Leitner, A., et al. (2004). Epoxycyclopentenone-containing oxidized phospholipids restore endothelial barrier function via Cdc42 and Rac. *Circulation Research*, *95*, 892–901.

Birukova, A. A., Tian, X., Cokic, I., Beckham, Y., Gardel, M. L., & Birukov, K. G. (2013). Endothelial barrier disruption and recovery is controlled by substrate stiffness. *Microvascular Research*, *87*, 50–57.

Brown, P. M., Kennedy, D. J., Morton, R. E., & Febbraio, M. (2015). CD36/SR-B2-TLR2 dependent pathways enhance Porphyromonas gingivalis mediated atherosclerosis in the Ldlr KO mouse model. *PLoS One*, *10*, e0125126.

Byfield, F. J., Aranda-Espinoza, H., Romanenko, V. G., Rothblat, G. H., & Levitan, I. (2004). Cholesterol depletion increases membrane stiffness of aortic endothelial cells. *Biophysical Journal*, *87*, 3336–3343.

Byfield, F. J., Reen, R. K., Shentu, T. P., Levitan, I., & Gooch, K. J. (2009). Endothelial actin and cell stiffness is modulated by substrate stiffness in 2D and 3D. *Journal of Biomechanics*, *42*, 1114–1119.

Byfield, F. J., Tikku, S., Rothblat, G. H., Gooch, K. J., & Levitan, I. (2006). OxLDL increases endothelial stiffness, force generation, and network formation. *Journal of Lipid Research*, *47*, 715–723.

Chai, C. K., Speelman, L., Oomens, C. W., & Baaijens, F. P. (2014). Compressive mechanical properties of atherosclerotic plaques—Indentation test to characterise the local anisotropic behaviour. *Journal of Biomechanics*, *47*, 784–792.

Cherfils, J., & Zeghouf, M. (2013). Regulation of small GTPases by GEFs, GAPs, and GDIs. *Physiological Reviews*, *93*, 269–309.

Chistiakov, D. A., Orekhov, A. N., & Bobryshev, Y. V. (2015). Endothelial barrier and its abnormalities in cardiovascular disease. *Frontiers in Physiology*, *6*, 365.

Chiu, J. J., & Chien, S. (2011). Effects of disturbed flow on vascular endothelium: Pathophysiological basis and clinical perspectives. *Physiological Reviews*, *91*, 327–387.

Claesson-Welsh, L. (2015). Vascular permeability—The essentials. *Upsala Journal of Medical Sciences*, *120*, 135–143.

Collins, S., Martin, T. L., Surwit, R. S., & Robidoux, J. (2004). Genetic vulnerability to diet-induced obesity in the C57BL/6J mouse: Physiological and molecular characteristics. *Physiology & Behavior*, *81*, 243–248.

Collins, C., Osborne, L. D., Guilluy, C., Chen, Z., O'Brien, E. T., 3rd, Reader, J. S., et al. (2014). Haemodynamic and extracellular matrix cues regulate the mechanical phenotype and stiffness of aortic endothelial cells. *Nature Communications, 5*, 3984.

Collins, C., & Tzima, E. (2011). Hemodynamic forces in endothelial dysfunction and vascular aging. *Experimental Gerontology, 46*, 185–188.

Daiber, A., Steven, S., Weber, A., Shuvaev, V. V., Muzykantov, V. R., Laher, I., et al. (2017). Targeting vascular (endothelial) dysfunction. *British Journal of Pharmacology, 174*, 1591–1619.

Dawson, D. W., Pearce, S. F., Zhong, R., Silverstein, R. L., Frazier, W. A., & Bouck, N. P. (1997). CD36 mediates the in vitro inhibitory effects of thrombospondin-1 on endothelial cells. *The Journal of Cell Biology, 138*, 707–717.

DePaola, N., Davies, P. F., Pritchard, W. F., Jr., Florez, L., Harbeck, N., & Polacek, D. C. (1999). Spatial and temporal regulation of gap junction connexin43 in vascular endothelial cells exposed to controlled disturbed flows in vitro. *Proceedings of the National Academy of Sciences of the United States of America, 96*, 3154–3159.

Discher, D. E., Janmey, P., & Wang, Y. L. (2005). Tissue cells feel and respond to the stiffness of their substrate. *Science, 310*, 1139–1143.

Ehara, S., Ueda, M., Naruko, T., Haze, K., Itoh, A., Otsuka, M., et al. (2001). Elevated levels of oxidized low density lipoprotein show a positive relationship with the severity of acute coronary syndromes. *Circulation, 103*, 1955–1960.

Elosegui-Artola, A., Oria, R., Chen, Y., Kosmalska, A., Perez-Gonzalez, C., Castro, N., et al. (2016). Mechanical regulation of a molecular clutch defines force transmission and transduction in response to matrix rigidity. *Nature Cell Biology, 18*, 540–548.

Endemann, G., Stanton, L. W., Madden, K. S., Bryant, C. M., White, R. T., & Protter, A. A. (1993). CD36 is a receptor for oxidized low density lipoprotein. *The Journal of Biological Chemistry, 268*, 11811–11816.

Essler, M., Retzer, M., Bauer, M., Heemskerk, J. W., Aepfelbacher, M., & Siess, W. (1999). Mildly oxidized low density lipoprotein induces contraction of human endothelial cells through activation of Rho/Rho kinase and inhibition of myosin light chain phosphatase. *The Journal of Biological Chemistry, 274*, 30361–30364.

Fan, J., Kitajima, S., Watanabe, T., Xu, J., Zhang, J., Liu, E., et al. (2015). Rabbit models for the study of human atherosclerosis: From pathophysiological mechanisms to translational medicine. *Pharmacology & Therapeutics, 146*, 104–119.

Febbraio, M., Hajjar, D. P., & Silverstein, R. L. (2001). CD36: A class B scavenger receptor involved in angiogenesis, atherosclerosis, inflammation, and lipid metabolism. *The Journal of Clinical Investigation, 108*, 785–791.

Febbraio, M., Podrez, E. A., Smith, J. D., Hajjar, D. P., Hazen, S. L., Hoff, H. F., et al. (2000). Targeted disruption of the class B scavenger receptor CD36 protects against atherosclerotic lesion development in mice. *The Journal of Clinical Investigation, 105*, 1049–1056.

Fessler, M. B., Arndt, P. G., Frasch, S. C., Lieber, J. G., Johnson, C. A., Murphy, R. C., et al. (2004). Lipid rafts regulate lipopolysaccharide-induced activation of Cdc42 and inflammatory functions of the human neutrophil. *The Journal of Biological Chemistry, 279*, 39989–39998.

Fleming, I., Mohamed, A., Galle, J., Turchanowa, L., Brandes, R. P., Fisslthaler, B., et al. (2005). Oxidized low-density lipoprotein increases superoxide production by endothelial nitric oxide synthase by inhibiting PKCalpha. *Cardiovascular Research, 65*, 897–906.

Gavara, N., & Chadwick, R. S. (2016). Relationship between cell stiffness and stress fiber amount, assessed by simultaneous atomic force microscopy and live-cell fluorescence imaging. *Biomechanics and Modeling in Mechanobiology, 15*, 511–523.

Getz, G. S., & Reardon, C. A. (2012). Animal models of atherosclerosis. *Arteriosclerosis, Thrombosis, and Vascular Biology, 32*, 1104–1115.

Gimbrone, M. A., Jr., & Garcia-Cardena, G. (2013). Vascular endothelium, hemodynamics, and the pathobiology of atherosclerosis. *Cardiovascular Pathology*, *22*, 9–15.

Goldstein, J. L., & Brown, M. S. (2009). The LDL receptor. *Arteriosclerosis, Thrombosis, and Vascular Biology*, *29*, 431–438.

Gray, K. M., & Stroka, K. M. (2017). Vascular endothelial cell mechanosensing: New insights gained from biomimetic microfluidic models. *Seminars in Cell & Developmental Biology*, *71*, 106–117.

Guy, E., Kuchibhotla, S., Silverstein, R., & Febbraio, M. (2007). Continued inhibition of atherosclerotic lesion development in long term Western diet fed CD36o/apoEo mice. *Atherosclerosis*, *192*, 123–130.

Hassanisaber, H., Rouleau, L., & Faucheux, N. (2019). Effect of BMP-9 on endothelial cells and its role in atherosclerosis. *Frontiers in Bioscience (Landmark Ed)*, *24*, 994–1023.

Hayashi, K., & Higaki, M. (2017). Stiffness of intact endothelial cells from fresh aortic bifurcations of atherosclerotic rabbits-atomic force microscopic study. *Journal of Cellular Physiology*, *232*, 7–13.

Henry, J. J. D., Yu, J., Wang, A., Lee, R., Fang, J., & Li, S. (2017). Engineering the mechanical and biological properties of nanofibrous vascular grafts for in situ vascular tissue engineering. *Biofabrication*, *9*, 035007.

Holinstat, M., Mehta, D., Kozasa, T., Minshall, R. D., & Malik, A. B. (2003). Protein kinase Calpha-induced p115RhoGEF phosphorylation signals endothelial cytoskeletal rearrangement. *The Journal of Biological Chemistry*, *278*, 28793–28798.

Huber, J., Boechzelt, H., Karten, B., Surboeck, M., Bochkov, V. N., Binder, B. R., et al. (2002). Oxidized cholesteryl linoleates stimulate endothelial cells to bind monocytes via the extracellular signal-regulated kinase 1/2 pathway. *Arteriosclerosis, Thrombosis, and Vascular Biology*, *22*, 581–586.

Huveneers, S., & Danen, E. H. (2009). Adhesion signaling—Crosstalk between integrins, Src and Rho. *Journal of Cell Science*, *122*, 1059–1069.

Huynh, J., Nishimura, N., Rana, K., Peloquin, J. M., Califano, J. P., Montague, C. R., et al. (2011). Age-related intimal stiffening enhances endothelial permeability and leukocyte transmigration. *Science Translational Medicine*, *3*, 112ra122.

Itabe, H. (2012). Oxidized low-density lipoprotein as a biomarker of in vivo oxidative stress: From atherosclerosis to periodontitis. *Journal of Clinical Biochemistry and Nutrition*, *51*, 1–8.

Jalali, S., Tafazzoli-Shadpour, M., Haghighipour, N., Omidvar, R., & Safshekan, F. (2015). Regulation of endothelial cell adherence and elastic modulus by substrate stiffness. *Cell Communication & Adhesion*, *22*, 79–89.

Karki, P., & Birukova, A. A. (2018). Substrate stiffness-dependent exacerbation of endothelial permeability and inflammation: Mechanisms and potential implications in ALI and PH (2017 Grover conference series). *Pulmonary Circulation*, *8*, 2045894018773044.

Kohn, J. C., Chen, A., Cheng, S., Kowal, D. R., King, M. R., & Reinhart-King, C. A. (2016). Mechanical heterogeneities in the subendothelial matrix develop with age and decrease with exercise. *Journal of Biomechanics*, *49*, 1447–1453.

Kohn, J. C., Lampi, M. C., & Reinhart-King, C. A. (2015). Age-related vascular stiffening: Causes and consequences. *Frontiers in Genetics*, *6*, 112.

Kothapalli, D., Liu, S. L., Bae, Y. H., Monslow, J., Xu, T., Hawthorne, E. A., et al. (2012). Cardiovascular protection by ApoE and ApoE-HDL linked to suppression of ECM gene expression and arterial stiffening. *Cell Reports*, *2*, 1259–1271.

Ku, D. N., Giddens, D. P., Zarins, C. K., & Glagov, S. (1985). Pulsatile flow and atherosclerosis in the human carotid bifurcation. Positive correlation between plaque location and low oscillating shear stress. *Arteriosclerosisk*, *5*, 293–302.

Kugelmann, D., Rotkopf, L. T., Radeva, M. Y., Garcia-Ponce, A., Walter, E., & Waschke, J. (2018). Histamine causes endothelial barrier disruption via Ca(2+)-mediated RhoA activation and tension at adherens junctions. *Scientific Reports*, *8*, 13229.

Kutys, M. L., & Chen, C. S. (2016). Forces and mechanotransduction in 3D vascular biology. *Current Opinion in Cell Biology*, *42*, 73–79.

Le Master, E., Fancher, I. S., Lee, J., & Levitan, I. (2018). Comparative analysis of endothelial cell and sub-endothelial cell elastic moduli in young and aged mice: Role of CD36. *Journal of Biomechanics*, *76*, 263–268.

Le Master, E., Huang, R. T., Zhang, C., Bogachkov, Y., Coles, C., Shentu, T. P., et al. (2018). Proatherogenic flow increases endothelial stiffness via enhanced CD36-mediated uptake of oxidized low-density lipoproteins. *Arteriosclerosis, Thrombosis, and Vascular Biology*, *38*, 64–75.

Lee, P. F., Bai, Y., Smith, R. L., Bayless, K. J., & Yeh, A. T. (2013). Angiogenic responses are enhanced in mechanically and microscopically characterized, microbial transglutaminase crosslinked collagen matrices with increased stiffness. *Acta Biomaterialia*, *9*, 7178–7190.

Levitan, I., & Shentu, T. P. (2011). Impact of oxLDL on cholesterol-rich membrane rafts. *Journal of Lipids*, *2011*, 730209.

Levitan, I., Volkov, S., & Subbaiah, P. V. (2010). Oxidized LDL: Diversity, patterns of recognition, and pathophysiology. *Antioxidants & Redox Signaling*, *13*, 39–75.

Li, X., Liu, J., Chen, B., & Fan, L. (2018). A positive feedback loop of profilin-1 and RhoA/ROCK1 promotes endothelial dysfunction and oxidative stress. *Oxidative Medicine and Cellular Longevity*, *2018*, 4169575.

Libby, P., Ridker, P. M., & Hansson, G. K. (2011). Progress and challenges in translating the biology of atherosclerosis. *Nature*, *473*, 317–325.

Lubrano, V., & Balzan, S. (2014). LOX-1 and ROS, inseparable factors in the process of endothelial damage. *Free Radical Research*, *48*, 841–848.

Martinelli, R., Zeiger, A. S., Whitfield, M., Sciuto, T. E., Dvorak, A., Van Vliet, K. J., et al. (2014). Probing the biomechanical contribution of the endothelium to lymphocyte migration: Diapedesis by the path of least resistance. *Journal of Cell Science*, *127*, 3720–3734.

Martorell, J., Santoma, P., Kolandaivelu, K., Kolachalama, V. B., Melgar-Lesmes, P., Molins, J. J., et al. (2014). Extent of flow recirculation governs expression of atherosclerotic and thrombotic biomarkers in arterial bifurcations. *Cardiovascular Research*, *103*, 37–46.

May, J. M., & Qu, Z. C. (2010). Ascorbic acid prevents increased endothelial permeability caused by oxidized low density lipoprotein. *Free Radical Research*, *44*, 1359–1368.

Miyazaki, H., & Hayashi, K. (1999). Atomic force microscopic measurement of the mechanical properties of intact endothelial cells in fresh arteries. *Medical & Biological Engineering & Computing*, *37*, 530–536.

Monson, K. L., Goldsmith, W., Barbaro, N. M., & Manley, G. T. (2003). Axial mechanical properties of fresh human cerebral blood vessels. *Journal of Biomechanical Engineering*, *125*, 288–294.

Monson, K. L., Goldsmith, W., Barbaro, N. M., & Manley, G. T. (2005). Significance of source and size in the mechanical response of human cerebral blood vessels. *Journal of Biomechanics*, *38*, 737–744.

Moore, K. J., Kunjathoor, V. V., Koehn, S. L., Manning, J. J., Tseng, A. A., Silver, J. M., et al. (2005). Loss of receptor-mediated lipid uptake via scavenger receptor A or CD36 pathways does not ameliorate atherosclerosis in hyperlipidemic mice. *The Journal of Clinical Investigation*, *115*, 2192–2201.

Murakami, M., & Simons, M. (2009). Regulation of vascular integrity. *Journal of Molecular Medicine (Berlin, Germany)*, *87*, 571–582.

Nakashima, Y., Plump, A. S., Raines, E. W., Breslow, J. L., & Ross, R. (1994). ApoE-deficient mice develop lesions of all phases of atherosclerosis throughout the arterial tree. *Arteriosclerosis and Thrombosis*, *14*, 133–140.

Needham, D., & Nunn, R. S. (1990). Elastic deformation and failure of lipid bilayer membranes containing cholesterol. *Biophysical Journal, 58*, 997–1009.

Oh, M. J., Zhang, C., LeMaster, E., Adamos, C., Berdyshev, E., Bogachkov, Y., et al. (2016). Oxidized LDL signals through Rho-GTPase to induce endothelial cell stiffening and promote capillary formation. *Journal of Lipid Research, 57*, 791–808.

Paigen, B. (1995). Genetics of responsiveness to high-fat and high-cholesterol diets in the mouse. *The American Journal of Clinical Nutrition, 62*, 458S–462S.

Palotie, A., Tryggvason, K., Peltonen, L., & Seppa, H. (1983). Components of subendothelial aorta basement membrane. Immunohistochemical localization and role in cell attachment. *Laboratory Investigation, 49*, 362–370.

Park, Y. M. (2014). CD36, a scavenger receptor implicated in atherosclerosis. *Experimental & Molecular Medicine, 46*, e99.

Park, Y., Booth, F. W., Lee, S., Laye, M. J., & Zhang, C. (2012). Physical activity opposes coronary vascular dysfunction induced during high fat feeding in mice. *The Journal of Physiology, 590*, 4255–4268.

Predescu, S. A., Predescu, D. N., & Malik, A. B. (2007). Molecular determinants of endothelial transcytosis and their role in endothelial permeability. *American Journal of Physiology. Lung Cellular and Molecular Physiology, 293*, L823–L842.

Rangaswamy, S., Penn, M. S., Saidel, G. M., & Chisolm, G. M. (1997). Exogenous oxidized low-density lipoprotein injures and alters the barrier function of endothelium in rats in vivo. *Circulation Research, 80*, 37–44.

Rebuffe-Scrive, M., Surwit, R., Feinglos, M., Kuhn, C., & Rodin, J. (1993). Regional fat distribution and metabolism in a new mouse model (C57BL/6J) of non-insulin-dependent diabetes mellitus. *Metabolism, 42*, 1405–1409.

Rohwedder, I., Montanez, E., Beckmann, K., Bengtsson, E., Duner, P., Nilsson, J., et al. (2012). Plasma fibronectin deficiency impedes atherosclerosis progression and fibrous cap formation. *EMBO Molecular Medicine, 4*, 564–576.

Sato, M., Levesque, M. J., & Nerem, R. M. (1987). Micropipette aspiration of cultured bovine aortic endothelial cells exposed to shear stress. *Arteriosclerosis, 7*, 276–286.

Sato, M., & Ohashi, T. (2005). Biorheological views of endothelial cell responses to mechanical stimuli. *Biorheology, 42*, 421–441.

Sato, M., Ohshima, N., & Nerem, R. M. (1996). Viscoelastic properties of cultured porcine aortic endothelial cells exposed to shear stress. *Journal of Biomechanics, 29*, 461–467.

Sato, M., Theret, D. P., Wheeler, L. T., Ohshima, N., & Nerem, R. M. (1990). Application of the micropipette technique to the measurement of cultured porcine aortic endothelial cell viscoelastic properties. *Journal of Biomechanical Engineering, 112*, 263–268.

Schaefer, A., & Hordijk, P. L. (2015). Cell-stiffness-induced mechanosignaling—A key driver of leukocyte transendothelial migration. *Journal of Cell Science, 128*, 2221–2230.

Schaefer, A., Te Riet, J., Ritz, K., Hoogenboezem, M., Anthony, E. C., Mul, F. P., et al. (2014). Actin-binding proteins differentially regulate endothelial cell stiffness, ICAM-1 function and neutrophil transmigration. *Journal of Cell Science, 127*, 4470–4482.

Shen, Q., Wu, M. H., & Yuan, S. Y. (2009). Endothelial contractile cytoskeleton and microvascular permeability. *Cell Health and Cytoskeleton, 2009*, 43–50.

Shentu, T. P., Singh, D. K., Oh, M. J., Sun, S., Sadaat, L., Makino, A., et al. (2012). The role of oxysterols in control of endothelial stiffness. *Journal of Lipid Research, 53*, 1348–1358.

Shentu, T. P., Titushkin, I., Singh, D. K., Gooch, K. J., Subbaiah, P. V., Cho, M., et al. (2010). oxLDL-induced decrease in lipid order of membrane domains is inversely correlated with endothelial stiffness and network formation. *American Journal of Physiology. Cell Physiology, 299*, C218–C229.

Shimokawa, H., Sunamura, S., & Satoh, K. (2016). RhoA/Rho-kinase in the cardiovascular system. *Circulation Research, 118*, 352–366.

Strassheim, D., Gerasimovskaya, E., Irwin, D., Dempsey, E. C., Stenmark, K., & Karoor, V. (2019). RhoGTPase in vascular disease. *Cell, 8*, 551.

Stroka, K. M., & Aranda-Espinoza, H. (2010). A biophysical view of the interplay between mechanical forces and signaling pathways during transendothelial cell migration. *The FEBS Journal, 277*, 1145–1158.

Sugimoto, K., Ishibashi, T., Sawamura, T., Inoue, N., Kamioka, M., Uekita, H., et al. (2009). LOX-1-MT1-MMP axis is crucial for RhoA and Rac1 activation induced by oxidized low-density lipoprotein in endothelial cells. *Cardiovascular Research, 84*, 127–136.

Sun, M., Northup, N., Marga, F., Huber, T., Byfield, F. J., Levitan, I., et al. (2007). The effect of cellular cholesterol on membrane-cytoskeleton adhesion. *Journal of Cell Science, 120*, 2223–2231.

Tracqui, P., Broisat, A., Toczek, J., Mesnier, N., Ohayon, J., & Riou, L. (2011). Mapping elasticity moduli of atherosclerotic plaque in situ via atomic force microscopy. *Journal of Structural Biology, 174*, 115–123.

Vandenbroucke, E., Mehta, D., Minshall, R., & Malik, A. B. (2008). Regulation of endothelial junctional permeability. *Annals of the New York Academy of Sciences, 1123*, 134–145.

VanderBurgh, J. A., & Reinhart-King, C. A. (2018). The role of age-related intimal remodeling and stiffening in atherosclerosis. *Advances in Pharmacology, 81*, 365–391.

Watson, A. D., Leitinger, N., Navab, M., Faull, K. F., Horkko, S., Witztum, J. L., et al. (1997). Structural identification by mass spectrometry of oxidized phospholipids in minimally oxidized low density lipoprotein that induce monocyte/endothelial interactions and evidence for their presence in vivo. *The Journal of Biological Chemistry, 272*, 13597–13607.

Woodrum, D. A., Romano, A. J., Lerman, A., Pandya, U. H., Brosh, D., Rossman, P. J., et al. (2006). Vascular wall elasticity measurement by magnetic resonance imaging. *Magnetic Resonance in Medicine, 56*, 593–600.

Wu, C., Huang, R. T., Kuo, C. H., Kumar, S., Kim, C. W., Lin, Y. C., et al. (2015). Mechanosensitive PPAP2B regulates endothelial responses to atherorelevant hemodynamic forces. *Circulation Research, 117*, e41–e53.

Yeh, Y. C., Ling, J. Y., Chen, W. C., Lin, H. H., & Tang, M. J. (2017). Mechanotransduction of matrix stiffness in regulation of focal adhesion size and number: Reciprocal regulation of caveolin-1 and beta1 integrin. *Scientific Reports*, 7, 15008.

Yeung, T., Georges, P. C., Flanagan, L. A., Marg, B., Ortiz, M., Funaki, M., et al. (2005). Effects of substrate stiffness on cell morphology, cytoskeletal structure, and adhesion. *Cell Motility and the Cytoskeleton, 60*, 24–34.

Yi, B., Shen, Y., Tang, H., Wang, X., & Zhang, Y. (2020). Stiffness of the aligned fibers affects structural and functional integrity of the oriented endothelial cells. *Acta Biomaterialia, 108*, 237–249.

Yla-Herttuala, S., Palinski, W., Rosenfeld, M. E., Parthasarathy, S., Carew, T. E., Butler, S., et al. (1989). Evidence for the presence of oxidatively modified low density lipoprotein in atherosclerotic lesions of rabbit and man. *The Journal of Clinical Investigation, 84*, 1086–1095.

Yoshimoto, R., Fujita, Y., Kakino, A., Iwamoto, S., Takaya, T., & Sawamura, T. (2011). The discovery of LOX-1, its ligands and clinical significance. *Cardiovascular Drugs and Therapy, 25*, 379–391.

Zhang, C., Adamos, C., Oh, M. J., Baruah, J., Ayee, M. A. A., Mehta, D., et al. (2017). oxLDL induces endothelial cell proliferation via rho/ROCK/Akt/p27(kip1) signaling: Opposite effects of oxLDL and cholesterol loading. *American Journal of Physiology. Cell Physiology, 313*, C340–C351.

CHAPTER SIX

Vascular smooth muscle stiffness and its role in aging

Andreea Trache[a,b,*], Michael P. Massett[c], and Christopher R. Woodman[d]

[a]Department of Medical Physiology, Texas A&M University Health Science Center, Bryan, TX, United States
[b]Department of Biomedical Engineering, Texas A&M University, College Station, TX, United States
[c]Department of Kinesiology and Sport Management, Texas Tech University, Lubbock, TX, United States
[d]Department of Health and Kinesiology, Texas A&M University, College Station, TX, United States
[*]Corresponding author: e-mail address: trache@tamu.edu

Contents

Abstract

Vascular smooth muscle cells (VSMC) are now considered important contributors to the pathophysiological and biophysical mechanisms underlying arterial stiffening in aging. Here, we review mechanisms whereby VSMC stiffening alters vascular function and contributes to the changes in vascular stiffening observed in aging and cardiovascular disease. Vascular stiffening in arterial aging was historically associated with changes in the

Current Topics in Membranes, Volume 86
ISSN 1063-5823
https://doi.org/10.1016/bs.ctm.2020.08.008

extracellular matrix; however, new evidence suggests that endothelial and vascular smooth muscle cell stiffness also contribute to overall blood vessel stiffness. Furthermore, VSMC play an integral role in regulating matrix deposition and vessel wall contractility via interaction between the actomyosin contractile unit and adhesion structures that anchor the cell within the extracellular matrix. Aged-induce phenotypic modulation of VSMC from a contractile to a synthetic phenotype is associated with decreased cellular contractility and increased cell stiffness. Aged VSMC also display reduced mechanosensitivity and adaptation to mechanical signals from their microenvironment due to impaired intracellular signaling. Finally, evidence for decreased contractility in arteries from aged animals demonstrate that changes at the cellular level result in decreased functional properties at the tissue level.

1. Introduction

Cardiovascular disease is the leading cause of death in the US. Aging is an established risk factor for the development of cardiovascular disease (Najjar, Scuteri, & Lakatta, 2005). By 2030 the US will have almost 73 million people or 20% of the population above the age of 65 (Ortman, Velkoff, & Hogan, 2014). At the same time, the annual cost of cardiovascular disease is expected to increase by 238% for people over 65 years old (Heidenreich et al., 2011). Slowing or preventing age-induced arterial vasomotor dysfunction, which is associated with an increased risk for cardiovascular diseases (CVDs), such as hypertension and atherosclerosis, remains a significant clinical challenge (Lakatta, 2015).

Arterial aging is directly linked with arterial stiffness associated with vessel wall remodeling, smooth muscle plasticity and endothelial dysfunction (Wang, Monticone, & Lakatta, 2010). Contributions to vascular stiffening along the vascular tree include material properties such as intrinsic wall stiffness, and functional arterial heterogeneity. Age-altered extracellular matrix architecture is well recognized as an important player in vascular stiffness (Dingemans, Teeling, Lagendijk, & Becker, 2000; Duca et al., 2016; Wagenseil & Mecham, 2009; Zhang et al., 2016) along with endothelial dysfunction (Barton et al., 1997; Celermajer et al., 1994; Herrera, Mingorance, Rodriguez-Rodriguez, & Alvarez de Sotomayor, 2010; Muller-Delp, Spier, Ramsey, & Delp, 2002; Muller-Delp, Spier, Ramsey, Lesniewski, et al., 2002). Despite wide recognition of age-associated arterial stiffness and progressive decline of arterial function with aging, limited research has been directed toward understanding the mechanistic role of vascular smooth muscle cell (VSMC) biomechanics on arterial stiffening. Recent studies uncovered important contributors to the pathophysiological and biophysical mechanisms underlying arterial stiffening in aging (Briones, Salaices, & Vila, 2007; Huveneers, Daemen, & Hordijk, 2015; Mitchell et al., 2004; Saphirstein et al., 2013; Zhu et al., 2012, 2018). Biophysical properties of vascular smooth muscle and endothelial cells have emerged as active players in regulating arterial stiffness (Kusche-Vihrog et al., 2011; Lang, 2011; Sehgel, Sun, et al., 2015; Sehgel, Vatner, & Meininger, 2015). In this chapter, we will discuss these new concepts with an emphasis on the contribution of VSMC stiffness to vascular aging.

2. Aging as an independent risk factor in cardiovascular disease

A recent area of interest in the link between aging and CVD is vascular aging—a change in the structure and function of the cardiovascular system leading to decreased function and increased susceptibility to CVD. Arterial aging results in progressive changes giving rise to a reduction in elasticity of the vessel wall and an impaired ability of aged blood vessels to control local blood flow and pressure. At the microscopic level, this translates to decreased VSMC responsivity to mechanical stimuli (i.e., mechanosensitivity). This impairment, in turn, induces compensatory hypertrophic or hyperplastic remodeling of aged arteries (Briones et al., 2007; Najjar et al., 2005; O'Rourke & Hashimoto, 2007). Two primary modifications that define vascular aging are endothelial dysfunction and arterial stiffening.

2.1 Endothelial dysfunction

As blood vessels age, the ability of endothelial cells to regulate vascular tone, respond to changes in blood flow and modulate antiatherogenic processes decline (Celermajer et al., 1994; Vita et al., 1990). Conversely, oxidative stress and inflammatory cytokines increase with age predisposing individuals to hypertension, atherosclerosis and peripheral vascular disease (Madamanchi, Vendrov, & Runge, 2005). Endothelial dysfunction with aging can be attributed to an overall decrease in nitric oxide (NO) bioavailability resulting from a combination of insufficient NO production and increased NO degradation by oxidative stress. NO maintains vascular homeostasis by regulating vascular tone and mediating antiatherogenic processes. The enzyme endothelial nitric oxide synthase (eNOS) converts L-arginine to L-citrulline and NO (Moncada & Higgs, 1993; Palmer, Ashton, & Moncada, 1988). NO production is dependent on the availability of substrate (L-arginine), cofactors involved in the reaction (e.g., tetrahydrobiopterin (BH4)), eNOS protein content, and eNOS enzyme activity. These factors have been shown to be altered in aged blood vessels. Evidence for the reduction of L-arginine is limited and inconsistent (Bode-Boger et al., 2003; Delp, Behnke, Spier, Wu, & Muller-Delp, 2008; Gates, Boucher, Silver, Monahan, & Seals, 2007). In contrast, a reduction in NO bioavailability due to BH4 deficiency is well-supported (Delp et al., 2008; Eskurza, Myerburgh, Kahn, & Seals, 2005; Heitzer et al., 2000, Vasquez-Vivar, Kalyanaraman, & Martasek, 2003). A reduction of BH4 can lead to eNOS uncoupling and production of superoxide at the cost of NO. Evidence for reduced eNOS protein content and activity with aging is mixed. Some evidence affirms a decrease in eNOS (Barton et al., 1997; Csiszar et al., 2002; Woodman, Price, & Laughlin, 2005). In contrast, other investigators have reported an increase or no change in eNOS protein or mRNA with aging (Donato et al., 2009; Goettsch et al., 2001; Rodriguez-Manas et al., 2009; Woodman, Price, & Laughlin, 2002). Van der Loo et al. (2000) reported that eNOS activity was sevenfold higher in aorta from aged rats when compared to young rats, whereas other investigators reported that eNOS activity declined with age (Berkowitz et al., 2003; White et al., 2006). It is important to note, however, that the age-related increase in eNOS activity reported by Van der Loo et al. (2000) was associated with increased production of superoxide and reduced NO availability indicating that the eNOS was uncoupled. In addition to eNOS uncoupling, NO bioavailability can be reduced through the interaction of NO with superoxide leading to the formation of the potent oxidant peroxynitrite (Francia et al., 2004; Rodriguez-Manas et al., 2009; Van der Loo et al., 2000). Reactive oxygen species can also be produced by vascular NADPH oxidase and the mitochondrial respiratory chain (Donato, Eskurza, et al., 2007; Rodriguez-Manas et al., 2009), further reducing the bioavailability of NO and causing oxidative damage. Thus, in all of these studies, the net effect of aging was a reduced NO bioavailability and endothelial dysfunction.

The loss of NO bioavailability, endothelial dysfunction and changes in arterial stiffness are interconnected. For example, inhibition of eNOS with N^G-monomethyl-L-arginine (L-NMMA) decreased brachial artery compliance and increased pulse wave velocity (Kinlay et al., 2001). In contrast, infusion of the NO donor nitroglycerin increased compliance. Thus, evidence supports the premise that NO regulates large artery distensibility (Kinlay et al., 2001; Schmitt et al., 2005; Wilkinson et al., 2002). Therefore, the loss of NO bioavailability with aging can contribute to increased arterial stiffness. This increase in stiffness can then increase the hemodynamic forces on the endothelium leading to further damage and endothelial dysfunction (Franklin et al., 1997).

2.2 Arterial stiffness

Large elastic arteries buffer the cyclical changes in arterial blood pressure. Arterial elasticity is determined by vascular smooth muscle tone and structural characteristics of the vessel wall. Arteries stiffen with age and disease, thereby reducing the ability of the vessel to dampen the arterial blood pressure wave as it travels toward the microcirculation (Sun, 2015). An increase in pulse pressure that is transmitted from large vessels to small vessels and an overall increase in pressure in small vessels can lead to end-organ damage (O'Rourke & Safar, 2005).

Arterial stiffness can be assessed *in vivo* by estimating arterial compliance from stroke volume and blood pressure or measuring pulse wave velocity (PWV). The rate at which arterial blood pressure waves are propagated through the arterial tree is higher in less compliant blood vessels (O'Rourke & Hashimoto, 2007). Measurement of carotid-femoral PWV is now considered the gold standard for clinical assessment of arterial stiffness (Van Bortel et al., 2012). Aortic stiffness measured as PWV is an independent risk factor and predictor of coronary heart disease, stroke, and CVD events (Ben-Shlomo et al., 2014). The predictive ability of PWV is higher in high-risk individuals (e.g., subjects with renal disease, diabetes, coronary artery disease, etc.) (Vlachopoulos, Aznaouridis, & Stefanadis, 2010). An increase of 1 m/s in aortic PWV is associated with a 14–15% increase in cardiovascular events and all-cause mortality (Vlachopoulos et al., 2010). Ben-Shlomo et al. (2014) estimated that a 1 m/s increase in PWV is associated with a 7% increase risk of a cardiovascular event in a 60-year-old man with no major risk factors. Collectively, these results suggest that understanding the mechanisms contributing to arterial stiffness is critical for developing strategies to reduce long-term CVD risk.

Clinical studies have shown that arterial stiffness is directly related to aging (AlGhatrif et al., 2013; Mitchell et al., 2004) and hypertension (Tomiyama et al., 2004). Vascular stiffening in arterial aging was historically associated with matrix dysregulation by an increase in collagen and elastin fragmentation (Fig. 1) (Duca et al., 2016; Wagenseil et al., 2005). Age-induced structural and functional changes are, generally, associated with an increase of collagen:elastin ratio and a decrease in the number of

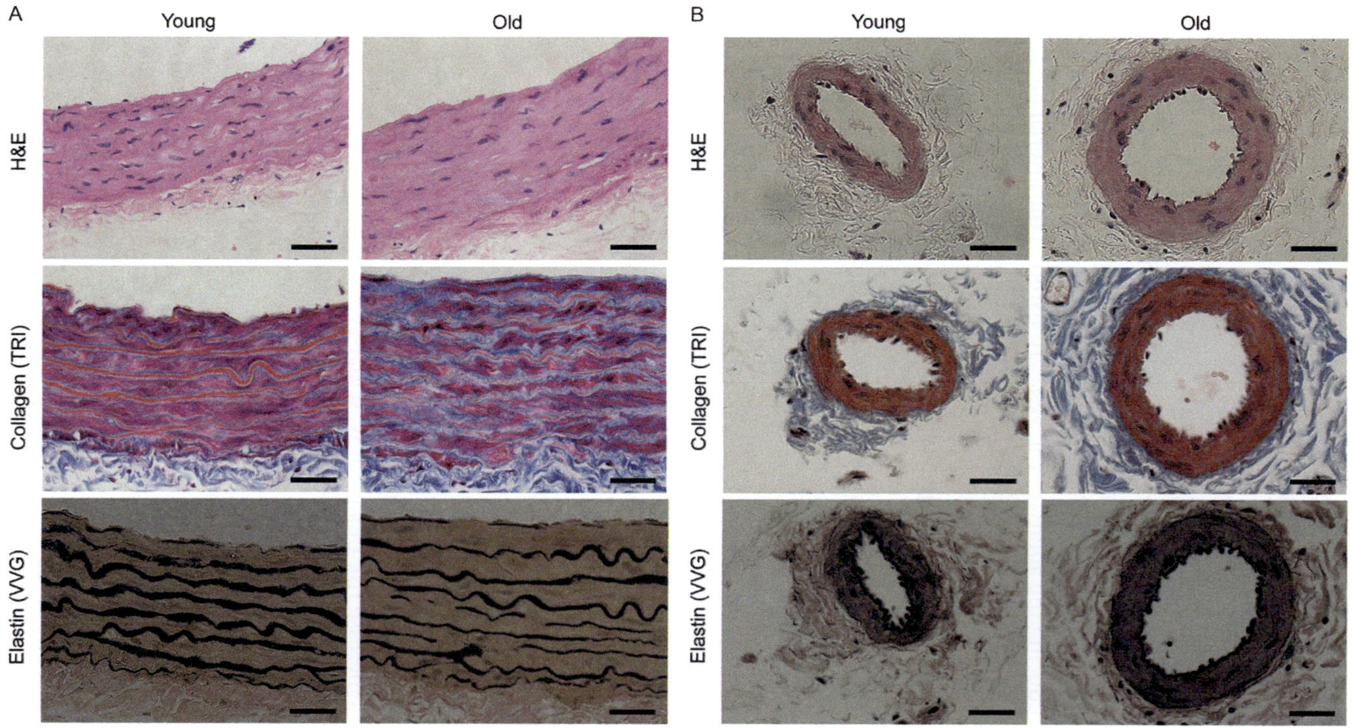

Fig. 1 Histological sections for (A) aorta, and (B) soleus resistance arteries isolated from young (3 months) and old (24 months) Fischer 344 rats. Histological sections 5 μm thick were stained through standard techniques for hematoxylin and eosin (H&E), trichrome stain (TRI, collagen), and Verhoeff-Van Gieson stain (VVG, elastin). Size bar is 10 μm.

VSMC that control the active tone of the vessels (Fritze et al., 2012; Yamada, Sakata, Wada, Tashiro, & Tayama, 2015). However, Sehgel, Vatner, and Meininger (2015) showed that in aortas from aged monkeys there was no change in collagen with aging, but there was a significant reduction in elastin, which was more pronounced in aged males than females. These structural changes related well with *in vivo* aortic stiffness measurements showing a greater increase in aged males than females. The structural contributions of collagen and elastin to the aortic stiffness directly relate to their mechanical properties. From the matrix side, at low wall stretch elastin fibers dominate the mechanical properties of the vessel wall, while at higher stresses, collagen fibers are activated, bearing the mechanical load (Kohn, Lampi, & Reinhart-King, 2015; Schriefl, Zeindlinger, Pierce, Regitnig, & Holzapfel, 2012).

Collagen and elastin are the main components of the extracellular matrix that have an important role in determining the biomechanical properties of blood vessels. Other extracellular matrix proteins such as fibronectin and laminin are also present in the vessel wall and might influence vessel biomechanics in health and disease. However, their contribution to age-related vascular stiffness is unclear. For example, isoform expression of laminin in the basal membrane of endothelial cells change with aging. Laminin β1 is upregulated and laminin β2 downregulated in the basal membrane of 18 months old mice vs 4 months old mice (Wagner et al., 2018). This switch was associated with decreased endothelial cell adhesion and migration, and could have implications for age-related fibrosis (Wagner et al., 2018). In the medial-adventitial border of rat tail vein and basilar and common carotid arteries laminin immunoreactivity is lower (Gavazzi, Boyle, Edgar, & Cowen, 1995). The level of immunoreactivity correlates with patterns of innervation, suggesting that laminin is related to age-related neurodegeneration. However, the impact of this decrease on vascular function or biomechanics have not been addressed.

Increases in fibronectin are more consistently associated with animal models of aging and hypertension (Bezie et al., 1998; Li, Froehlich, Galis, & Lakatta, 1999; Takasaki, Chobanian, Sarzani, & Brecher, 1990; Wang et al., 2006; Yoon et al., 2016). The increase in collagen:elastin ratio observed in aging and hypertension can also be accompanied by elevated levels of fibronectin (Lakatta, 2013). However, it is unclear whether this change is related to aging, *per se*, or the combination of increasing age and coincident hypertension. In addition to an increase in the amount of fibronectin, evidence supports a role for increased adhesion to fibronectin as a contributing factor to arterial stiffening in aging and hypertension

(Qiu et al., 2010; Sehgel, Sun, et al., 2015). Sehgel, Sun, et al. (2015) showed that the probability of VSMC adhesion via integrins to fibronectin and integrin-fibronectin adhesion force are increased in VSMC from aged rats. These changes were not dependent on blood pressure but exacerbated in VSMC of aged hypertensive rats. These results suggest that extracellular matrix remodeling as well as the interaction of VSMC with the extracellular matrix are altered with aging and likely contribute to increases in arterial stiffness.

In addition to traditional extracellular matrix components, recent work has focused on the role of lipids in regulating vascular stiffness. Although not specific to aging, VSMC exhibit decreased cholesterol efflux and changes in extracellular matrix composition in atherosclerosis. Sanyour et al. (2019, 2020) recently demonstrated an interaction between cell membrane cholesterol and substrate stiffness. These factors act synergistically to affect VSMC cytoskeleton and biomechanics. Depletion of membrane cholesterol decreased VSMC adhesion force, stiffness, and reduced vasoconstriction in rat aorta. In contrast, increasing cholesterol tended to increase adhesion and VSMC stiffness. These results suggest that modulating lipid levels might impact VSMC and vascular stiffness. The role of lipids in regulating vascular stiffness is extensively reviewed in the chapter "The interplay of membrane cholesterol and substrate on vascular smooth muscle biomechanics" by Hanna et al., in this volume.

3. Arterial stiffness and its contribution to aging

Arterial stiffening involves both arterial hemodynamics and cellular mechanosensing. Blood vessel wall stresses (i.e., axial wall stress, circumferential wall stretch) and intraluminal shear stress are the most important biomechanical forces in aging and cardiovascular disease (Fig. 2). The increase in matrix deposition and/or VSMC hyperplasia that leads to intima thickening in aging have detrimental consequences for maintaining the cellular phenotype and vascular function (Lacolley, Regnault, Nicoletti, Li, & Michel, 2012; Lakatta, 2003; Lundberg & Crow, 1999). Increasing evidence supports that changes in the extracellular matrix alone are insufficient to fully account for increased vascular stiffness and that VSMC are important contributors to the mechanisms underlying arterial stiffening and vasomotor dysfunction in aging. Here, emphasis will be placed on discussing structural and functional adaptations of VSMC in aging.

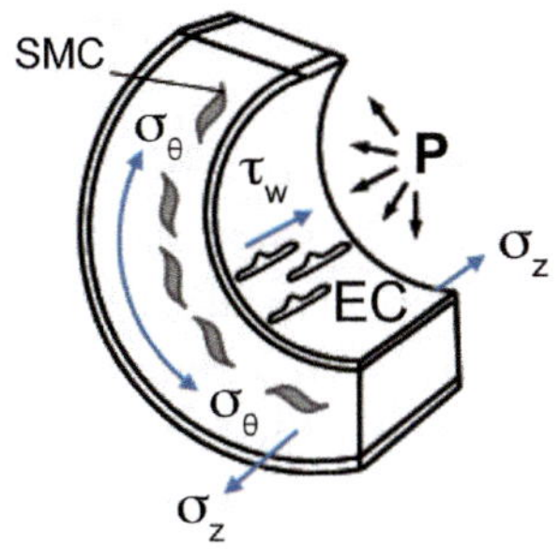

Fig. 2 Distribution of main mechanical stresses in the vascular wall. Axial wall stress (σ_z) represents the longitudinal blood vessel tension that exerts a force along the vessel wall length; circumferential stretch (σ_θ) is due to increased intraluminal pressure of the blood that exerts a force perpendicular to vessel wall surface; flow induced wall shear stress (τ_w) is generated due to friction of the blood against the vessel wall exerting a force parallel with the vessel wall surface. P, intraluminal pressure; EC, endothelial cells; SMC, smooth muscle cells.

3.1 Vessel wall mechanical properties present regional and local heterogeneity

Identification of signals essential for maintaining healthy vascular structure and function throughout the lifespan remains a significant challenge. In large conduit arteries (e.g., aorta and carotid), aging is associated with wall stiffening due to an increase in collagen and fragmentation of elastin, as well as VSMC hyperplasia. In contrast, small resistance arteries contain fewer layers of VSMC and have less matrix than conduit vessels. Some results have shown that age-induced adaptation of resistance arteries and modulation of their vascular structure and function differ from that of conduit vessels (Hanna et al., 2014; Luttrell et al., 2020; Mitchell et al., 2004; Muller-Delp, Spier, Ramsey, Lesniewski, et al., 2002). Moreover, aging is differentially regulated along the vascular tree, which contributes to anatomic heterogeneity of aged vasculature (Barton et al., 1997; Kohn et al., 2016). These findings indicate that vascular aging throughout the arterial network occurs in a non-uniform manner.

VSMC play an important role in regulating matrix deposition in the vessel wall (Lim & Park, 2014; Wagenseil & Mecham, 2009). VSMC in the vessel wall wrap-around the vessel lumen and are sandwiched between layers of extracellular matrix, mainly collagen and elastin. While in conduit vessels (i.e., aorta) there are several distinct layers of VSMC and well-organized collagen and elastin fibers (Fig. 1), in microcirculation (i.e., distal arteries with a diameter less than 100 μm) there are only a few layers of VSM cell and a much-reduced collagen-elastin matrix, with an important decrease in the matrix/cell ratio (Dinardo et al., 2014; Greenwald, 2007). This heterogeneous property of the vessel walls is remarkable among the different segments of the vascular tree (Barton et al., 1997; Augustin & Koh, 2017). Furthermore, there is a relative heterogeneity within each particular segment of the vasculature that is necessary to accommodate local longitudinal changes in length (i.e., tensile/axial stress) as well as pre-existing stress conditions of the respective vascular beds (Humphrey, 2008; Humphrey, Eberth, Dye, Gleason, 2009; Lawrence & Gooch, 2009). Thus, vessel wall mechanical properties present both regional and local heterogeneity.

Understanding the relationship between local tissue passive properties (i.e., stiffness) and their correlation with structural components of the vessel wall is key in elucidating how matrix stiffness affects transmural stresses and VSMC compliance (Hayenga, Trache, Trzeciakowski, & Humphrey, 2011). Local stiffness properties of the aortic wall measured by atomic force microscopy (AFM) in cross-sections of the aortic rings from ApoE−/− mice, an animal model used to study atherosclerotic plaque formation, were paired with histological tissue staining on the same dissection. Local wall stiffness measurements in cross-sections of aortic rings maintained at normal wall tension showed two distinct populations at 9.8 and 76.7 kPa (Fig. 3). Correlating the matrix protein location with stiffness measurements showed that higher stiffness values corresponded to elastin fibers, while the lower values corresponded with the plaque areas. This combined approach represents a valuable tool in elucidating the coupling of discrete material properties with the respective elements that comprise arterial tissue at the microscopic level.

3.2 Vascular cell stiffness contributes to overall vessel stiffness

Previous studies demonstrated differential regulation of the vascular aging process in large and small arteries. In an established animal model of aging in the absence of hypertension, Fischer 344 rats, there is a decrease in cytoskeletal proteins with aging that may play a role in remodeling of the aortic wall (Li, Cheng, Lederer, Froehlich, & Lakatta, 1997). In contrast, groups studying the interactive effects of aging and hypertension on mechanisms of arterial stiffening showed an increase in VSMC stiffness in large conduit vessels (e.g., aorta) in animal models for hypertension (monkeys and spontaneously hypertensive rats) (Sehgel, Sun, et al., 2015; Sehgel, Vatner, & Meininger, 2015). Moreover, the role of RhoA upregulation in arterial

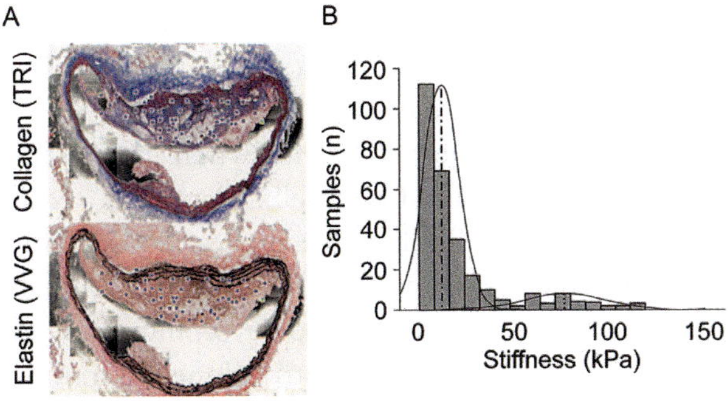

Fig. 3 Histology combined with AFM stiffness measurements. (A) Representative images of histological aorta sections from ApoE mice stained for collagen (TRI) and elastin (VVG) overlaid with brightfield images of the same sections, showing locations (+) for AFM stiffness measurements. (B) Histogram of axial stiffness measurements in ApoE−/− aortic rings. Normal distributions constructed from the mean and standard deviations are outlined by solid curves for the softer and stiffer populations of the experimental stiffness measurements. *Reproduced with permission from Hayenga, H. N., Trache, A., Trzeciakowski, J., & Humphrey, J. D. (2011). Regional atherosclerotic plaque properties in ApoE−/− mice quantified by atomic force, immunofluorescence, and light microscopy.* Journal of Vascular Research, *48(6), 495–504.*

stiffening that precedes hypertension is well recognized (Kaess et al., 2012), and efforts have been made to develop ROCK inhibitors (e.g., fasudil) as pharmacological treatments of hypertension in the clinic (Fujita et al., 2010; Shimokawa & Rashid, 2007). Endothelial cells contribute to the regulation of VSMC contractile function by secreting vasoactive factors (Vanhoutte, Shimokawa, Feletou, & Tang, 2017). It has also been shown that NO production by endothelium, as well as NO-mediated signaling to VSMC, is inhibited by RhoA/ROCK pathway activation (Satoh, Fukumoto, & Shimokawa, 2011; Takemoto, Sun, Hiroki, Shimokawa, & Liao, 2002). Additionally, it has been shown in rat aorta that NO donors inhibit phenylephrine-induced activation of RhoA (Sauzeau et al., 2000), while NO-induced relaxation was related in part to downregulation of RhoA signaling (Chitaley & Webb, 2002). Current treatments in hypertension are directed toward decreasing blood pressure (Dhaun et al., 2008; Schiffrin, 1999), rather than modulating vessel wall compliance (Lakatta & Levy, 2003). The endogenous production of NO by endothelial cells has an important impact on blood pressure regulation (Fels, Oberleithner, & Kusche-Vihrog, 2010). Stiff endothelial cells release less NO. This leads to increased vascular resistance (Oberleithner et al., 2007) and increased blood pressure (Fels et al., 2010).

The term "*stiff endothelial cell syndrome*" was introduced recently in respect to this decrease in production of vasorelaxing factors that follows endothelium stiffening (Lang, 2011; Oberleithner et al., 2007).

VSMC plasticity reflects their ability to switch from a contractile to a synthetic phenotype (Chamley-Campbell, Campbell, & Ross, 1979; Martinez-Lemus, Hill, Bolz, Pohl, & Meininger, 2004). This phenotypic modulation induces changes in matrix composition, eventually leading to arterial stiffening in aging. Importantly, however, increasing evidence supports that changes in the extracellular matrix alone are insufficient to fully account for increased vascular stiffness and that VSMC are important contributors to arterial stiffening and vasomotor dysfunction in aging (Sehgel, Vatner, & Meininger, 2015). Increased vascular stiffness is also accompanied by a significant increase in the VSMC stiffness and adhesion to the extracellular matrix. Thus, it was proposed that VSMC stiffening alters vascular function and contributes to the changes in vascular stiffening observed in aging and cardiovascular disease, which was termed the "*smooth muscle cell stiffness syndrome.*" (Sehgel, Vatner, & Meininger, 2015). Elucidating how VSMC stiffness occurs in the context of vascular wall stiffness and its underlying mechanisms will provide new knowledge necessary for identifying future therapeutic targets to relieve or reverse vascular wall stiffness in aging or disease.

3.3 Actomyosin apparatus and cell adhesion properties coordinate cellular adaptation to external force

VSMC regulate vessel wall contractility via interactions between the actomyosin contractile unit and adhesions structures formed at the cell membrane that mechanically link the cell to the extracellular matrix. Since VSMC are anchored to the extracellular matrix, their homeostatic state is characterized by a pre-existing force equilibrium between the actomyosin contractility and extracellular mechanical forces (Gunst, Tang, & Opazo Saez, 2003; Ingber, 2002; Stamenovic, 2008; Worth, Rolfe, Song, & Campbell, 2001). *In vitro* studies conducted using cells isolated from microvasculature and further embedded in fibrous collagen matrices showed that VSMC (red) from young (3 months) Fischer 344 rats exert force on the collagen matrix (green) inducing extensive collagen fiber remodeling (Fig. 4A, see arrows), whereas no significant fiber remodeling was observed in collagen gels containing cells from aged (24 months) rats. Quantitative assessment of collagen matrix remodeling assessed by matrix fibers alignment within the image frame increased for hydrogels containing young cells, while no

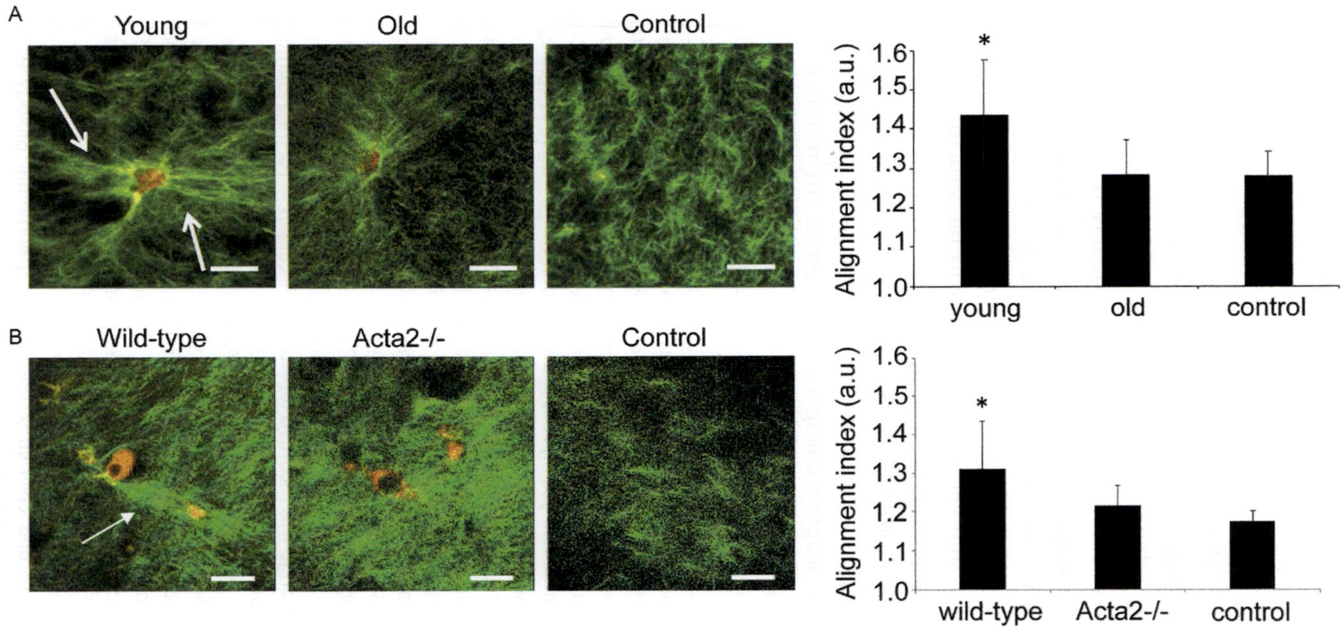

Fig. 4 See figure legend on next page.

change was measured for hydrogels containing aged cells with respect to control (i.e., no cells). These results suggest that aged VSMC are not able to generate the force needed to induce matrix remodeling and maintain arterial contractility. Similar experiments performed with VSMC lacking smooth muscle alpha-actin (Acta2−/−) showed that alignment of collagen fibers was no different with respect to control (i.e., no cells) while the wild-type cells were able to generate the force needed to induce matrix remodeling necessary to maintain cellular contractility (Fig. 4B) (Massett et al., 2020). Our data support a synthetic phenotype for aged cells with decreased contractility that involves a reduced presence of the main players of the contractile apparatus (i.e., smooth muscle alpha-actin) (Seawright et al., 2018).

Cytoskeleton and cell adhesions are dynamic structures that assemble, disperse and turn over as cells respond to mechanical force. External mechano-chemical stresses alter the structural and functional properties of the cells, leading to rapid responses in the adjacent tissue or neighboring cells that induce adaptive changes to the external environment. The extracellular matrix provides the physical microenvironment in which cells live, representing the substrate for cell adhesion. Thus, extracellular matrix

Fig. 4 VSMC-induced matrix remodeling assay. (A) Representative projections of the 3D volumetric images and their alignment index show that young VSMC exert force on the collagen matrix inducing higher remodeling of the collagen fibers (see arrows). In contrast, old VSMC have reduced local matrix remodeling. The distribution of collagen fiber angle alignment index for young cells shows a significant increase compared with control (collagen gel without cells), while no change was recorded for old cells. $N = 8$–10 tri-dimensional volumetric images per condition. Scale bar represents 50 μm. Data shown as mean ± SD. Significance was evaluated at $p < 0.05$. (B) Representative projections of the 3D volumetric images and their alignment index show that wild-type cells exert force on the collagen matrix inducing higher remodeling of the collagen fibers (see arrow), while VSMC lacking smooth muscle alpha-actin (Acta2−/−) have reduced local matrix remodeling. The distribution of collagen fiber angle alignment index for wild-type cells shows a significant increase compared with control (collagen gel without cells), while no change was recorded for Acta2 −/− cells. $N = 10$–14 tri-dimensional volumetric images per condition. Scale bar represents 50 μm. Data shown as mean ± SD. Significance was evaluated at $p < 0.05$. *Panel (A) Reproduced with permission from Seawright, J. W., Sreenivasappa, H., Gibbs, H. C., Padgham, S., Shin, S. Y., Chaponnier, C., Yeh, A. T., Trzeciakowski, J. P., Woodman, C. R., & Trache, A. (2018). Vascular smooth muscle contractile function declines with age in skeletal muscle feed arteries.* Frontiers in Physiology, 9, *856, Panel (B) Reproduced with permission from Massett, M. P., Bywaters, B. C., Gibbs, H. C., Trzeciakowski, J. P., Padgham, S., Chen, J., Rivera, G., Yeh, A. T., Milewicz, D. M., & Trache, A. (2020). Loss of smooth muscle alpha-actin effects on mechanosensing and cell-matrix adhesions.* Experimental Biology and Medicine (Maywood, N.J.), 245*(4), 374–384.*

topography, molecular composition and mechanical properties are of great importance to the cell-matrix cross-talk needed for transmitting environmental signals to the cell through the adhesion proteins as mediators of the process. It has been suggested that cells probe the chemical nature of the matrix through integrin binding, and determine the physical rigidity of the matrix through contractile forces (An et al., 2009; Ingber, 2002; Wei, Lin, Shen, & Tang, 2008).

Cell-matrix adhesion structures contain clusters of integrins, heterodimeric transmembrane adhesion proteins (Arnaout, Goodman, & Xiong, 2007; Legate & Fässler, 2009), that connect to the actin cytoskeleton via a number of specific proteins. Talin, vinculin, α-actinin, paxillin and focal adhesion kinase (FAK) are just a few of the ~905 proteins that localize at cell-matrix adhesions (Kuo, Han, Hsiao, Yates, & Waterman, 2011). During cellular contraction, specific proteins undergo orientation changes, become phosphorylated (Opazo-Saez et al., 2004; Tang, Mehta, & Gunst, 1999; Turner, 2000; Zaidel-Bar, Milo, Kam, & Geiger, 2007) or translocate away from the cell-matrix adhesions (Cattaruzza, Lattrich, & Hecker, 2004; Wang & Gilmore, 2003). Which proteins are absolutely required for the formation of cell-matrix adhesions is still under investigation, and the order these proteins bind integrins to actin or to each other is only partially understood (Kanchanawong et al., 2010).

Intracellular signaling pathways of VSMC mechanotransduction have been extensively studied and determined that integrins have mechanosensory contributions in VSMC (Davis et al., 2001; Li & Xu, 2000). Integrins transfer the mechanical forces from the extracellular matrix to the cytoskeleton. The most prevalent integrins on VSMC are α5β1, α4β1, α1β1, α2β1 and αvβ3 (Weis, 2007). It was shown that α5β1 (Hein et al., 2001; Mogford, Davis, & Meininger, 1997) and α4β1 have an important role in arteriole constriction through upregulation of the L-type calcium channel dependent mechanism (Waitkus-Edwards et al., 2002), whereas αvβ3 has a dilatory effect through a downregulation of L-type calcium channels (Mogford, Davis, Platts, & Meininger, 1996). Also, the link between the L-type calcium channel and integrin signaling appears to involve proteins from the cell-matrix adhesions such as vinculin, paxillin and FAK (Boudreau & Jones, 1999; Wu, Davis, Meininger, Wilson, & Davis, 2001).

Integrins α5β1 and αvβ3 have distinct roles in regulating VSMC contractile function (Martinez-Lemus, Hill, & Meininger, 2009; Martinez-Lemus, Sun, Trache, Trzciakowski, & Meininger, 2005; Martinez-Lemus et al., 2003). Integrin activation by RGD binding showed a role for integrin α5β1 in regulating vascular contraction (Wu et al., 1998, 2001), while integrin αvβ3 binding to the same ligand regulates vascular relaxation (D'Angelo, Mogford, Davis, Davis, & Meininger, 1997). However, vasoconstrictor response to increased pressure requires activation of both integrins (Martinez-Lemus, Crow, Davis, & Meininger, 2005). Mechanical stretch alters the conformation of integrins (Katsumi, Naoe, Matsushita, Kaibuchi, & Schwartz, 2005) such that their β cytoplasmic tails become phosphorylated to provide binding sites for c-Src and FAK (Woodside et al., 2001). The phosphorylation of FAK exposes binding sites for vinculin and paxillin (Mitra, Hanson, & Schlaepfer, 2005; Pasapera, Schneider, Rericha, Schlaepfer, & Waterman, 2010; Tang et al., 1999; Webb et al., 2004) to assemble an adhesion complex at the site of force application (Alenghat & Ingber, 2002; Geiger & Bershadsky, 2001). This process is followed by actin stress fiber remodeling, which is necessary to redistribute physical forces needed for cell contraction and to enable cell adaptation to the extracellular microenvironment (Chen, Pasapera, Koretsky, & Waterman, 2013; Zhang & Gunst, 2006). Conversely, intracellular stresses may be transmitted to the underlying matrix via cell-matrix adhesions. Thus, forces are transmitted bidirectionally across the plasma membrane through outside-in and inside-out integrin signaling (Campbell & Humphries, 2011; Hynes, 2002; Luo, Carman, & Springer, 2007).

Recent studies have shown that expression and function of integrins is altered with aging. In a study using 8-month-old and 30-month-old Fisher 344 crossbred Brown Norway rats (F344XBN), Zhu et al. (2018) showed that even though old VSMC showed an increase in the mRNA levels of α3, α7, β3 and β5 integrin subunits vs young VSMC, there were no differences in the transcript levels of collagen binding integrin dimers α1β1 and α2β1 or fibronectin binding integrin dimers α5β1 and αvβ3. However, primary VSMC in culture stained for integrin α5β1 and αvβ3 showed higher overall fluorescence intensities in old VSMC by comparison with young VSMC. By using the biophysical protocol of Fourier transform traction microscopy, they also showed that old VSMC were larger in size and presented significant increases in traction forces to the substrate than young VSMC. These results are also supported by an elegant spectral analysis of real-time measurements of integrin α5β1 adhesion to fibronectin performed using the AFM (Zhu et al., 2012). Complementary to stiffness measurements, the AFM can be also used to measure single ligand-receptor (i.e., integrin-matrix) adhesion strength (Trache & Meininger, 2008).

Thus, adhesion force and binding probability was increased in old VSMC isolated from old monkeys (24–25 year old) by comparison with young monkeys (5–8 year old). In addition, local cell stiffness measured in the same VSMC has been found to be significantly higher in old VSMC than young ones.

The mechanotransduction process involves tension-dependent conformational changes of integrin mechanotransducers but also activation of the RhoA/Rho Kinase (ROCK) pathway that functions as a molecular switch to induce actomyosin contractility (Huveneers & Danen, 2009; Lim, Trzeciakowski, Sreenivasappa, Dangott, & Trache, 2012; Schiller et al., 2013). In addition, serum response factor (SRF), a transcription factor which regulates contractile genes, is in turn controlled by RhoA. Previous studies showed that αv integrin activates the RhoA pathway, which couples with downstream effector formin mDia1 to induce stress fiber formation, but not myosin II activation. In contrast, α5β1 integrin activation mediates force generation by increasing myosin II activity via RhoA/ROCK pathway (Schiller et al., 2013). Integrin-mediated activation of RhoA affects nuclear myocardin-related transcription factor-A (MRTF-A) levels with the highest level in cells expressing both β1 and αv integrins and lower in cells expressing only β1 integrins (Hermann et al., 2016). In the cytoplasm, MRTF-A is sequestered by monomeric G-actin. Reduced MRTF-A activity promotes proliferation of VSMC, as occurs during atherosclerosis and restenosis (Owens, Kumar, & Wamhoff, 2004). Actin polymerization stimulates MRTF-A nuclear translocation where it interacts with SRF enhancing transcription of contractile proteins (Lacolley, Li, Challande, & Regnault, 2017). Thus, MRTF-A/SRF complex functions as a central regulator of VSMC phenotypic changes, and its dysfunction contributes to pathological VSMC proliferation in the vessel wall. These findings indicate that α5β1 and αvβ3 integrin expression differentially regulate the RhoA pathway, which in turn modulates MRTF-A/SRF activation and determines contractile state of VSMC.

Cell stiffness is linked to the physical and functional integrity of the actomyosin apparatus, and therefore, can be considered a measure of cytoskeletal tension (Sreenivasappa et al., 2014). Using fibronectin functionalized AFM probes (Trache & Meininger, 2008), we showed that differential modulation of cytoskeletal tension by RhoA and c-Src alters functional coupling between adhesion force and cell stiffness in VSMC. The pre-stress level of the cytoskeletal tension (i.e., cell stiffness) determines integrin affinity for the matrix. Adhesion force of α5β1 integrin to fibronectin directly increased with RhoA activity, while expression of constitutively active c-Src induced only a modest increase in adhesion force. Thus, above a cell stiffness threshold of 17 kPa that corresponded to an integrin-fibronectin adhesion force of ~37 pN, the force increased rapidly with the increase in cell stiffness. Soft cells exhibiting a cytoskeletal tension below the threshold value presented no significant variation in the adhesion force to the matrix, while stiffer cells exhibiting a cytoskeletal tension above threshold presented a significant increase in adhesion strength, presumably, through a mechanism that involves integrin activation. Taken together, these results suggest that RhoA and c-Src activation have different effects on cytoskeletal tension development, inducing two distinct force–stiffness functional regimes for α5β1-integrin binding to fibronectin.

3.4 Vascular smooth muscle cell adaptation to mechanical stresses in the vessel wall

Mechanical stresses in the vessel wall can be mimicked *in vitro* with the aim to dissect discrete protein contributions to the cellular adaptation to force. Tensile stress induced by AFM mechanical stimulation at the apical cell surface was shown to induce significantly different cell responses that depend on cytoskeletal tension and myosin function (Lim et al., 2012). Modulation of intracellular tension can be achieved by expression of RhoA-mutants or myosin function inhibition, while tensile stress was applied by upward movement of fibronectin functionalized AFM cantilever in discrete force steps inducing low tensile stress (0.5 nN) followed by higher tensile stress (1 nN). By tracking overall probe displacement as a function of time for the duration of the experiment (Fig. 5) we showed that for cells expressing RhoA-T19N (dominant negative), the AFM probe detaches from the cell surface at the first mechanical stimulation with high level force (1 nN). A decrease in RhoA activation reduces formation of cortical actin fibers and myosin activity. Thus, cells appear softer, have reduced contractility, and the overall probe displacement vs time exhibits a linear dependence. Expression of RhoA-Q63L (constitutively active) induces strong actin fibers formation and cells exhibit high contractile activity. Thus, these cells were also elastic and able to pull back after each tensile force application, exhibiting a high reactive response of the overall probe displacement vs time. In contrast, cells treated with ML-7, a potent smooth muscle myosin inhibitor, presented an overall probe displacement vs time that plateaus at high forces. These cells were significantly softer than RhoA-Q63L expressing cells, however, they have an intact actin fiber network, but decreased contractility due to reduced myosin activity. Taken together, these results

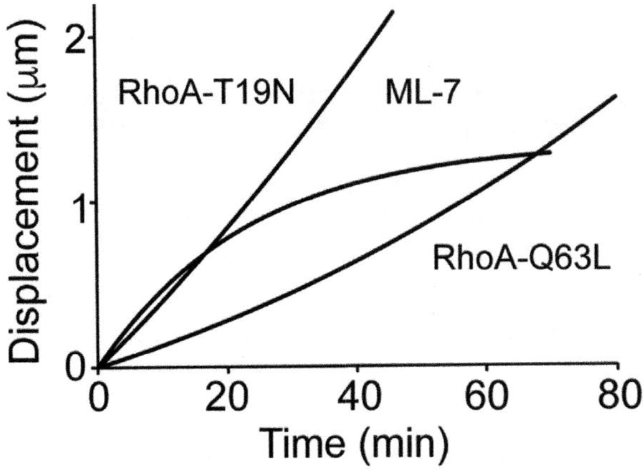

Fig. 5 Cellular adaptive response to AFM induced tensile stress. Cells expressing RhoA-T19N (dominant negative) are detaching from the fibronectin functionalized probe at the first mechanical stimulation with 1 nN force, and the overall curve exhibits a linear dependence with a maximal displacement. RhoA-Q63L (constitutively active) expressing cells were found to resist to the applied tensile stress and exhibit high contractility power showing a reactive response to the applied stress. In contrast, cells treated with ML-7 display an overall dependence that plateaus at high force stimulation. *Adapted with permission from Lim, S. M., Trzeciakowski, J. P., Sreenivasappa, H., Dangott, L. J., & Trache, A. (2012). RhoA-induced cytoskeletal tension controls adaptive cellular remodeling to mechanical signaling.* Integrative Biology, *4(6), 615–627.*

suggest that pre-existing cytoskeletal tension modulates VSMC-adaptive response to external mechanical stimuli. Moreover, actomyosin driven cellular adaptation to their mechanical microenvironment is a characteristic of the integrated cell system as a whole.

By combining a biaxial cell-stretching device with both AFM and confocal microscopy (Na, Trache, Meininger, & Humphrey, 2008) we were able to correlate VSMC stiffness and remodeling of cell-matrix adhesions as a function of the cyclic stretch applied. Results showed that cyclic 10% equibiaxial stretching at 0.25 Hz significantly and rapidly alters both cell stiffness and recruitment of adhesion proteins, paxillin and vinculin, at cell-matrix adhesions, followed by a return toward baseline after 5 min of stretching. These results suggest that cyclic stretch causes significant transient alterations in cell stiffness that correlate well with adaptive structural changes involving remodeling of cell-matrix adhesions which further regulate stress fiber formation (Na, Meininger, & Humphrey, 2007) in order to maintain a homeostatic biomechanical state when perturbed above a threshold value (Chiquet, 1999; Hinz, 2006).

The age-induced progressive changes in the mechanical properties of the blood vessel wall induce an increase in wall stiffness. These blood vessel wall changes correlate well with its impaired ability to control local blood flow and pressure, which translate to a decreased mechanosensitivity of the vascular cells (i.e., decreased response to mechanical stimuli). Thus, extracellular matrix stiffening induces alteration of cell-matrix adhesion and further activation of intracellular pathways leading to actomyosin contraction and an increased cytoskeletal tension. While adhesion to the matrix and cell stiffness increase with age (Fig. 6A), upregulation of Rho-kinase in its active

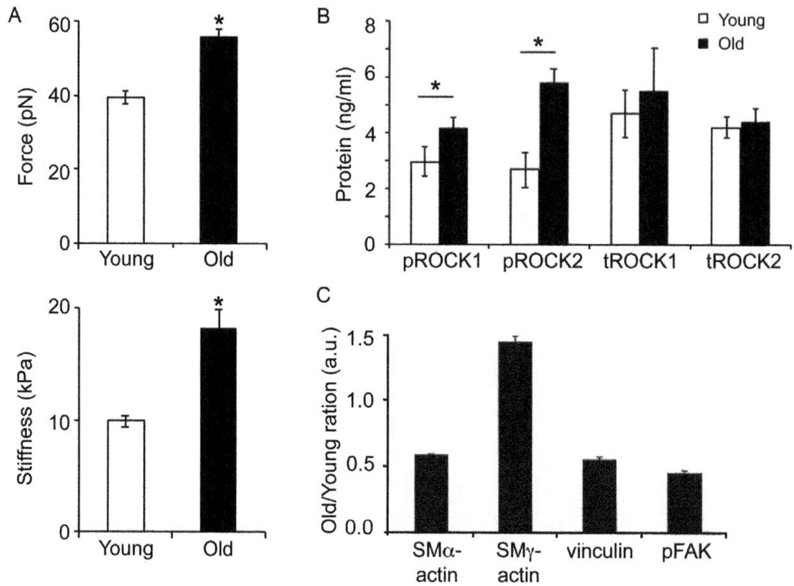

Fig. 6 (A) Integrin α5β1-fibronectin adhesion force spectroscopy measurements. Adhesion force of integrin α5β1 binding to fibronectin and cell stiffness significantly increased with age. $N \sim 4000$ individual measurements for each condition. Significance was evaluated at $p < 0.05$. (B) Elisa assay values for total and phospho-ROCK proteins. While total ROCK (tROCK) does not change with age, there was a significant increase in old cells for both isoforms of the active form pROCK1 and pROCK2. $N = 4$. Data shown as mean ± SD. Significance was evaluated at $p < 0.05$. (C) Age effects on VSMC morphology. Quantitative analysis of fluorescence images is presented as old:young ratios of relative protein area to respective cell area. $N = 7$–11 cells per condition. Values different from a ratio of 1 are statistically significant ($p < 0.05$). *Reproduced with permission from Seawright, J. W., Sreenivasappa, H., Gibbs, H. C., Padgham, S., Shin, S. Y., Chaponnier, C., Yeh, A. T., Trzeciakowski, J. P., Woodman, C. R., & Trache, A. (2018). Vascular smooth muscle contractile function declines with age in skeletal muscle feed arteries.* Frontiers in Physiology, *9, 856.*

form (Fig. 6B) is not able to rescue the reduced contractility state of the aged cells (Seawright et al., 2018). This is due, in part, to reduced presence of the essential contractile and adhesion proteins (Fig. 6C). These data suggest that age-induced changes at the molecular level support a VSMC synthetic phenotype that plays a central role in reduced arterial contractile function resulting in reduced mechanosensitivity and adaptation to mechanical signals from their microenvironment. Further studies are needed to elucidate the physiological and pathophysiological effects of endothelial and smooth muscle cells stiffness, and their contribution to the overall vascular stiffness.

4. Vascular smooth muscle function in aged arteries

Building on age-induced structural and functional adaptations of VSMC we will discuss here the effects of aging on the ability of VSM to regulate vascular tone in intact arteries. Emphasis will be placed on resistance arteries and arterioles, because these vessels play an integral role in controlling blood pressure and skeletal muscle blood flow at rest and during exercise (Williams & Segal, 1993).

4.1 Aging induces changes in smooth muscle contractile function in resistance arteries

4.1.1 Myogenic constrictor responses

In skeletal muscle, proper distribution of blood flow depends, in part, on regulation of VSM tone in blood vessels. One mechanism by which resistance arteries and arterioles regulate VSM tone is through the myogenic response. Specifically, VSM in resistance arteries and arterioles respond to increased or decreased transmural pressure by contracting or relaxing, respectively (Davis & Hill, 1999). Consequently, an aging-induced impairment of VSM myogenic responses of blood vessels impairs the ability to regulate skeletal muscle perfusion and blood pressure.

To evaluate the effects of aging on VSM myogenic responses, Muller-Delp, Spier, Ramsey, Lesniewski, et al. (2002) utilized a cannulated artery preparation in which skeletal muscle resistance arteries were isolated *in vitro*, and then pressurized to *in vivo* pressure. VSM contractile responses were assessed by exposing the arteries to graded increases or decreases in intraluminal pressure and measuring the ability of VSM to contract or relax. Using this technique, these investigators examined the effects of aging on myogenic responses in first-order arterioles from the soleus muscle and superficial white portion of the gastrocnemius muscle, representing oxidative and glycolytic muscle, respectively. Results revealed that myogenic constrictor responses were significantly impaired in first-order arterioles from soleus and gastrocnemius muscles from aged Fischer 344 rats when compared to myogenic responses in corresponding arterioles from young rats.

In a follow up study, the importance of large-conductance Ca^{2+} activated K^+ (BK_{Ca}) channels and voltage-dependent K^+ (K_V) channels in myogenic responses was assessed in arterioles from young and old Fischer 344 rats (Kang et al., 2009). Results showed that myogenic responses in soleus muscle first-order arterioles were impaired in arterioles from aged rats due to increased modulation of myogenic constriction by both K_V and BK_{Ca} channels in aged arterioles. In contrast, the decline in myogenic constriction in gastrocnemius arterioles from aged rats was attributed solely to increased modulation by K_V channels (Kang et al., 2009). These results indicate that age-related decrements in myogenic constriction differentially occur in arterioles from oxidative and glycolytic muscle. Moreover, impaired myogenic responses have also been documented in resistance arteries in the cerebral microcirculation (Springo et al., 2015).

4.1.2 Agonist induced constrictor responses

In addition to age-related decrements in smooth muscle myogenic constrictor responses, previous studies indicate that agonist-induced smooth muscle contractile responses decline with age in skeletal muscle resistance arteries. Sinkler and Segal (2014) reported that smooth muscle constrictor responses to both α1 and α2 receptor agonists were impaired in feed arteries and arterioles from the gluteus maximus muscle of old mice when compared to arteries isolated from young mice. Similarly, alpha adrenergic-mediated constrictor responses were reported to be attenuated in soleus muscle feed arteries isolated from aged Fischer 344 rats (Seawright, Trache, Wilson, & Woodman, 2016).

In a related study, Gittemeier et al. (2017) assessed the effect of aging on the rate of VSM-meditated constriction in resistance arteries. Importantly, these investigators reported that aging was associated with a threefold increase in the time to reach stable diameter following stimulation with norepinephrine, a non-selective α adrenergic receptor agonist. The detrimental effect of aging on VSM contractile function was attributed to impaired calcium release in VSM of aged arteries. Interestingly, VSM constrictor responses to adrenergic receptor-independent agonists have been reported to be enhanced with aging in some vascular beds (Donato, Lesniewski, &

Delp, 2005, 2007; Park et al., 2012). Specifically, VSM constrictor responses to endothelin-1 (ET-1) were reported to be augmented with age in first-order arterioles from white gastrocnemius muscle (Donato et al., 2005). The greater VSM constrictor responses were attributed to enhanced signaling through VSMC ET_A receptors. Similarly, VSM constrictor responses to Ang II were reported to be enhanced in skeletal muscle first-order arterioles (Donato, Lesniewski, & Delp, 2007), and were attributed to a greater pro-inflammatory state mediated by TNFα (Park et al., 2012). Thus, enhanced constrictor responses to Ang II and ET-1 may help compensate for impaired adrenergic signaling mechanisms in aged VSM of resistance arteries.

4.2 Exercise training improves smooth muscle contractile function in aged resistance arteries

Based on previous observation (Kang et al., 2009) of age-related decline of the myogenic responses in the first-order arterioles from oxidative (soleus) and glycolytic (gastrocnemius) muscles, Ghosh et al. (2015) completed a study to determine whether aerobic exercise training attenuates or reverses the detrimental effects of aging on myogenic responses. To evaluate the effects of exercise training, young and old Fischer 344 rats remained sedentary or completed a 10-week endurance exercise training program. Rats in the exercise training group ran on a motorized treadmill 5 days per week, 60 min per day. At the end of the 10-week intervention, myogenic constrictor responses were less in arterioles isolated from old sedentary rats than young sedentary rats. Exercise training improved myogenic constrictor responses in aged arterioles such that responses were not different from young sedentary arterioles. In addition, results revealed that the training-induced improvement of myogenic responses in aged arteries was mediated, in part, by training-induced adaptations in voltage gated potassium channels K_V1 in VSMC.

Endurance exercise training has also been shown to be an effective intervention to improve agonist-induced vasoconstrictor responses in aged resistance arteries. Donato, Lesniewski, and Delp (2007) reported that enhanced α-adrenergic-mediated constrictor responses in aged resistance arterioles were reversed following completion of an endurance training program. In addition, endurance exercise-training reversed the detrimental effects of aging on Ang II-induced constriction in skeletal muscle arterioles (Park et al., 2012). By comparison, ET-1 mediated constrictor responses in first-order arterioles from white gastrocnemius and soleus muscles were not altered by training (Donato et al., 2005). Collectively, results from these studies indicate that endurance exercise training improves adrenergic-mediated VSM contractility in aged skeletal muscle resistance arteries.

Exercise training not only improves vascular function, but also has an important role in remodeling vessel wall structure. Hanna et al. (2014) showed that 10 weeks of exercise training also reduced collagen:elastin ratio in coronary resistance arteries from old, but not young Fischer 344 rats (Fig. 7A). The percent elastin, as a function of cross-sectional area,

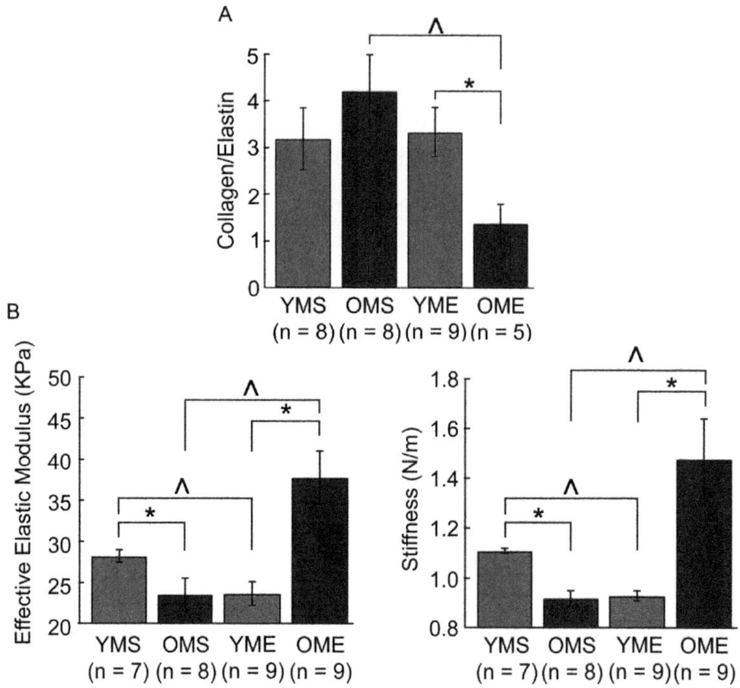

Fig. 7 (A) Collagen-to-elastin ratio in exercised and sedentary Fischer 344 rats. The collagen:elastin is most uniform for old male exercised (OME) group, which is significantly different from its old-matched sedentary group (OMS) and corresponding young exercised group (YME). (B) Effective elastic modulus (Eeff (Kpa)) and calculated stiffness (S (N/m)). Exercise training reduced the Eeff and S in the young group to a level comparable to OMS, whereas increases in Eeff and S were observed with exercise training in the old group (OME). Values are means ± SE; n, number of rats. $^*p \leq 0.05$ vs corresponding old group. $\hat{}p \leq 0.05$ vs age-matched sedentary group. Y/OMS, young/old male sedentary; Y/OME, young/old male exercised. *Reproduced with permission from Hanna, M. A., Taylor, C. R., Chen, B., La, H. S., Maraj, J. J., Kilar, C. R., Behnke, B. J., Delp, M. D., & Muller-Delp, J. M. (2014). Structural remodeling of coronary resistance arteries: Effects of age and exercise training.* Journal of Applied Physiology (Bethesda, MD: 1985), 117*(6), 616–623.*

was higher in old exercised animals compared with old sedentary, or young sedentary and young exercised animals, while no significant difference in percent collagen as a function of cross-sectional area was detected among the animal groups. However, the increase in elastin content did not prevent stiffening of the aged vascular wall with exercise training. Nanoindentation experiments carried out on an *en face* vessel preparation using resistance arteries allowed *ex vivo* measurements of vessel wall stiffness and effective elastic modulus. Fig. 7B shows that both parameters were increased in old exercised trained rats. This could be eventually explained by the increase in collagen cross-linking with age, that may have altered mechanical properties of the vessel wall to the extent that the increase in elastin could not compensate and maintain vessel wall compliance. In addition, exercise training may induce a transition to a contractile VSM phenotype that can be potentially stiffer. This hypothesis is supported by the increase in myogenic response found in aged coronary resistance arteries post-exercise training.

4.3 Mechanical stimulation improves smooth muscle contractile function in aged resistance arteries

Age-related decrements in VSM contractile responses are attenuated or reversed following a program of endurance exercise training in skeletal muscle resistance arteries and arterioles. During exercise, blood pressure (i.e., perfusion pressure) increases to facilitate the delivery of oxygen rich blood to the active muscle. Therefore, it is conceivable that increased intraluminal pressure during exercise provides an important mechanical signal promoting healthy VSM function. Thus, we proposed to identify the signals associated with exercise responsible for the improvement in VSM function by pressurizing soleus muscle resistance arteries to the pressure believed to be present in these arteries at rest (90 cm H_2O) or during exercise (130 cm H_2O) for 1 h (Seawright et al., 2016). At the end of the 1-h treatment period, pressure in all arteries was lowered to resting pressure conditions and VSM constrictor responses were assessed. Following the 1-h pressure treatment, constrictor responses to norepinephrine (NE), phenylephrine (PE), and angiotensin II (Ang II) were improved in the aged resistance arteries such that responses were not different from the young arteries. In addition, the beneficial effect of pressure treatment was abolished in the presence of Y27632, a Rho Kinase inhibitor (Fig. 8). Based on these findings, we concluded that the beneficial effect of pressure treatment was mediated, in part, through the Rho Kinase signaling pathway (Seawright et al., 2016).

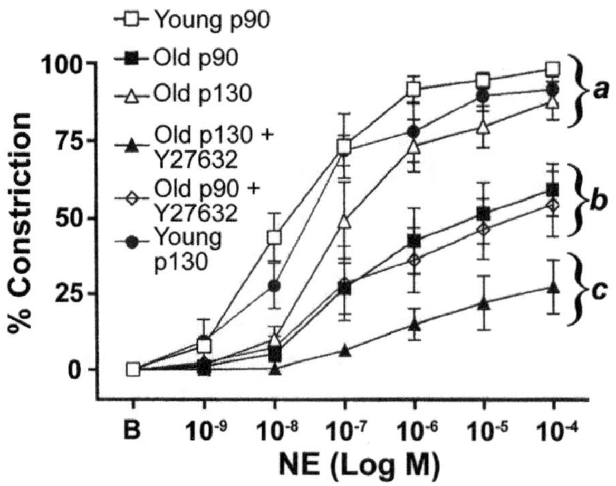

Fig. 8 Pressure pre-treatment of aged arteries improves VSM contraction via RhoA pathway. The effect of mechanical stimulation tested by increased intraluminal pressure on VSM contractility to norepinephrine (NE) was performed on *ex vivo* cannulated arteries. Results showed that pre-treatment of aged resistance arteries with increased pressure (Old p130) improved VSM constrictor responses to NE. In the presence of ROCK inhibitor (Y27632), constrictor responses were reduced in Old p130 arteries whereas, there was no effect on Old p90. Significance was considered at $p < 0.05$. Curves with different letters are significantly different vs all other curves. B represents the baseline diameter at a pressure of 90 cm H_2O before the first dose of NE. *Adapted from Seawright, J. W., Sreenivasappa, H., Gibbs, H. C., Padgham, S., Shin, S. Y., Chaponnier, C., Yeh, A. T., Trzeciakowski, J. P., Woodman, C. R., & Trache, A. (2018). Vascular smooth muscle contractile function declines with age in skeletal muscle feed arteries.* Frontiers in Physiology, *9, 856 and Seawright, J. W., Trache, A., Wilson, E., & Woodman, C. R. (2016). Short-duration increases in intraluminal pressure improve vasoconstrictor responses in aged skeletal muscle feed arteries.* European Journal of Applied Physiology, 116*(5), 931–937.*

These studies are consistent with a decline in VSM function in skeletal muscle resistance arteries with age. The detrimental effect of aging is characterized, in part, by impaired vasoconstrictor responses in skeletal muscle feed arteries and arterioles. Reduced VSM contractility has been proposed to contribute to orthostatic intolerance and reduced exercise capacity in the elderly (Davy, Seals, & Tanaka, 1998; Musch, Eklund, Hageman, & Poole, 2004). Endurance exercise training attenuates or reverses the detrimental effects of aging on VSM function in some vascular beds. One signal associated with exercise responsible for promoting healthy VSM function in aged arteries likely involves mechanical stimulation of the arterial wall through short duration increases in intraluminal pressure during individual exercise bouts.

5. Conclusions

Age-induced arterial vasomotor dysfunction is associated with an increased risk for cardiovascular disease. Endothelial dysfunction and arterial stiffening are recognized contributors to impaired vascular function with aging. This review focuses on arterial stiffening with an emphasis on the contribution of VSMC stiffness to vascular aging. Our work using AFM and that of others demonstrate that stiffness properties of the vessel wall vary throughout the vascular tree and is dependent, in part, on the interaction between VMSC and the surrounding matrix. Furthermore, we detail the role of changes in the actomyosin contractile unit and cell-matrix adhesions of aged VSMC that contribute to their phenotypic switch from a contractile to synthetic phenotype that associates with increased vascular stiffness. This phenotypic switch is characterized by the inability of aged VSMC to generate force to induce matrix remodeling, and such, impairs their ability to maintain arterial contractility. Moreover, changes in matrix composition eventually lead to arterial stiffening in aging. Our recent work showed that despite having increased Rho-kinase activity, contractility is not restored in aged VSMC due to reduced presence of the essential contractile and adhesion proteins. These impairments in contractility of VSMC have functional consequences by decreasing vasoconstrictor function and myogenic response in resistance arteries with aging. Exercise training can improve contractile function in aged resistance arteries. To address one potential mechanism underlying this improvement, we showed that short duration increases in intraluminal pressure also improved contractile function. We conclude that mechanical stimulation has important implications in improving contractile function in aged resistance arteries by promoting healthy vascular smooth muscle function in aged blood vessels.

Acknowledgments

Authors thank Lin Bustamante, Veterinary Integrative Biosciences Histology Service Lab, Texas A&M University for her contribution to the preparation of histological samples. The work was supported by the National Institutes of Health (NIA) grant 1R03AG06455 to A.T.

References

Alenghat, F. J., & Ingber, D. E. (2002). Mechanotransduction: All signals point to cytoskeleton, matrix, and integrins. *Science's STKE, 2002*(119), pe6.

AlGhatrif, M., Strait, J. B., Morrell, C. H., Canepa, M., Wright, J., Elango, P., et al. (2013). Longitudinal trajectories of arterial stiffness and the role of blood pressure: The Baltimore Longitudinal Study of Aging. *Hypertension, 62*(5), 934–941.

An, S. S., Kim, J., Ahn, K., Trepat, X., Drake, K. J., Kumar, S., et al. (2009). Cell stiffness, contractile stress and the role of extracellular matrix. *Biochemical and Biophysical Research Communications, 382*(4), 697–703.

Arnaout, M. A., Goodman, S. L., & Xiong, J. P. (2007). Structure and mechanics of integrin-based cell adhesion. *Current Opinion in Cell Biology, 19*(5), 495–507.

Augustin, H. G., & Koh, G. Y. (2017). Organotypic vasculature: From descriptive heterogeneity to functional pathophysiology. *Science, 357*(6353), eaal2379.

Barton, M., Cosentino, F., Brandes, R. P., Moreau, P., Shaw, S., & Luscher, T. F. (1997). Anatomic heterogeneity of vascular aging: Role of nitric oxide and endothelin. *Hypertension, 30*(4), 817–824.

Ben-Shlomo, Y., Spears, M., Boustred, C., May, M., Anderson, S. G., Benjamin, E. J., et al. (2014). Aortic pulse wave velocity improves cardiovascular event prediction: An individual participant meta-analysis of prospective observational data from 17, 635 subjects. *Journal of the American College of Cardiology, 63*(7), 636–646.

Berkowitz, D. E., White, R., Li, D., Minhas, K. M., Cernetich, A., Kim, S., et al. (2003). Arginase reciprocally regulates nitric oxide synthase activity and contributes to endothelial dysfunction in aging blood vessels. *Circulation, 108*(16), 2000–2006.

Bezie, Y., Lamaziere, J. M., Laurent, S., Challande, P., Cunha, R. S., Bonnet, J., et al. (1998). Fibronectin expression and aortic wall elastic modulus in spontaneously hypertensive rats. *Arteriosclerosis, Thrombosis, and Vascular Biology, 18*(7), 1027–1034.

Bode-Boger, S. M., Muke, J., Surdacki, A., Brabant, G., Boger, R. H., & Frolich, J. C. (2003). Oral L-arginine improves endothelial function in healthy individuals older than 70 years. *Vascular Medicine, 8*(2), 77–81.

Boudreau, N. J., & Jones, P. L. (1999). Extracellular matrix and integrin signalling: The shape of things to come. *Biochemical Journal, 339*(Pt. 3), 481–488. Retrieved from https://www.ncbi.nlm.nih.gov/pubmed/10215583.

Briones, A. M., Salaices, M., & Vila, E. (2007). Mechanisms underlying hypertrophic remodeling and increased stiffness of mesenteric resistance arteries from aged rats. *The Journals of Gerontology. Series A, Biological Sciences and Medical Sciences, 62*(7), 696–706.

Campbell, I. D., & Humphries, M. J. (2011). Integrin structure, activation, and interactions. *Cold Spring Harbor Perspectives in Biology, 3*(3), a004994.

Cattaruzza, M., Lattrich, C., & Hecker, M. (2004). Focal adhesion protein zyxin is a mechanosensitive modulator of gene expression in vascular smooth muscle cells. *Hypertension, 43*(4), 726–730.

Celermajer, D. S., Sorensen, K. E., Spiegelhalter, D. J., Georgakopoulos, D., Robinson, J., & Deanfield, J. E. (1994). Aging is associated with endothelial dysfunction in healthy men years before the age-related decline in women. *Journal of the American College of Cardiology, 24*(2), 471–476.

Chamley-Campbell, J., Campbell, G. R., & Ross, R. (1979). The smooth muscle cell in culture. *Physiological Reviews, 59*(1), 1–61.

Chen, Y., Pasapera, A. M., Koretsky, A. P., & Waterman, C. M. (2013). Orientation-specific responses to sustained uniaxial stretching in focal adhesion growth and turnover. *Proceedings of the National Academy of Sciences of the United States of America, 110*(26), E2352–E2361.

Chiquet, M. (1999). Regulation of extracellular matrix gene expression by mechanical stress. *Matrix Biology, 18*(5), 417–426.

Chitaley, K., & Webb, R. C. (2002). Nitric oxide induces dilation of rat aorta via inhibition of rho-kinase signaling. *Hypertension, 39*(2 Pt. 2), 438–442. Retrieved from http://www.ncbi.nlm.nih.gov/pubmed/11882586.

Csiszar, A., Ungvari, Z., Edwards, J. G., Kaminski, P., Wolin, M. S., Koller, A., et al. (2002). Aging-induced phenotypic changes and oxidative stress impair coronary arteriolar function. *Circulation Research, 90*(11), 1159–1166.

D'Angelo, G., Mogford, J. E., Davis, G. E., Davis, M. J., & Meininger, G. A. (1997). Integrin-mediated reduction in vascular smooth muscle [Ca2+]i induced by RGD-containing peptide. *The American Journal of Physiology*, *272*(4 Pt. 2), H2065–H2070.

Davis, M. J., & Hill, M. A. (1999). Signaling mechanisms underlying the vascular myogenic response. *Physiological Reviews*, *79*(2), 387–423.

Davis, M. J., Wu, X., Nurkiewicz, T. R., Kawasaki, J., Davis, G. E., Hill, M. A., et al. (2001). Integrins and mechanotransduction of the vascular myogenic response. *American Journal of Physiology. Heart and Circulatory Physiology*, *280*(4), H1427–H1433.

Davy, K. P., Seals, D. R., & Tanaka, H. (1998). Augmented cardiopulmonary and integrative sympathetic baroreflexes but attenuated peripheral vasoconstriction with age. *Hypertension*, *32*(2), 298–304.

Dhaun, N., Goddard, J., Kohan, D. E., Pollock, D. M., Schiffrin, E. L., & Webb, D. J. (2008). Role of endothelin-1 in clinical hypertension: 20 years on. *Hypertension*, *52*(3), 452–459.

Delp, M. D., Behnke, B. J., Spier, S. A., Wu, G., & Muller-Delp, J. M. (2008). Ageing diminishes endothelium-dependent vasodilatation and tetrahydrobiopterin content in rat skeletal muscle arterioles. *The Journal of Physiology*, *586*(4), 1161–1168.

Dinardo, C. L., Venturini, G., Zhou, E. H., Watanabe, I. S., Campos, L. C., Dariolli, R., et al. (2014). Variation of mechanical properties and quantitative proteomics of VSMC along the arterial tree. *American Journal of Physiology. Heart and Circulatory Physiology*, *306*(4), H505–H516.

Dingemans, K. P., Teeling, P., Lagendijk, J. H., & Becker, A. E. (2000). Extracellular matrix of the human aortic media: An ultrastructural histochemical and immunohistochemical study of the adult aortic media. *The Anatomical Record*, *258*(1), 1–14.

Donato, A. J., Eskurza, I., Silver, A. E., Levy, A. S., Pierce, G. L., Gates, P. E., et al. (2007). Direct evidence of endothelial oxidative stress with aging in humans: Relation to impaired endothelium-dependent dilation and upregulation of nuclear factor-kappaB. *Circulation Research*, *100*(11), 1659–1666.

Donato, A. J., Gano, L. B., Eskurza, I., Silver, A. E., Gates, P. E., Jablonski, K., et al. (2009). Vascular endothelial dysfunction with aging: Endothelin-1 and endothelial nitric oxide synthase. *American Journal of Physiology. Heart and Circulatory Physiology*, *297*(1), H425–H432.

Donato, A. J., Lesniewski, L. A., & Delp, M. D. (2005). The effects of aging and exercise training on endothelin-1 vasoconstrictor responses in rat skeletal muscle arterioles. *Cardiovascular Research*, *66*(2), 393–401.

Donato, A. J., Lesniewski, L. A., & Delp, M. D. (2007). Ageing and exercise training alter adrenergic vasomotor responses of rat skeletal muscle arterioles. *The Journal of Physiology*, *579*(Pt. 1), 115–125.

Duca, L., Blaise, S., Romier, B., Laffargue, M., Gayral, S., El Btaouri, H., et al. (2016). Matrix ageing and vascular impacts: Focus on elastin fragmentation. *Cardiovascular Research*, *110*(3), 298–308.

Eskurza, I., Myerburgh, L. A., Kahn, Z. D., & Seals, D. R. (2005). Tetrahydrobiopterin augments endothelium-dependent dilatation in sedentary but not in habitually exercising older adults. *The Journal of Physiology*, *568*(Pt. 3), 1057–1065.

Fels, J., Oberleithner, H., & Kusche-Vihrog, K. (2010). Menage a trois: Aldosterone, sodium and nitric oxide in vascular endothelium. *Biochimica et Biophysica Acta*, *1802*(12), 1193–1202.

Francia, P., Delli Gatti, C., Bachschmid, M., Martin-Padura, I., Savoia, C., Migliaccio, E., et al. (2004). Deletion of p66shc gene protects against age-related endothelial dysfunction. *Circulation*, *110*(18), 2889–2895.

Franklin, S. S., Gustin, W., 4th, Wong, N. D., Larson, M. G., Weber, M. A., Kannel, W. B., et al. (1997). Hemodynamic patterns of age-related changes in blood pressure. The Framingham Heart Study. *Circulation*, *96*(1), 308–315.

Fritze, O., Romero, B., Schleicher, M., Jacob, M. P., Oh, D. Y., Starcher, B., et al. (2012). Age-related changes in the elastic tissue of the human aorta. *Journal of Vascular Research, 49*(1), 77–86.

Fujita, H., Fukumoto, Y., Saji, K., Sugimura, K., Demachi, J., Nawata, J., et al. (2010). Acute vasodilator effects of inhaled fasudil, a specific Rho-kinase inhibitor, in patients with pulmonary arterial hypertension. *Heart and Vessels, 25*(2), 144–149.

Gates, P. E., Boucher, M.-L., Silver, A. E., Monahan, K. D., & Seals, D. R. (2007). Impaired flow-mediated dilation with age is not explained by L-arginine bioavailability or endothelial asymmetric dimethylarginine protein expression. *Journal of Applied Physiology (Bethesda, MD: 1985), 102*(1), 63–71.

Gavazzi, I., Boyle, K. S., Edgar, D., & Cowen, T. (1995). Reduced laminin immunoreactivity in the blood vessel wall of ageing rats correlates with reduced innervation in vivo and following transplantation. *Cell and Tissue Research, 281*(1), 23–32.

Geiger, B., & Bershadsky, A. (2001). Assembly and mechanosensory function of focal contacts. *Current Opinion in Cell Biology, 13*(5), 584–592.

Ghosh, P., Mora Solis, F. R., Dominguez, J. M., 2nd, Spier, S. A., Donato, A. J., Delp, M. D., et al. (2015). Exercise training reverses aging-induced impairment of myogenic constriction in skeletal muscle arterioles. *Journal of Applied Physiology (Bethesda, MD: 1985), 118*(7), 904–911.

Gittemeier, E. M., Ericson, T., Ghosh, P., Copp, S. W., Opoku-Acheampong, A. B., & Behnke, B. J. (2017). Effects of aging and exercise training on the dynamics of vasoconstriction in skeletal muscle resistance vessels. *European Journal of Applied Physiology, 117*(3), 397–407.

Goettsch, W., Lattmann, T., Amann, K., Szibor, M., Morawietz, H., Munter, K., et al. (2001). Increased expression of endothelin-1 and inducible nitric oxide synthase isoform II in aging arteries in vivo: Implications for atherosclerosis. *Biochemical and Biophysical Research Communications, 280*(3), 908–913.

Greenwald, S. E. (2007). Ageing of the conduit arteries. *The Journal of Pathology, 211*(2), 157–172.

Gunst, S. J., Tang, D. D., & Opazo Saez, A. (2003). Cytoskeletal remodeling of the airway smooth muscle cell: A mechanism for adaptation to mechanical forces in the lung. *Respiratory Physiology & Neurobiology, 137*(2–3), 151–168.

Hanna, M. A., Taylor, C. R., Chen, B., La, H. S., Maraj, J. J., Kilar, C. R., et al. (2014). Structural remodeling of coronary resistance arteries: Effects of age and exercise training. *Journal of Applied Physiology (Bethesda, MD: 1985), 117*(6), 616–623.

Hayenga, H. N., Trache, A., Trzeciakowski, J., & Humphrey, J. D. (2011). Regional atherosclerotic plaque properties in Apo E−/− mice quantified by atomic force, immunofluorescence, and light microscopy. *Journal of Vascular Research, 48*(6), 495–504.

Heidenreich, P. A., Trogdon, J. G., Khavjou, O. A., Butler, J., Dracup, K., Ezekowitz, M. D., et al. (2011). Forecasting the future of cardiovascular disease in the United States: A policy statement from the American Heart Association. *Circulation, 123*(8), 933–944.

Hein, T. W., Platts, S. H., Waitkus-Edwards, K. R., Kuo, L., Mousa, S. A., & Meininger, G. A. (2001). Integrin-binding peptides containing RGD produce coronary arteriolar dilation via cyclooxygenase activation. *American Journal of Physiology. Heart and Circulatory Physiology, 281*(6), H2378–H2384.

Heitzer, T., Brockhoff, C., Mayer, B., Warnholtz, A., Mollnau, H., Henne, S., et al. (2000). Tetrahydrobiopterin improves endothelium-dependent vasodilation in chronic smokers: Evidence for a dysfunctional nitric oxide synthase. *Circulation Research, 86*(2), E36–E41.

Hermann, M. R., Jakobson, M., Colo, G. P., Rognoni, E., Jakobson, M., Kupatt, C., et al. (2016). Integrins synergise to induce expression of the MRTF-A-SRF target gene ISG15 for promoting cancer cell invasion. *Journal of Cell Science, 129*(7), 1391–1403.

Herrera, M. D., Mingorance, C., Rodriguez-Rodriguez, R., & Alvarez de Sotomayor, M. (2010). Endothelial dysfunction and aging: An update. *Ageing Research Reviews*, *9*(2), 142–152.

Hinz, B. (2006). Masters and servants of the force: The role of matrix adhesions in myofibroblast force perception and transmission. *European Journal of Cell Biology*, *85*(3–4), 175–181.

Humphrey, J. D. (2008). Mechanisms of arterial remodeling in hypertension: Coupled roles of wall shear and intramural stress. *Hypertension.*, *52*, 195–200.

Humphrey, J. D., Eberth, J. F., Dye, W. W., & Gleason, R. L. (2009). Fundamental role of axial stress incompensatory adaptations by arteries. *The Journal of Biomechanics*, *42*, 1–8.

Huveneers, S., Daemen, M. J., & Hordijk, P. L. (2015). Between Rho(k) and a hard place: The relation between vessel wall stiffness, endothelial contractility, and cardiovascular disease. *Circulation Research*, *116*(5), 895–908.

Huveneers, S., & Danen, E. H. (2009). Adhesion signaling—Crosstalk between integrins, Src and Rho. *Journal of Cell Science*, *122*(Pt. 8), 1059–1069.

Hynes, R. O. (2002). Integrins: Bidirectional, allosteric signaling machines. *Cell*, *110*(6), 673–687.

Ingber, D. E. (2002). Mechanical signaling and the cellular response to extracellular matrix in angiogenesis and cardiovascular physiology. *Circulation Research*, *91*(10), 877–887.

Kaess, B. M., Rong, J., Larson, M. G., Hamburg, N. M., Vita, J. A., Levy, D., et al. (2012). Aortic stiffness, blood pressure progression, and incident hypertension. *JAMA*, *308*(9), 875–881.

Kanchanawong, P., Shtengel, G., Pasapera, A. M., Ramko, E. B., Davidson, M. W., Hess, H. F., et al. (2010). Nanoscale architecture of integrin-based cell adhesions. *Nature*, *468*(7323), 580–584.

Kang, L. S., Kim, S., Dominguez, J. M., 2nd, Sindler, A. L., Dick, G. M., & Muller-Delp, J. M. (2009). Aging and muscle fiber type alter K+ channel contributions to the myogenic response in skeletal muscle arterioles. *Journal of Applied Physiology (Bethesda, MD: 1985)*, *107*(2), 389–398.

Katsumi, A., Naoe, T., Matsushita, T., Kaibuchi, K., & Schwartz, M. A. (2005). Integrin activation and matrix binding mediate cellular responses to mechanical stretch. *The Journal of Biological Chemistry*, *280*(17), 16546–16549.

Kinlay, S., Creager, M. A., Fukumoto, M., et al. (2001). Endothelium-derived nitric oxide regulates arterial elasticity in human arteries in vivo. *Hypertension*, *38*(5), 1049–1053.

Kohn, J. C., Chen, A., Cheng, S., Kowal, D. R., King, M. R., & Reinhart-King, C. A. (2016). Mechanical heterogeneities in the subendothelial matrix develop with age and decrease with exercise. *Journal of Biomechanics*, *49*(9), 1447–1453.

Kohn, J. C., Lampi, M. C., & Reinhart-King, C. A. (2015). Age-related vascular stiffening: Causes and consequences. *Frontiers in Genetics*, *6*, 112.

Kuo, J. C., Han, X., Hsiao, C. T., Yates, J. R., 3rd, & Waterman, C. M. (2011). Analysis of the myosin-II-responsive focal adhesion proteome reveals a role for β-Pix in negative regulation of focal adhesion maturation. *Nature Cell Biology*, *13*(4), 383–393.

Kusche-Vihrog, K., Urbanova, K., Blanque, A., Wilhelmi, M., Schillers, H., Kliche, K., et al. (2011). C-reactive protein makes human endothelium stiff and tight. *Hypertension*, *57*(2), 231–237.

Lacolley, P., Li, Z., Challande, P., & Regnault, V. (2017). SRF/myocardin: A novel molecular axis regulating vascular smooth muscle cell stiffening in hypertension. *Cardiovascular Research*, *113*(2), 120–122.

Lacolley, P., Regnault, V., Nicoletti, A., Li, Z., & Michel, J. B. (2012). The vascular smooth muscle cell in arterial pathology: A cell that can take on multiple roles. *Cardiovascular Research*, *95*(2), 194–204.

Lakatta, E. G. (2003). Arterial and cardiac aging: Major shareholders in cardiovascular disease enterprises: Part III: Cellular and molecular clues to heart and arterial aging. *Circulation, 107*(3), 490–497.

Lakatta, E. G. (2013). The reality of aging viewed from the arterial wall. *Artery Research*, 7(2), 73–80.

Lakatta, E. G. (2015). So! What's aging? Is cardiovascular aging a disease? *Journal of Molecular and Cellular Cardiology, 83*, 1–13.

Lakatta, E. G., & Levy, D. (2003). Arterial and cardiac aging: Major shareholders in cardiovascular disease enterprises: Part I: Aging arteries: A "set up" for vascular disease. *Circulation, 107*(1), 139–146.

Lang, F. (2011). Stiff endothelial cell syndrome in vascular inflammation and mineralocorticoid excess. *Hypertension, 57*(2), 146–147.

Lawrence, A. R., & Gooch, K. J. (2009). Transmural pressure and axial loading interactively regulate arterial remodeling ex vivo. *American Journal of Physiology. Heart and Circulatory Physiology, 297*(1), H475–H484.

Legate, K. R., & Fässler, R. (2009). Mechanisms that regulate adaptor binding to β-integrin cytoplasmic tails. *Journal of Cell Science, 122*(2), 187–198.

Li, Z., Cheng, H., Lederer, W. J., Froehlich, J., & Lakatta, E. G. (1997). Enhanced proliferation and migration and altered cytoskeletal proteins in early passage smooth muscle cells from young and old rat aortic explants. *Experimental and Molecular Pathology, 64*(1), 1–11.

Li, Z., Froehlich, J., Galis, Z. S., & Lakatta, E. G. (1999). Increased expression of matrix metalloproteinase-2 in the thickened intima of aged rats. *Hypertension, 33*(1), 116–123.

Li, C., & Xu, Q. (2000). Mechanical stress-initiated signal transductions in vascular smooth muscle cells. *Cellular Signalling, 12*(7), 435–445.

Lim, S., & Park, S. (2014). Role of vascular smooth muscle cell in the inflammation of atherosclerosis. *BMB Reports, 47*(1), 1–7.

Lim, S. M., Trzeciakowski, J. P., Sreenivasappa, H., Dangott, L. J., & Trache, A. (2012). RhoA-induced cytoskeletal tension controls adaptive cellular remodeling to mechanical signaling. *Integrative Biology, 4*(6), 615–627.

Lundberg, M. S., & Crow, M. T. (1999). Age-related changes in the signaling and function of vascular smooth muscle cells. *Experimental Gerontology, 34*(4), 549–557.

Luo, B. H., Carman, C. V., & Springer, T. A. (2007). Structural basis of integrin regulation and signaling. *Annual Review of Immunology, 25*, 619–647.

Luttrell, M., Kim, H., Shin, S. Y., Holly, D., Massett, M. P., & Woodman, C. R. (2020). Heterogeneous effect of aging on vasorelaxation responses in large and small arteries. *Physiological Reports, 8*(1), e14341.

Madamanchi, N. R., Vendrov, A., & Runge, M. S. (2005). Oxidative stress and vascular disease. *Arteriosclerosis, Thrombosis, and Vascular Biology, 25*(1), 29–38.

Martinez-Lemus, L. A., Crow, T., Davis, M. J., & Meininger, G. A. (2005). alphavbeta3- and alpha5beta1-integrin blockade inhibits myogenic constriction of skeletal muscle resistance arterioles. *American Journal of Physiology. Heart and Circulatory Physiology, 289*(1), H322–H329.

Martinez-Lemus, L. A., Hill, M. A., Bolz, S. S., Pohl, U., & Meininger, G. A. (2004). Acute mechanoadaptation of vascular smooth muscle cells in response to continuous arteriolar vasoconstriction: Implications for functional remodeling. *The FASEB Journal, 18*(6), 708–710.

Martinez-Lemus, L. A., Hill, M. A., & Meininger, G. A. (2009). The plastic nature of the vascular wall: A continuum of remodeling events contributing to control of arteriolar diameter and structure. *Physiology (Bethesda), 24*, 45–57.

Martinez-Lemus, L. A., Sun, Z., Trache, A., Trzciakowski, J. P., & Meininger, G. A. (2005). Integrins and regulation of the microcirculation: From arterioles to molecular studies using atomic force microscopy. *Microcirculation, 12*(1), 99–112.

Martinez-Lemus, L. A., Wu, X., Wilson, E., Hill, M. A., Davis, G. E., Davis, M. J., et al. (2003). Integrins as unique receptors for vascular control. *Journal of Vascular Research, 40*(3), 211–233.

Massett, M. P., Bywaters, B. C., Gibbs, H. C., Trzeciakowski, J. P., Padgham, Ṡ., Chen, J., et al. (2020). Loss of smooth muscle alpha-actin effects on mechanosensing and cell-matrix adhesions. *Experimental Biology and Medicine (Maywood, N.J.), 245*(4), 374–384.

Mitchell, G. F., Parise, H., Benjamin, E. J., Larson, M. G., Keyes, M. J., Vita, J. A., et al. (2004). Changes in arterial stiffness and wave reflection with advancing age in healthy men and women: The Framingham Heart Study. *Hypertension, 43*(6), 1239–1245.

Mitra, S. K., Hanson, D. A., & Schlaepfer, D. D. (2005). Focal adhesion kinase: In command and control of cell motility. *Nature Reviews. Molecular Cell Biology, 6*(1), 56–68.

Mogford, J. E., Davis, G. E., & Meininger, G. A. (1997). RGDN peptide interaction with endothelial alpha5beta1 integrin causes sustained endothelin-dependent vasoconstriction of rat skeletal muscle arterioles. *The Journal of Clinical Investigation, 100*(6), 1647–1653.

Mogford, J. E., Davis, G. E., Platts, S. H., & Meininger, G. A. (1996). Vascular smooth muscle alpha v beta 3 integrin mediates arteriolar vasodilation in response to RGD peptides. *Circulation Research, 79*(4), 821–826.

Moncada, S., & Higgs, A. (1993). The L-arginine-nitric oxide pathway. *The New England Journal of Medicine, 329*(27), 2002–2012.

Muller-Delp, J. M., Spier, S. A., Ramsey, M. W., & Delp, M. D. (2002). Aging impairs endothelium-dependent vasodilation in rat skeletal muscle arterioles. *American Journal of Physiology. Heart and Circulatory Physiology, 283*(4), H1662–H1672.

Muller-Delp, J., Spier, S. A., Ramsey, M. W., Lesniewski, L. A., Papadopoulos, A., Humphrey, J. D., et al. (2002). Effects of aging on vasoconstrictor and mechanical properties of rat skeletal muscle arterioles. *American Journal of Physiology. Heart and Circulatory Physiology, 282*(5), H1843–H1854.

Musch, T. I., Eklund, K. E., Hageman, K. S., & Poole, D. C. (2004). Altered regional blood flow responses to submaximal exercise in older rats. *Journal of Applied Physiology (Bethesda, MD: 1985), 96*(1), 81–88.

Na, S., Meininger, G. A., & Humphrey, J. D. (2007). A theoretical model for F-actin remodeling in vascular smooth muscle cells subjected to cyclic stretch. *Journal of Theoretical Biology, 246*(1), 87–99.

Na, S., Trache, A., Trzeciakowski, J., Sun, Z., Meininger, G. A., & Humphrey, J. D. (2008). Time-dependent changes in smooth muscle cell stiffness and focal adhesion area in response to cyclic equibiaxial stretch. *Annals of Biomedical Engineering, 36*(3), 369–380.

Najjar, S. S., Scuteri, A., & Lakatta, E. G. (2005). Arterial aging: Is it an immutable cardiovascular risk factor? *Hypertension, 46*(3), 454–462.

Oberleithner, H., Riethmuller, C., Schillers, H., MacGregor, G. A., de Wardener, H. E., & Hausberg, M. (2007). Plasma sodium stiffens vascular endothelium and reduces nitric oxide release. *Proceedings of the National Academy of Sciences of the United States of America, 104*(41), 16281–16286.

Opazo-Saez, A., Zhang, W., Wu, Y., Turner, C. E., Tang, D. D., & Gunst, S. J. (2004). Tension development during contractile stimulation of smooth muscle requires recruitment of paxillin and vinculin to the membrane. *American Journal of Physiology-Cell Physiology, 286*(2), C433–C447.

O'Rourke, M. F., & Hashimoto, J. (2007). Mechanical factors in arterial aging: A clinical perspective. *Journal of the American College of Cardiology, 50*(1), 1–13.

O'Rourke, M. F., & Safar, M. E. (2005). Relationship between aortic stiffening and microvascular disease in brain and kidney: Cause and logic of therapy. *Hypertension, 46*(1), 200–204.

Ortman, J. M., Velkoff, V. A., & Hogan, H. (2014, May). *An aging nation: The older population in the United States*. US Census Bureau. https://www.census.gov/library/publications/2014/demo/p25-1140.html.

Owens, G. K., Kumar, M. S., & Wamhoff, B. R. (2004). Molecular regulation of vascular smooth muscle cell differentiation in development and disease. *Physiological Reviews*, *84*(3), 767–801.

Palmer, R. M., Ashton, D. S., & Moncada, S. (1988). Vascular endothelial cells synthesize nitric oxide from L-arginine. *Nature*, *333*(6174), 664–666.

Park, Y., Prisby, R. D., Behnke, B. J., Dominguez, J. M., 2nd, Lesniewski, L. A., Donato, A. J., et al. (2012). Effects of aging, TNF-α, and exercise training on angiotensin II-induced vasoconstriction of rat skeletal muscle arterioles. *Journal of Applied Physiology (Bethesda, MD: 1985)*, *113*(7), 1091–1100.

Pasapera, A. M., Schneider, I. C., Rericha, E., Schlaepfer, D. D., & Waterman, C. M. (2010). Myosin II activity regulates vinculin recruitment to focal adhesions through FAK-mediated paxillin phosphorylation. *The Journal of Cell Biology*, *188*(6), 877–890.

Qiu, H., Zhu, Y., Sun, Z., Trzeciakowski, J. P., Gansner, M., Depre, C., et al. (2010). Short communication: Vascular smooth muscle cell stiffness as a mechanism for increased aortic stiffness with aging. *Circulation Research*, *107*(5), 615–619.

Rodriguez-Manas, L., El-Assar, M., Vallejo, S., Lopez-Doriga, P., Solis, J., Petidier, R., et al. (2009). Endothelial dysfunction in aged humans is related with oxidative stress and vascular inflammation. *Aging Cell*, *8*(3), 226–238.

Sanyour, H. J., Li, N., Rickel, A. P., Childs, J. D., Kinser, C. N., & Hong, Z. (2019). Membrane cholesterol and substrate stiffness co-ordinate to induce the remodelling of the cytoskeleton and the alteration in the biomechanics of vascular smooth muscle cells. *Cardiovascular Research*, *115*(8), 1369–1380.

Sanyour, H. J., Li, N., Rickel, A. P., Torres, H. M., Anderson, R. H., Miles, M. R., et al. (2020). Statin-mediated cholesterol depletion exerts coordinated effects on the alterations in rat vascular smooth muscle cell biomechanics and migration. *The Journal of Physiology*, *598*(8), 1505–1522.

Saphirstein, R. J., Gao, Y. Z., Jensen, M. H., Gallant, C. M., Vetterkind, S., Moore, J. R., et al. (2013). The focal adhesion: A regulated component of aortic stiffness. *PLoS One*, *8*(4), e62461.

Satoh, K., Fukumoto, Y., & Shimokawa, H. (2011). Rho-kinase: Important new therapeutic target in cardiovascular diseases. *American Journal of Physiology. Heart and Circulatory Physiology*, *301*(2), H287–H296.

Sauzeau, V., Le Jeune, H., Cario-Toumaniantz, C., Smolenski, A., Lohmann, S. M., Bertoglio, J., et al. (2000). Cyclic GMP-dependent protein kinase signaling pathway inhibits RhoA-induced Ca2+ sensitization of contraction in vascular smooth muscle. *The Journal of Biological Chemistry*, *275*(28), 21722–21729.

Schiffrin, E. L. (1999). State-of-the-Art lecture. Role of endothelin-1 in hypertension. *Hypertension*, *34*(4 Pt. 2), 876–881. Retrieved from http://www.ncbi.nlm.nih.gov/pubmed/10523377.

Schiller, H. B., Hermann, M. R., Polleux, J., Vignaud, T., Zanivan, S., Friedel, C. C., et al. (2013). β1- and αv-class integrins cooperate to regulate myosin II during rigidity sensing of fibronectin-based microenvironments. *Nature Cell Biology*, *15*(6), 625–636.

Schmitt, M., Avolio, A., Qasem, A., McEniery, C. M., Butlin, M., Wilkinson, I. B., et al. (2005). Basal NO locally modulates human iliac artery function in vivo. *Hypertension*, *46*(1), 227–231.

Schriefl, A. J., Zeindlinger, G., Pierce, D. M., Regitnig, P., & Holzapfel, G. A. (2012). Determination of the layer-specific distributed collagen fibre orientations in human thoracic and abdominal aortas and common iliac arteries. *Journal of the Royal Society, Interface*, *9*(71), 1275–1286.

Seawright, J. W., Sreenivasappa, H., Gibbs, H. C., Padgham, S., Shin, S. Y., Chaponnier, C., et al. (2018). Vascular smooth muscle contractile function declines with age in skeletal muscle feed arteries. *Frontiers in Physiology*, *9*, 856.

Seawright, J. W., Trache, A., Wilson, E., & Woodman, C. R. (2016). Short-duration increases in intraluminal pressure improve vasoconstrictor responses in aged skeletal muscle feed arteries. *European Journal of Applied Physiology*, *116*(5), 931–937.

Sehgel, N. L., Sun, Z., Hong, Z., Hunter, W. C., Hill, M. A., Vatner, D. E., et al. (2015). Augmented vascular smooth muscle cell stiffness and adhesion when hypertension is superimposed on aging. *Hypertension*, *65*(2), 370–377.

Sehgel, N. L., Vatner, S. F., & Meininger, G. A. (2015). "smooth muscle cell stiffness syndrome"—Revisiting the structural basis of arterial stiffness. *Frontiers in Physiology*, *6*, 335.

Shimokawa, H., & Rashid, M. (2007). Development of Rho-kinase inhibitors for cardiovascular medicine. *Trends in Pharmacological Sciences*, *28*(6), 296–302.

Sinkler, S. Y., & Segal, S. S. (2014). Aging alters reactivity of microvascular resistance networks in mouse gluteus maximus muscle. *American Journal of Physiology. Heart and Circulatory Physiology*, *307*(6), H830–H839.

Springo, Z., Toth, P., Tarantini, S., Ashpole, N. M., Tucsek, Z., Sonntag, W. E., et al. (2015). Aging impairs myogenic adaptation to pulsatile pressure in mouse cerebral arteries. *Journal of Cerebral Blood Flow and Metabolism*, *35*(4), 527–530.

Sreenivasappa, H., Chaki, S. P., Lim, S. M., Trzeciakowski, J. P., Davidson, M. W., Rivera, G. M., et al. (2014). Selective regulation of cytoskeletal tension and cell-matrix adhesion by RhoA and Src. *Integrative Biology*, *6*(8), 743–754.

Stamenovic, D. (2008). Cytoskeletal mechanics in airway smooth muscle cells. *Respiratory Physiology & Neurobiology*, *163*(1–3), 25–32.

Sun, Z. (2015). Aging, arterial stiffness, and hypertension. *Hypertension*, *65*(2), 252–256.

Takasaki, I., Chobanian, A. V., Sarzani, R., & Brecher, P. (1990). Effect of hypertension on fibronectin expression in the rat aorta. *The Journal of Biological Chemistry*, *265*(35), 21935–21939.

Takemoto, M., Sun, J., Hiroki, J., Shimokawa, H., & Liao, J. K. (2002). Rho-kinase mediates hypoxia-induced downregulation of endothelial nitric oxide synthase. *Circulation*, *106*(1), 57–62.

Tang, D., Mehta, D., & Gunst, S. J. (1999). Mechanosensitive tyrosine phosphorylation of paxillin and focal adhesion kinase in tracheal smooth muscle. *The American Journal of Physiology*, *276*(1), C250–C258.

Tomiyama, H., Arai, T., Koji, Y., Yambe, M., Motobe, K., Zaydun, G., et al. (2004). The age-related increase in arterial stiffness is augmented in phases according to the severity of hypertension. *Hypertension Research*, *27*(7), 465–470.

Trache, A., & Meininger, G. A. (2008). Atomic force microscopy (AFM). *Current Protocols in Microbiology*. Chapter 2, Unit 2C 2.

Turner, C. E. (2000). Paxillin and focal adhesion signalling. *Nature Cell Biology*, *2*(12), E231–E236.

Van Bortel, L. M., Laurent, S., Boutouyrie, P., Chowienczyk, P., Cruickshank, J. K., De Backer, T., et al. (2012). Expert consensus document on the measurement of aortic stiffness in daily practice using carotid-femoral pulse wave velocity. *Journal of Hypertension*, *30*(3), 445–448.

Van der Loo, B., Labugger, R., Skepper, J. N., Bachschmid, M., Kilo, J., Powell, J. M., et al. (2000). Enhanced peroxynitrite formation is associated with vascular aging. *The Journal of Experimental Medicine*, *192*(12), 1731–1744.

Vanhoutte, P. M., Shimokawa, H., Feletou, M., & Tang, E. H. (2017). Endothelial dysfunction and vascular disease—A 30th anniversary update. *Acta Physiologica (Oxford, England)*, *219*(1), 22–96.

Vasquez-Vivar, J., Kalyanaraman, B., & Martasek, P. (2003). The role of tetrahydrobiopterin in superoxide generation from eNOS: Enzymology and physiological implications. *Free Radical Research*, *37*(2), 121–127.

Vita, J. A., Treasure, C. B., Nabel, E. G., McLenachan, J. M., Fish, R. D., Yeung, A. C., et al. (1990). Coronary vasomotor response to acetylcholine relates to risk factors for coronary artery disease. *Circulation, 81*(2), 491–497.

Vlachopoulos, C., Aznaouridis, K., & Stefanadis, C. (2010). Prediction of cardiovascular events and all-cause mortality with arterial stiffness: A systematic review and meta-analysis. *Journal of the American College of Cardiology, 55*(13), 1318–1327.

Wagenseil, J. E., & Mecham, R. P. (2009). Vascular extracellular matrix and arterial mechanics. *Physiological Reviews, 89*(3), 957–989.

Wagenseil, J. E., Nerurkar, N. L., Knutsen, R. H., Okamoto, R. J., Li, D. Y., & Mecham, R. P. (2005). Effects of elastin haploinsufficiency on the mechanical behavior of mouse arteries. *American Journal of Physiology. Heart and Circulatory Physiology, 289*(3), H1209–H1217.

Wagner, J. U. G., Chavakis, E., Rogg, E. M., Muhly-Reinholz, M., Glaser, S. F., Gunther, S., et al. (2018). Switch in laminin β2 to laminin β1 isoforms during aging controls endothelial cell functions—Brief report. *Arteriosclerosis, Thrombosis, and Vascular Biology, 38*(5), 1170–1177.

Waitkus-Edwards, K. R., Martinez-Lemus, L. A., Wu, X., Trzeciakowski, J. P., Davis, M. J., Davis, G. E., et al. (2002). alpha(4)beta(1) Integrin activation of L-type calcium channels in vascular smooth muscle causes arteriole vasoconstriction. *Circulation Research, 90*(4), 473–480.

Wang, Y., & Gilmore, T. D. (2003). Zyxin and paxillin proteins: Focal adhesion plaque LIM domain proteins go nuclear. *Biochimica et Biophysica Acta, 1593*(2–3), 115–120.

Wang, M., Monticone, R. E., & Lakatta, E. G. (2010). Arterial aging: A journey into subclinical arterial disease. *Current Opinion in Nephrology and Hypertension, 19*(2), 201–207.

Wang, M., Zhao, D., Spinetti, G., Zhang, J., Jiang, L. Q., Pintus, G., et al. (2006). Matrix metalloproteinase 2 activation of transforming growth factor-beta1 (TGF-beta1) and TGF-beta1-type II receptor signaling within the aged arterial wall. *Arteriosclerosis, Thrombosis, and Vascular Biology, 26*(7), 1503–1509.

Webb, D. J., Donais, K., Whitmore, L. A., Thomas, S. M., Turner, C. E., Parsons, J. T., et al. (2004). FAK-Src signalling through paxillin, ERK and MLCK regulates adhesion disassembly. *Nature Cell Biology, 6*(2), 154–161.

Wei, W. C., Lin, H. H., Shen, M. R., & Tang, M. J. (2008). Mechanosensing machinery for cells under low substratum rigidity. *American Journal of Physiology. Cell Physiology, 295*(6), C1579–C1589.

Weis, S. M. (2007). Evaluating integrin function in models of angiogenesis and vascular permeability. *Methods in Enzymology, 426*, 505–528.

White, A. R., Ryoo, S., Li, D., Champion, H. C., Steppan, J., Wang, D., et al. (2006). Knockdown of arginase I restores NO signaling in the vasculature of old rats. *Hypertension, 47*(2), 245–251.

Wilkinson, I. B., Qasem, A., McEniery, C. M., Webb, D. J., Avolio, A. P., & Cockcroft, J. R. (2002). Nitric oxide regulates local arterial distensibility in vivo. *Circulation, 105*(2), 213–217.

Williams, D. A., & Segal, S. S. (1993). Feed artery role in blood flow control to rat hindlimb skeletal muscles. *The Journal of Physiology, 463*, 631–646.

Woodman, C. R., Price, E. M., & Laughlin, M. H. (2002). Aging induces muscle-specific impairment of endothelium-dependent dilation in skeletal muscle feed arteries. *Journal of Applied Physiology (Bethesda, MD: 1985), 93*(5), 1685–1690.

Woodman, C. R., Price, E. M., & Laughlin, M. H. (2005). Shear stress induces eNOS mRNA expression and improves endothelium-dependent dilation in senescent soleus muscle feed arteries. *Journal of Applied Physiology (Bethesda, MD: 1985), 98*(3), 940–946.

Woodside, D. G., Obergfell, A., Leng, L., Wilsbacher, J. L., Miranti, C. K., Brugge, J. S., et al. (2001). Activation of Syk protein tyrosine kinase through interaction with integrin beta cytoplasmic domains. *Current Biology*, *11*(22), 1799–1804.

Worth, N. F., Rolfe, B. E., Song, J., & Campbell, G. R. (2001). Vascular smooth muscle cell phenotypic modulation in culture is associated with reorganisation of contractile and cytoskeletal proteins. *Cell Motility and the Cytoskeleton*, *49*(3), 130–145.

Wu, X., Davis, G. E., Meininger, G. A., Wilson, E., & Davis, M. J. (2001). Regulation of the L-type calcium channel by alpha 5beta 1 integrin requires signaling between focal adhesion proteins. *The Journal of Biological Chemistry*, *276*(32), 30285–30292.

Wu, X., Mogford, J. E., Platts, S. H., Davis, G. E., Meininger, G. A., & Davis, M. J. (1998). Modulation of calcium current in arteriolar smooth muscle by alphav beta3 and alpha5 beta1 integrin ligands. *The Journal of Cell Biology*, *143*(1), 241–252.

Yamada, H., Sakata, N., Wada, H., Tashiro, T., & Tayama, E. (2015). Age-related distensibility and histology of the ascending aorta in elderly patients with acute aortic dissection. *Journal of Biomechanics*, *48*(12), 3267–3273.

Yoon, H. E., Kim, E. N., Kim, M. Y., Lim, J. H., Jang, I. A., Ban, T. H., et al. (2016). Age-associated changes in the vascular renin-angiotensin system in mice. *Oxidative Medicine and Cellular Longevity*, *2016*, 6731093.

Zaidel-Bar, R., Milo, R., Kam, Z., & Geiger, B. (2007). A paxillin tyrosine phosphorylation switch regulates the assembly and form of cell-matrix adhesions. *Journal of Cell Science*, *120*(Pt. 1), 137–148.

Zhang, W., & Gunst, S. J. (2006). Dynamic association between α-actinin and β-integrin regulates contraction of canine tracheal smooth muscle. *The Journal of Physiology*, *572*(Pt. 3), 659–676.

Zhang, J., Zhao, X., Vatner, D. E., McNulty, T., Bishop, S., Sun, Z., et al. (2016). Extracellular matrix disarray as a mechanism for greater abdominal versus thoracic aortic stiffness with aging in primates. *Arteriosclerosis, Thrombosis, and Vascular Biology*, *36*(4), 700–706.

Zhu, W., Kim, B. C., Wang, M., Huang, J., Isak, A., Bexiga, N. M., et al. (2018). TGFβ1 reinforces arterial aging in the vascular smooth muscle cell through a long-range regulation of the cytoskeletal stiffness. *Scientific Reports*, *8*, 2668.

Zhu, Y., Qiu, H., Trzeciakowski, J. P., Sun, Z., Li, Z., Hong, Z., et al. (2012). Temporal analysis of vascular smooth muscle cell elasticity and adhesion reveals oscillation waveforms that differ with aging. *Aging Cell*, *11*, 741–750.

CHAPTER SEVEN

Mechanobiology of microvesicle release, uptake, and microvesicle-mediated activation

Samantha C. Schwager and Cynthia A. Reinhart-King*
Department of Biomedical Engineering, Vanderbilt University, Nashville, TN, United States
*Corresponding author: e-mail address: cynthia.reinhart-king@vanderbilt.edu

Contents

Abstract

Microvesicles are small, membrane-bound vesicles that are shed from the plasma membrane of cells into the extracellular space. Microvesicles contain a variety of cargo not typically thought to be released from cells, including receptor tyrosine kinases, cytosolic signaling proteins, and microRNAs, which are transferred from donor cells to recipient cells. The transfer of microvesicle cargo can result in the transformation of recipient cells thereby supporting disease progression, including modified fibroblast metabolism, epithelial cell contractility, vascular remodeling, and immune cell inflammatory signaling. Additionally, microvesicles are believed to play prominent roles in cell–cell communication and disease progression as they are detected at elevated concentrations in diseased tissues. As microvesicle uptake by recipient cells can modulate cell function to promote disease progression, understanding the mechanisms and mechanosensitivity of microvesicle release, internalization, and the resulting signaling is crucial to fully comprehend their functions in disease. Here, we review recent advances in the understanding of actomyosin-regulated microvesicle biogenesis, microvesicle uptake via pinocytosis, and the resulting cellular transformation. We discuss the effects of altered cell contractility, mode of cell migration, and extracellular matrix compliance on microvesicle signaling, with direct implications in disease progression and identifying future therapeutic targets.

Current Topics in Membranes, Volume 86
ISSN 1063-5823
https://doi.org/10.1016/bs.ctm.2020.08.004

1. Introduction

Extracellular vesicles (EVs) have recently emerged as important mediators of cell communication and disease progression. The term "extracellular vesicles" refers to lipid bilayer-bound vesicles released from cells and include exosomes, microvesicles (MVs), large oncosomes, and apoptotic bodies (Vader, Breakefield, & Wood, 2014; Wendler et al., 2017). Exosomes, 50–150 nm in diameter, are formed via the endocytic pathway as secretory multi-vesicular bodies (MVBs) fuse with the plasma membrane (Wendler et al., 2017). Exosomes are distinguished from other EVs by the presence of endosomal sorting complex required for transport (ESCRT) proteins that regulate MVB transportation, including Alix, TSG101, HSP70, and HSP90 (Doyle & Wang, 2019). Additionally, exosome release relies on the neutral sphingomyelinase pathway, resulting in the enrichment of CD63, CD9, and CD81 tetraspanins in exosomes (Doyle & Wang, 2019). MVs, 100 or 200 nm– 1 μm in diameter, are formed via the budding and subsequent shedding of the plasma membrane (Wendler et al., 2017). As a result, MVs primarily contain cytosolic and plasma membrane-associated proteins, including tetraspanins (Doyle & Wang, 2019). Cytoskeletal proteins, heat shock proteins, integrins, and glycosylated and phosphorylated proteins are also commonly identified in MVs (Doyle & Wang, 2019). Large oncosomes, classified as EVs larger than 1 μm in diameter, are produced by plasma membrane blebbing but differ in their buoyant density from other EVs (Colombo, Raposo, & Théry, 2014; Wendler et al., 2017). Large oncosomes are believed to only be released from cancer and cancer-associated cells, with the number of large oncosomes released correlating with cancer aggressiveness (Minciacchi, Freeman, & Di Vizio, 2015). As such, oncosomes primarily contain oncogenic proteins (Minciacchi et al., 2015). Lastly, apoptotic bodies are approximately 50 nm – 5 μm in diameter and are shed from the plasma membrane of apoptotic cells (Vader et al., 2014). Unlike other EVs, apoptotic bodies contain complete organelles and higher levels of proteins associated with the nucleus, mitochondria, Golgi apparatus, and endoplasmic reticulum (Doyle & Wang, 2019).

The role of MVs in disease progression are newly emerging. MVs were first described in 1967 after particulate material, described as "dust," was observed to be released from platelets (Wolf, 1967). It has now been determined that this "dust" is plasma membrane bound vesicles that are shed from the plasma membrane of both healthy and diseased cells. Now termed microvesicles, these vesicles contain a variety of cargo not typically thought to be released from cells, including receptor tyrosine kinases, cytosolic signaling proteins, and microRNAs. MVs are believed to play a prominent role in disease progression as they have been detected in body fluids including blood, urine, and synovial fluid from diseased patients, and elevated MV counts have been observed in cardiovascular diseases and cancerous tissue (Dickhout & Koenen, 2018; György et al., 2011; Menck et al., 2017). The transfer of cargo via MVs can result in the transformation of recipient cells to support disease progression. For example, MV signaling can promote extracellular matrix and vascular remodeling, recruit tumor-promoting stromal cells, and induce transformed phenotypes within neighboring cells (Antonyak et al., 2011; Bordeleau et al., 2016; Huaitong et al., 2017; Jiang et al., 2019; Kawamoto et al., 2012; Menck et al., 2015; Schwager et al., 2019).

The mechanobiology and membrane dynamics of MV release and uptake are under investigation. As MV uptake by recipient cells can modulate cell function, understanding the mechanisms of MV release, internalization, and the resulting signaling is crucial to fully comprehend the roles of MVs in disease progression. Here, we review recent advances in the understanding of actomyosin-regulated MV biogenesis, MV uptake via pinocytosis, and subsequent MV signaling. We additionally discuss the roles of altered cell contractility, mode of cell migration, and matrix compliance in regulating MV signaling.

2. Mechanobiology of microvesicle biogenesis

2.1 Intracellular contractility and MV biogenesis

The release of MVs from the plasma membrane of cells is integral to MV-mediated communication during disease progression. While the complete mechanism of MV release is not defined, MV biogenesis is believed to be a multistep process that includes the trafficking of cellular cargo to the plasma membrane, a rearrangement of membrane lipids, and vesicle pinching mediated by contractile machinery (Tricarico, Clancy, & D'Souza-Schorey, 2016). As contractile machinery is essential for MVs release, altered cell contractility likely affects MV release and is an area currently being investigated (Li, Antonyak, Zhang, & Cerione, 2012; Sedgwick, Clancy, Balmert, & D'Souza-Schorey, 2015).

Recent work has outlined a contractility-based MV release pathway regulated by ARF6, a small GTP-binding protein that regulates

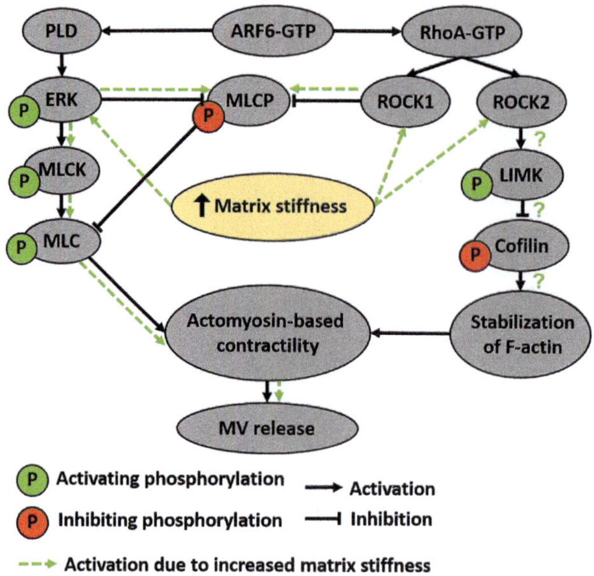

Fig. 1 Proposed pathway of intracellular contractility and matrix stiffness regulation of MV release. MV release is regulated by ARF6-GTP signaling through ERK and RhoA-GTP pathways. Increased matrix stiffness may regulate MV release through a variety of pathways. Green dashed lines represent pathways activated by matrix stiffness.

membrane trafficking and actin cytoskeleton remodeling (Li et al., 2012; Muralidharan-Chari et al., 2009; Sedgwick et al., 2015) (Fig. 1). Briefly, activation of phospholipase D (PLD) by ARF6-GTP promotes ERK activation. ERK subsequently activates myosin light chain kinase (MLCK) which signals through myosin light chain (MLC) to induce the actomyosin-based contraction required for MV release (Muralidharan-Chari et al., 2009).

Downstream of ARF6 signaling, a separate RhoA GTPase signaling pathway highly regulates MV release (Li et al., 2012; Sedgwick et al., 2015) (Fig. 1). RhoA activation targets both Rho-associated protein kinase 1 (ROCK1) and Rho-associated protein kinase 2 (ROCK2). Together, ROCK1 and ERK inhibit myosin light chain phosphatase (MLCP) which reduces the inactivation of MLC, further inducing actomyosin-based contractility and MV formation (Sedgwick et al., 2015). Alternatively, ROCK2 activates Lim kinase (LIMK), resulting in the inactivation of the actin-binding disassembly factor cofilin, resulting in the stabilization of F-actin necessary for MV generation (Li et al., 2012).

Importantly, a variety of external stimuli that alter cellular contractility also affect MV release dynamics (Li et al., 2012). Inhibition of ROCK1 and ROCK2 in cancer cells using the small molecule Y-27632 inhibitor resulted in decreased cell contractility and eliminated the appearance of MVs on the cell surface (Li et al., 2012), suggesting that cells with reduced ROCK-mediated contractility have decreased MV release. Additionally, it is well-established that contractile forces can be modulated by MLCK, which promotes MLC phosphorylation and the assembly of actin-myosin filaments (Gallagher, Herring, & Stull, 1997). Overexpression of MLCK in HeLa cells increased MV formation (Li et al., 2012). Similarly, treatment of melanoma cells with blebbistatin, a small molecule inhibitor of myosin II, or with latrunculin A, an actin-binding toxin that inhibits actin polymerization, both reduced MV release (Muralidharan-Chari et al., 2009). Altogether, these results suggest that external stimuli that alter actomyosin dynamics have profound effects on MV release. Further work is required to determine whether additional external stimuli, such as growth factors or cytokines, released by surrounding cells regulate MV release.

2.2 Amoeboid migration promotes MV release

Cell migration is a fundamental behavior in a variety of tissue processes and disease states, including development, wound healing, diabetes, and cancer (Bear & Haugh, 2014; Franz, Jones, & Ridley, 2002; Xuan et al., 2014). During cell migration, cells primarily display two distinct modes of motility: amoeboid and mesenchymal. Amoeboid migration is classically defined as cell migration that exhibits fast speeds, high levels of actomyosin contractility, round cell shapes, and little to no matrix adhesion (O'Neill et al., 2018). Amoeboid migration is protease-independent; cells squeeze through preexisting gaps or tracks in the extracellular matrix (ECM) (Charras & Paluch, 2008) (Fig. 2A). Alternatively, cells undergoing mesenchymal migration move much slower, are weakly polarized, and exhibit multiple types of protrusions, including lamellipodia and filopodia (Bear & Haugh, 2014). Mesenchymal migration relies on strong integrin-mediated adhesions to the matrix and proteases to degrade the matrix (Bear & Haugh, 2014) (Fig. 2B). Amoeboid and mesenchymal migration are highly regulated by the Rho GTPases, RhoA and Rac1 GTPase, respectively. As intracellular actin dynamics and Rho GTPase activity are required for both MV release and regulation of cell migration mode, mode of cell migration appears to be linked to MV release.

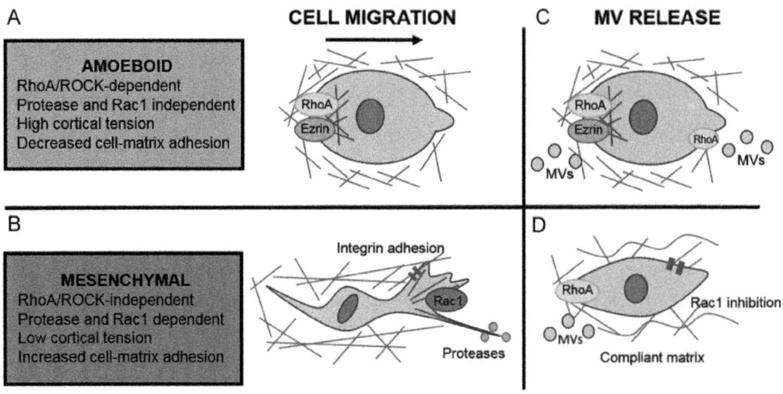

Fig. 2 RhoA regulates MV release (A) In amoeboid migration, actomyosin contractility and cytoplasmic flow form membrane blebs at the leading edge. RhoA and ezrin are polarized to the rear edge. (B) In mesenchymal migration, Rac1 expression at the leading edge drives actin polymerization in lamellipodia and filopodia protrusions which, along with increased cell-matrix integrin adhesion and protease activity, drive migration. (C) In amoeboid migration, MVs are likely released from both the leading edge, with short-lived RhoA expression, and the trailing edge where RhoA is constitutively expressed. (D) In mesenchymal migration, evidence suggests that to release substantial quantities of MVs, Rac1 inhibition is required. Inhibition of Rac1 increases RhoA activation, resulting in increased MV release on compliant matrices, but not stiff matrices.

In amoeboid migration, activation of RhoA GTPase at the cell rear leads to the assembly of contractile actin and myosin filaments. Actomyosin-based contractility at the cell rear generates hydrostatic pressure and cytoplasmic flow to form plasma membrane blebs at the leading edge, which are believed to be due to a decrease in membrane-actin cortex attachment or a local rupture of the actin cortex (Caswell & Zech, 2018; Charras & Paluch, 2008; Paluch & Raz, 2013). Blebs are distinct from other cellular protrusions as their formation is pressure-driven rather than actin polymerization-driven (Pollard & Borisy, 2003). For blebbing to induce amoeboid cell migration, blebs need to be formed at the leading edge of the cell, resulting in a transfer of cytoplasm in the direction of migration (Paluch & Raz, 2013) (Fig. 2A). As previously described, RhoA GTPase has been implicated in contractility-based MV formation through ROCK1 and ROCK2 signaling. Thus, RhoA GTPase activation may link amoeboid migration with increased MV release.

Whether MVs are released from the leading edge or trailing edge of migrating amoeboid cells is not currently defined and evidence suggests both scenarios may be possible. During amoeboid migration, directional membrane flow is generated by RhoA polarization to the cell rear, which creates force and pressure for membrane blebbing at the leading edge (Nguyen, Kholodenko, & von Kriegsheim, 2016; O'Neill et al., 2018) (Fig. 2A). As RhoA expression is enriched in MVs and required for contractility-based MV release (Li et al., 2012), RhoA polarization to the cell rear suggests that MV release is enhanced at the cell rear during amoeboid migration. Live cell imaging of migrating amoeboid cells has also revealed a trail of MVs in the wake of the migrating cells (Sedgwick et al., 2015). To promote their directional migration, polarized amoeboid cells also express ezrin, an actin binding protein that links the plasma membrane to the actin cytoskeleton, at the cell rear (Lorentzen, Bamber, Sadok, Elson-Schwab, & Marshall, 2011). Conflicting evidence exists as to whether ezrin expression prevents or enhances MV release. Because ezrin links the plasma membrane to the cytoskeleton, it is speculated that ezrin prevents the membrane and actin cortex detachment required for blebbing at the cell rear, thereby promoting directional migration (Lorentzen et al., 2011). This implies that MVs are not released from the cell rear during active amoeboid cell migration but rather from RhoA rich regions of the membrane during moments of reduced cell polarity. Interestingly, a smaller amount of RhoA is expressed at the leading edge of a cell during amoeboid migration (Pertz, Hodgson, Klemke, & Hahn, 2006). While RhoA at the rear edge appears to be constitutively active, RhoA expression at the leading edge is highly variable and may only last for seconds (Pertz et al., 2006). It is perhaps during these moments of leading edge RhoA expression that MVs are released from the front edge of a migrating cell (Fig. 2C).

Importantly, ezrin has been identified in MVs released from a variety of cell types, including T-lymphocytes and cancer cells (Clancy, Tricarico, Marous, & D'Souza-Schorey, 2019; Tucher et al., 2018), highlighting that ezrin might actually contribute to MV release. Modulation of ezrin expression and localization has direct impacts on MV release (Huang, Hays, Tomasek, Benyajati, & Zhang, 2020; Tucher et al., 2018). Expression of CD82, a metastasis-suppressing tetraspanin that inhibits tumor cell migration, resulted in ezrin expression in membrane bleb-like structures at the cell periphery and promoted ezrin release to the extracellular space via MVs (Huang et al., 2020). While CD82 expression impairs cell migration, it may also promote cancer progression due to increased MV release and signaling. Similarly, induction of an amoeboid phenotype resulted in the redistribution of fascin, an actin-binding protein, ezrin, and podocalyxin, a cell-adhesion glycoprotein, to the neck of budding MVs at the cell periphery to promote MV release (Clancy et al., 2019). Importantly, during plasma membrane blebbing, ezrin is recruited to plasma membrane blebs in a spatially and temporally-dependent fashion, potentially explaining the conflicting role of ezrin in blebbing (Charras, Hu, Coughlin, & Mitchison, 2006). While ezrin is the first protein recruited to plasma membrane blebs, it is not enriched in the membrane during bleb expansion (Charras et al., 2006). It later re-localizes to the bleb as expansion slows (Charras et al., 2006), potentially to reattach the plasma membrane to the cytoskeleton as the MV is released. In summary, RhoA GTPase regulation of amoeboid migration links amoeboid migration with MV release. Further work is necessary to determine the relationship between amoeboid cell polarization, ezrin localization, and MV release.

In mesenchymal migration, Rac1 GTPase is exclusively active at the leading edge to promote invadopodia formation (Nguyen et al., 2016) (Fig. 2B). In contrast, RhoA GTPase is expressed in lower amounts at the cell rear (Nguyen et al., 2016). Rac1 and RhoA GTPases have a complex, balanced relationship in cell migration (Chauhan, Lou, Zheng, & Lang, 2011; Kim et al., 2017; Moorman, Luu, Wickham, Bobak, & Hahn, 1999). Inhibition of one GTPase leads to activation of the other (Moorman et al., 1999). Recent work has studied the direct link between Rac1 GTPase expression and MV release (Sedgwick et al., 2015). Sustained Rac1 activation abrogated MV release from cancer cells on compliant matrices, supporting that Rac1 expression does not promote MV release. Treatment of Rac1-expressing cells with the Rac1 small molecule inhibitor NSC23766 or transfection with the dominant-negative Rac1 mutant Rac1(T17N) resulted in decreased invadopodia and invadopodia-mediated proteolysis, larger focal adhesions, enhanced RhoA activation, and increased MV release on compliant matrices, but not on stiff matrices (Sedgwick et al., 2015). This implies that Rac1 inhibition promotes MV release on compliant matrices by a switch to RhoA GTPase expression (Fig. 2D). As Rac1 inhibition could not promote MV release on stiff matrices, a mechanosensitive pathway independent from RhoA/Rac1 signaling also likely mediates MV release.

Importantly, a third Rho GTPase, Cdc42, is also involved in the formation of invadopodia and appears linked to MV shedding from the cell surface. Constitutive expression of Cdc42 in HeLa cells did not induce MV generation (Li et al., 2012), suggesting that Cdc42 expression does not drive MV formation. Interestingly, very few MVs were observed on the surface of diffuse B-lymphoma (Dbl)-transformed fibroblasts (Hart et al., 1994). As Dbl is a known activator of Cdc42 (Olivo et al., 2000), Cdc42 likely either inhibits MV formation or stimulates MV shedding from the cell surface. An activated-mutant of Cdc42 capable of constitutive GDP-GTP exchange increased the amount of MVs shed by HeLa cells into surrounding culture medium (Antonyak, Wilson, & Cerione, 2012), suggesting that Cdc42 actually stimulates MV shedding from the cell surface. As a result, invadopodia formation and MV shedding may be linked through Cdc42 expression.

Importantly, cancer cells exhibit plasticity in which they can switch between amoeboid and mesenchymal motility depending on their environment (Caswell & Zech, 2018). Recently, mesenchymal-to-amoeboid transition (MAT) has been described in which cells exhibit a motility shift due to environmental signals, including matrix metalloproteinase (MMP) activity and cytokine signaling (Taddei et al., 2014). This suggests that both RhoA/Rac1 signaling and MV release may be altered by environmental cues that alter mode of cell migration. Altogether, MV release appears linked to RhoA GTPase activity in amoeboid migration, with cells undergoing RhoA-mediated amoeboid migration releasing increased numbers of MVs compared to cells undergoing Rac1-mediated mesenchymal migration. While the relationship between mode of cell migration and MV release is highly regulated by RhoA/Rac1 expression, a separate mechanosensitive pathway may also regulate this relationship due to Rac1 inhibition not promoting MV release on stiff matrices.

2.3 Matrix stiffness alters cell contractility and cell migration to regulate MV release

Altered ECM mechanical properties occur in a variety of disease states, including pulmonary disease, Alzheimer's disease, diabetes, and cancer (Sonbol, 2018). While changes to the ECM are known to directly impact cell contractility and cell migration (Gkretsi & Stylianopoulos, 2018), the direct relationship between ECM mechanics and MV release have yet to be thoroughly investigated.

Altered ECM mechanical properties directly affect the intracellular contractile machinery previously implicated in MV release. Increased matrix stiffness increases cell spreading and cell contractility in a variety of cell types, including fibroblasts, epithelial cells, and endothelial cells (Califano & Reinhart-King, 2010; Kraning-Rush, Califano, & Reinhart-King, 2012; Schwager et al., 2019). As increased cell spreading and cytoskeletal tension promote MLC phosphorylation in vascular smooth muscle cells (Polte, Eichler, Wang, & Ingber, 2004), matrix stiffness likely additionally regulates the MLCK-dependent step of MV release (Fig. 1). Additionally, ECM stiffening induces integrin clustering which enhances ERK activation (Paszek et al., 2005) (Fig. 1). This suggests that stiffening of the microenvironment in various disease states may promote MV release from resident cells through MLCK- and ERK-mediated cell contractility. This enhanced MV release from diseased cells may work to further promote disease progression by promoting MV signaling to healthy cells. Increased substrate stiffness also increases RhoA and ROCK1 mRNA and protein in breast cancer cells (Peng et al., 2019). Integrin clustering associated with increased matrix stiffness increases Rho-mediated, ROCK1-generated myosin contractility and force generation (Paszek et al., 2005). As a result, increased matrix stiffness may also enhance MV release through ROCK1 signaling (Fig. 1). While increased ECM stiffness activates a variety of pathways associated with MV release, including MLCK, ERK, RhoA, and ROCK1 signaling, the direct effect of increased matrix stiffness on contractility-based MV release is not yet known.

While ERK and ROCK1 signaling are upregulated by increased matrix stiffness, the impact of increased matrix stiffness on ROCK2 signaling is less clear. Similarly to RhoA and ROCK1, increased substrate stiffness increases ROCK2 mRNA and protein in breast cancer cells (Ebata et al., 2017; Peng et al., 2019; Sugimoto et al., 2018) which may also promote contractility-based MV release (Fig. 1). However, the impact of increased matrix stiffness on downstream ROCK2 signaling is unclear. It was recently shown that cofilin phosphorylation, required for F-actin stabilization in MV release, does not appear to be altered by substrate stiffness. However, only a range of 10–57 kPa was tested, ignoring the effects of softer tissues on cofilin phosphorylation (Peng et al., 2019). Additional work is required to reveal whether ROCK2 signaling in MV release is regulated by matrix stiffness.

The role of matrix stiffness in RhoA-dependent MV release additionally suggests a relationship between cell differentiation and MV release, as substrate stiffness is known to regulate stem cell differentiation (Hoon, Tan, & Koh, 2016). The expression of smooth muscle-specific differentiation proteins was found to be dependent on matrix rigidity in 3D when RhoA is constitutively activated (Peyton, Kim, Ghajar, Seliktar, & Putnam, 2008). This implies that different amounts of MVs and MV cargo may be released by cells differentiated on compliant and stiff ECM.

Changes in matrix stiffness can also alter mode of cell migration to impact MV release. It was recently shown that cancer cells on compliant matrices adopted a rounded, blebbing, amoeboid morphology associated with RhoA GTPase activity (Sedgwick et al., 2015). Conversely, cancer cells on stiff matrices exhibited a flattened, spread, mesenchymal morphology associated with Rac1 GTPase activity (Sedgwick et al., 2015). These stiffness-mediated morphologies were determined to be associated with different MV release properties, with cells on soft matrices releasing more MVs than cells on stiff matrices (Sedgwick et al., 2015). While consistent with previous findings that RhoA signaling is required for MV release, these results suggest that decreased matrix stiffness is correlated with increased MV release due to the switch to an amoeboid phenotype. Importantly, matrix stiffness in this system was modulated by changing gelatin thickness and gelatin percentage in 2D (Sedgwick et al., 2015). Investigation of 3D environments and physiologically relevant biomaterials and stiffness would further enhance our understanding of stiffness-mediated MV release.

Altered matrix mechanics may also induce MAT. Increased cell confinement due to decreased pore size in the ECM results in cells that exhibit traits of both mesenchymal and amoeboid migration, with rounded morphology, increased migration speed, and large vinculin-adhesion sites (Mosier et al., 2019). While MV release in this intermediate mode of migration has yet to be investigated, the rounded cell morphology paired with large matrix adhesion sites may increase the effects of matrix stiffness on MV release.

In summary, MV release is highly dependent upon intracellular actin dynamics, mediated by ARF6 signaling through ERK and RhoA GTPase. The pathways downstream of ARF6 are highly dependent on altered intracellular contractility and matrix stiffness, indicating that soluble and mechanical signals in the tumor microenvironment may influence MV release and contribute to disease progression. Lastly, mode of cell migration, regulated by environmental signals and mechanical properties, regulates MV release through RhoA GTPase activity.

3. Mechanobiology of microvesicle transport and uptake

As MV uptake by recipient cells is known to modulate cell function, understanding the mechanisms of MV transport through the extracellular space and internalization by recipient cells are crucial to fully understand their function in disease progression. MVs have been detected in a variety of bodily fluids from cardiovascular disease, rheumatoid arthritis, and cancer patients (Dickhout & Koenen, 2018; György et al., 2011; Menck et al., 2017), suggesting that MVs are transported away from diseased tissue where they may interact with healthy cells. Mechanical signals generated by the diseased tissue likely play a role in regulating MV transport away from diseased tissue. First, decreased epithelial barrier integrity, induced by increased epithelial cell contractility and decreased cell adhesion, may regulate the transport of MVs in a variety of diseases, including cancers, gastrointestinal diseases, and pulmonary fibrosis (Brennan, Offiah, McSherry, & Hopkins, 2009; Cui et al., 2010; Goh et al., 2011; Groschwitz & Hogan, 2009; Ivanov, Parkos, & Nusrat, 2010; Nishida et al., 2017; Soler et al., 1999). The increased epithelial permeability induced by these diseases may promote MV transport away from the diseased tissue to signal to healthy cells. However, the direct relationship between MV transport and epithelial barrier function have yet to be investigated. Additionally, altered vascular permeability may similarly regulate MV transport into the blood stream. Increased vascular permeability, common in both atherosclerosis and cancer, is induced by altered tissue mechanics and arterial stiffening (Bordeleau et al., 2017; Huynh et al., 2011; VanderBurgh, Hotchkiss, Potharazu, Taufalele, & Reinhart-King, 2018). Increased vascular permeability is known to enhance leukocyte extravasation and cancer cell intravasation (Harney et al., 2015; Huynh et al., 2011; Reid et al., 2017). It is likely that MVs similarly exhibit enhanced transport into the vasculature in conditions of increased vascular permeability. As a result, mechanics of the vasculature likely regulate transport of MVs away from diseased tissues but have yet to be directly investigated.

The mechanisms of EV uptake by recipient cells have been thoroughly researched and reviewed in recent years (Costa Verdera, Gitz-Francois, Schiffelers, & Vader, 2017; Durak-Kozica, Baster, Kubat, & Stępień, 2018; French, Antonyak, & Cerione, 2017; Mulcahy, Pink, & Carter, 2014; Ofir-Birin et al., 2018; Toribio et al., 2019); however, these studies refer to their vesicle population as the generic class of EVs and do not distinguish based on vesicle type. As a result, the specific mechanisms of MV uptake are largely undefined. Current evidence suggests that while EVs can be taken up by direct fusion with the plasma membrane (Parolini et al., 2009), they are primarily taken up by endocytosis (Escrevente, Keller, Altevogt, & Costa, 2011; Montecalvo et al., 2012; Morelli et al., 2004; Mulcahy et al., 2014). EV uptake via endocytosis can occur through a variety of endocytic pathways, including clathrin-mediated endocytosis, lipid raft-mediated endocytosis, caveolin-mediated endocytosis, pinocytosis, and phagocytosis (Mulcahy et al., 2014). As these studies focused on uptake of the generic EV population, it is unclear how MVs are internalized by recipient cells and whether MV internalization is a mechanosensitive process.

Uptake of MVs released from alveolar macrophages by alveolar epithelial cells was shown to more closely resemble pinocytosis, or the non-specific uptake of extracellular fluid, rather than receptor-mediated endocytosis (Schneider et al., 2017). This result is significant in suggesting that MVs likely do not use specific transmembrane proteins to dock on recepient cells. Instead, uptake of MVs by alveolar epithelial cells is dependent upon recipient cell dynamin function (Schneider et al., 2017). Dynamin, a GTPase that is required for clathrin-dependent endocytosis, is recruited and assembled at the neck of clathrin-coated pits. Activated dynamin then catalyzes membrane fission to induce endocytosis (Mettlen, Pucadyil, Ramachandran, & Schmid, 2009; Sundborger & Hinshaw, 2014). Importantly, while MV uptake was greatly inhibited by the dynamin inhibitor dynasore, clathrin expression by MVs was not required for dynamin-mediated endocytosis (Schneider et al., 2017), suggesting that MV uptake does not depend upon clathrin-mediated endocytosis. This reveals a dynamin-dependent clathrin-independent endocytic pathway that regulates MV uptake. Interestingly, membrane tension, regulated by matrix stiffness and cellular contractility, is known to affect endocytosis (Dai, Ting-Beall, & Sheetz, 1997; Mandal et al., 2019; Nambiar, McConnell, & Tyska, 2009; Raucher & Sheetz, 1999). Decreased membrane tension is associated with increased endocytosis while increased membrane tension decreases endocytosis (Dai et al., 1997; Raucher & Sheetz, 1999). While certain endocytic responses correlate with a reduction in membrane tension, these responses are dynamin-independent (Thottacherry et al., 2018). This suggests that dynamin-dependent MV endocytosis may be independent of recipient cell membrane tension. However, other methods of MV endocytosis not currently defined may be regulated by membrane tension.

MV uptake has also been found to be dependent upon recipient cell actin polymerization (Schneider et al., 2017). MV uptake by alveolar epithelial cells was disrupted using the actin polymerization inhibitor cytochalasin D. As actin polymerization is required for endocytosis, the interaction between actin and dynamin likely regulates MV uptake.

While the relationship between cell contractility and MV uptake has not yet been investigated, this evidence suggests that increased cell contractility, which is associated with increased actin polymerization and actin remodeling, may correlate with increased MV uptake.

Actin dynamics in cell migration may additionally affect MV uptake but have yet to be investigated. RhoA regulates actin polymerization in amoeboid cells through mammalian homolog of Drosophila diaphanous 1 (mDia1) signaling, which catalyzes actin nucleation and polymerization (Narumiya, Tanji, & Ishizaki, 2009). Together, mDia1 and ROCK interact downstream of RhoA to induce actomyosin bundling (Narumiya et al., 2009). As a result, increased RhoA signaling in amoeboid cells may enhance their ability to uptake MVs. Conversely, Rac1 and Cdc42, associated with invadopodia formation, control actin polymerization through Wiskott-Aldrich syndrome protein (WASP) and WASP-family verprolin-homologous protein (WAVE) regulation (Paňková, Rösel, Novotný, & Brábek, 2010). Specifically, the interactions of WASP and WAVE with Arp2/3 promotes formation of the actin network at the leading edge of mesenchymal cells (Paňková et al., 2010). Thus, the leading edge of cells undergoing mesenchymal migration, rather than the trailing edge, may be more equipped for MV uptake due to increased actin polymerization.

In summary, the mechanisms of MV transport and internalization are vastly understudied. Increased epithelial and vascular permeability likely enhance transport of MVs away from diseased tissue and into the vasculature. However, the relationship between barrier integrity and MV transport have not yet been investigated. Additionally, recipient cell mechanics regulate MV uptake through actin polymerization (Schneider et al., 2017). Currently, the only known mechanism of MV uptake is via dynamin-dependent endocytosis which may not be affected by matrix stiffness and cell contractility. However, MV uptake does appear to be dependent on recipient cell actin polymerization, potentially through a mechanism other than the previously described dynamin-dependent endocytosis. As a result, recipient cell RhoA and Rac1 GTPase dynamics may regulate the ability of cells to uptake MVs via actin polymerization.

4. Mechanobiology of MV-mediated cell activation

Uptake of MVs can differentiate and transform recipient cells leading to altered cellular function. It is known that MVs can signal to and transform epithelial cells, endothelial cells, immune cells, and stromal fibroblasts (Antonyak et al., 2011; Bordeleau et al., 2016; Huaitong et al., 2017; Jiang et al., 2019; Kawamoto et al., 2012; Kreger, Dougherty, Greene, Cerione, & Antonyak, 2016; Lee, Zhang, Zhu, Dela Cruz, & Jin, 2016; Schwager et al., 2019; Zhao et al., 2015). Importantly, we have recently shown that MV-mediated transformation is mechanosensitive, as MV-mediated fibroblast activation was enhanced on stiff matrices mimicking the breast tumor periphery, compared to soft matrices mimicking healthy breast tissue (Schwager et al., 2019). This finding emphasizes that stiffer tumorigenic matrices may further promote cancer progression through enhanced MV signaling. However, the mechanism behind this enhanced MV signaling on stiff matrices is unclear. We speculate that it may be due to a heightened response to the MV cargo on stiff matrices or increased MV uptake, potentially due to increased fibroblast contractility on stiff matrices. It is additionally unclear whether this phenomenon is fibroblast-specific or whether matrix stiffness regulates MV-induced transformation in a variety of cell types. Further investigation is required to determine whether stiffened matrices associated with diseases such as lung fibrosis, atherosclerosis, or diabetes will enhance MV signaling to epithelial, endothelial, or immune cells.

Previous research has suggested that matrix stiffness sensitizes cells to various chemical cues. For example, increased matrix stiffness enhances vascular endothelial growth factor receptor-2 (VEGFR-2) internalization in endothelial cells and sensitizes epithelial cells to epidermal growth factor (EGF), resulting in increased epithelial cell proliferation (Kim & Asthagiri, 2011; LaValley et al., 2017). Together, these results suggest that matrix stiffness may increase cell sensitivity to MV cargo. This may be cargo-specific and recipient cells may only exhibit increased sensitivity to select MV cargo, such as growth factors, cytokines, or other soluble factors. Future investigation into the sensitivity of fibroblasts to specific MV cargo on a variety of substrate stiffness will elucidate the mechanism of this stiffness-dependent MV-mediated fibroblast activation.

Alternatively, an enhanced response to MVs on stiff matrices may be a result of increased MV uptake via increased actin polymerization and cell contractility, as previously described. As fibroblasts exhibit increased contractility on stiff matrices (Schwager et al., 2019), this may provide them with the necessary actin polymerization required for MV uptake. Similarly, increased substrate stiffness may alter recipient cell membrane tension to increase MV uptake via a form of endocytosis not yet implicated in MV uptake.

MVs may also be involved in a mechanosensitive feedback loop in the tumor microenvironment to promote cancer progression. As previously described, increased matrix stiffness enhances fibroblast response to MV cargo (Schwager et al., 2019). This results in increased fibroblast activation to further remodel and stiffen the tumor microenvironment (Schwager et al., 2019), which may ultimately further amplify fibroblast response to MVs, both through increased MV release by resident cancer cells, increased MV uptake by resident fibroblasts, and increased remodeling by activated fibroblasts. Matrix remodeling by cancer-associated fibroblasts has significant effects on cancer progression by inducing increased cancer cell invasion away from primary tumors and promoting pre-metastatic niche formation (Attieh et al., 2017; Erdogan et al., 2017; Fang et al., 2018). As a result, this mechanosensitive MV signaling feedback loop may further promote cancer metastasis.

MVs may also alter the mode of migration in recipient cells. Metastatic amoeboid cancer cells transferred their amoeboid phenotype to other cancer cells via exosomes to induce a more aggressive cancer phenotype (Schillaci et al., 2017). This signaling was shown to be a result of cargo transfer that activated the RhoA/ROCK signaling pathway (Schillaci et al., 2017). While MVs have not yet been studied in this context, components of the RhoA signaling pathway in MV cargo may also induce amoeboid signaling in recipient cells. If true, MV-induced MAT may enhance MV release from transformed recipient cells to further promote disease progression.

The effects of matrix stiffness and other environmental factors on MV signaling are poorly understood. While recent work suggests that increased matrix stiffness may enhance MV-mediated fibroblast activation, the mechanism behind this phenomenon is unknown. Further work is necessary to understand the role of matrix stiffness in the delivery of specific MV cargo to specific cell types and the effects on their downstream signaling. Additionally, investigation of MV regulation of MAT will further elucidate the roles of MVs in cancer progression.

5. Conclusions

The role of MVs in cell signaling and disease progression are newly emerging. The transfer of cellular cargo via MVs promotes extracellular matrix and vascular remodeling, recruits tumor-promoting stromal cells, and induces transformed phenotypes within recipient cells

(Antonyak et al., 2011; Bordeleau et al., 2016; Huaitong et al., 2017; Jiang et al., 2019; Kawamoto et al., 2012; Menck et al., 2015; Schwager et al., 2019). However, the roles of altered cell contractility, cell migration, and matrix stiffness in MV release, MV uptake, and downstream MV signaling have largely been ignored. In this review, we summarize recent literature suggesting that altered cellular behaviors and microenvironmental factors associated with disease progression can regulate MV signaling. Specifically, increased cell contractility and a shift to amoeboid migration enhance MV signaling through RhoA and ERK signaling. Although little is known about the mechanisms of MV uptake by recipient cells, it appears to be dependent upon recipient cell actin polymerization, suggesting a role for intracellular mechanics in MV uptake as well. In fact, our own data show that increased matrix stiffness enhances the effects of MVs on recipient cells (Schwager et al., 2019), potentially through increased sensitivity to MV cargo or increased MV uptake. As such, there is increasing evidence to show that the mechanical state of cells and the microenvironment are mediators of MV release, uptake, and sensitivity.

The mechanosensitivity of MV signaling has important implications in disease progression but is largely understudied, partially due to limitations in MV research. Current obstacles in MV research include differentiation between EV subtypes, MV labeling, and MV detection via microscopy. As a result, the majority of novel EV research has focused on exosomes as they possess distinct markers. As additional techniques and methods are developed for MV analysis, our understanding of MV dynamics and MV-mediated signaling will grow significantly. Identification of unique MV markers and microscopy techniques for live imaging of MV release and uptake would greatly enhance MV research. Recently, optogenetic activation of RhoA was shown to drive amoeboid cell migration (O'Neill et al., 2018). Optogenetic activation of RhoA to control mode of cell migration paired with a fluorescent MV marker could be used to visualize and investigate MV release during cell migration (O'Neill et al., 2018). Additionally, fluorescent RhoA and ezrin probes could be used to further determine their localization during MV release in amoeboid migration.

The majority of MV signaling research has been conducted *in vitro* as *in vivo* research poses increased difficulties in MV labeling and signaling. Importantly, these *in vitro* systems possess inherent limitations as they fail to accurately recapitulate human tissues and diseased states. As previously mentioned, the effects of ECM stiffness on MV release have only been investigated in 2D on gelatin matrices (Sedgwick et al., 2015). The use of engineered systems that accurately recapitulate the stiffness, matrix components, and ECM architecture found *in vivo* will better elucidate the effects of the disease microenvironment on MV release. Additionally, *in vivo* labeling mechanisms that selectively label and track MVs will be essential to investigate the interplay between cell and microenvironmental mechanics and MV release *in vivo*.

Acknowledgments

This work was supported by funding from the NIH (GM131178) to C.A.R.-K. and by the National Science Foundation Graduate Research Fellowship Program under Grant No. 1937963 to S.C.S.

References

Antonyak, M. A., Li, B., Boroughs, L. K., Johnson, J. L., Druso, J. E., Bryant, K. L., et al. (2011). Cancer cell-derived microvesicles induce transformation by transferring tissue transglutaminase and fibronectin to recipient cells. *Proceedings of the National Academy of Sciences of the United States of America, 108*(12), 4852–4857. https://doi.org/10.1073/pnas.1017667108.

Antonyak, M. A., Wilson, K. F., & Cerione, R. A. (2012). R(h)oads to microvesicles. *Small GTPases, 3*(4), 219–224. https://doi.org/10.4161/sgtp.20755.

Attieh, Y., Clark, A. G., Grass, C., Richon, S., Pocard, M., Mariani, P., et al. (2017). Cancer-associated fibroblasts lead tumor invasion through integrin-β3–dependent fibronectin assembly. *The Journal of Cell Biology, 216*(11), 3509–3520. https://doi.org/10.1083/jcb.201702033.

Bear, J. E., & Haugh, J. M. (2014). Directed migration of mesenchymal cells: Where signaling and the cytoskeleton meet. *Current Opinion in Cell Biology, 30*, 74–82. https://doi.org/10.1016/j.ceb.2014.06.005.

Bordeleau, F., Chan, B. S., Antonyak, M. A., Lampi, M. C., Cerione, R. A., & Reinhart-King, C. A. (2016). Microvesicles released from tumor cells disrupt epithelial cell morphology and contractility. *Journal of Biomechanics, 49*(8), 1272–1279. https://doi.org/10.1016/j.jbiomech.2015.10.003.

Bordeleau, F., Mason, B. N., Lollis, E. M., Mazzola, M., Zanotelli, M. R., Somasegar, S., et al. (2017). Matrix stiffening promotes a tumor vasculature phenotype. *Proceedings of the National Academy of Sciences, 114*(3), 492–497. https://doi.org/10.1073/pnas.1613855114.

Brennan, K., Offiah, G., McSherry, E. A., & Hopkins, A. M. (2009). Tight Junctions: A barrier to the initiation and progression of breast cancer? *Journal of Biomedicine and Biotechnology*. https://doi.org/10.1155/2010/460607.

Califano, J. P., & Reinhart-King, C. A. (2010). Substrate stiffness and cell area predict cellular traction stresses in single cells and cells in contact. *Cellular and Molecular Bioengineering, 3*(1), 68–75. https://doi.org/10.1007/s12195-010-0102-6.

Caswell, P. T., & Zech, T. (2018). Actin-based cell protrusion in a 3D matrix. *Trends in Cell Biology, 28*(10), 823–834. https://doi.org/10.1016/j.tcb.2018.06.003.

Charras, G., Hu, C.-K., Coughlin, M., & Mitchison, T. (2006). Reassembly of contractile actin cortex in cell blebs. *The Journal of Cell Biology, 175*(3), 477–490. https://doi.org/10.1083/jcb.200602085.

Charras, G., & Paluch, E. (2008). Blebs lead the way: How to migrate without lamellipodia. *Nature Reviews Molecular Cell Biology, 9*(9), 730–736. https://doi.org/10.1038/nrm2453.

Chauhan, B. K., Lou, M., Zheng, Y., & Lang, R. A. (2011). Balanced Rac1 and RhoA activities regulate cell shape and drive invagination morphogenesis in epithelia. *Proceedings of the National Academy of Sciences*, *108*(45), 18289–18294. https://doi.org/10.1073/pnas.1108993108.

Clancy, J. W., Tricarico, C. J., Marous, D. R., & D'Souza-Schorey, C. (2019). Coordinated regulation of intracellular fascin distribution governs tumor microvesicle release and invasive cell capacity. *Molecular and Cellular Biology*, *39*(3). https://doi.org/10.1128/MCB.00264-18.

Colombo, M., Raposo, G., & Théry, C. (2014). Biogenesis, secretion, and intercellular interactions of exosomes and other extracellular vesicles. *Annual Review of Cell and Developmental Biology*, *30*, 255–289. https://doi.org/10.1146/annurev-cellbio-101512-122326.

Costa Verdera, H., Gitz-Francois, J. J., Schiffelers, R. M., & Vader, P. (2017). Cellular uptake of extracellular vesicles is mediated by clathrin-independent endocytosis and macropinocytosis. *Journal of Controlled Release: Official Journal of the Controlled Release Society*, *266*, 100–108. https://doi.org/10.1016/j.jconrel.2017.09.019.

Cui, W., Li, L. X., Sun, C. M., Wen, Y., Zhou, Y., Dong, Y. L., et al. (2010). Tumor necrosis factor alpha increases epithelial barrier permeability by disrupting tight junctions in Caco-2 cells. *Brazilian Journal of Medical and Biological Research*, *43*(4), 330–337. https://doi.org/10.1590/S0100-879X2010007500020.

Dai, J., Ting-Beall, H. P., & Sheetz, M. P. (1997). The secretion-coupled endocytosis correlates with membrane tension changes in RBL 2H3 cells. *The Journal of General Physiology*, *110*(1), 1–10. https://doi.org/10.1085/jgp.110.1.1.

Dickhout, A., & Koenen, R. R. (2018). Extracellular vesicles as biomarkers in cardiovascular disease; chances and risks. *Frontiers in Cardiovascular Medicine, 5*. https://doi.org/10.3389/fcvm.2018.00113.

Doyle, L. M., & Wang, M. Z. (2019). Overview of extracellular vesicles, their origin, composition, purpose, and methods for exosome isolation and analysis. *Cells*, *8*(7). https://doi.org/10.3390/cells8070727.

Durak-Kozica, M., Baster, Z., Kubat, K., & Stępień, E. (2018). 3D visualization of extracellular vesicle uptake by endothelial cells. *Cellular & Molecular Biology Letters, 23*. https://doi.org/10.1186/s11658-018-0123-z.

Ebata, T., Mitsui, Y., Sugimoto, W., Maeda, M., Araki, K., Machiyama, H., et al. (2017). Substrate stiffness influences doxorubicin-induced p53 activation via ROCK2 Expression. *BioMed Research International*. https://doi.org/10.1155/2017/5158961.

Erdogan, B., Ao, M., White, L. M., Means, A. L., Brewer, B. M., Yang, L., et al. (2017). Cancer-associated fibroblasts promote directional cancer cell migration by aligning fibronectin. *The Journal of Cell Biology*, *216*(11), 3799–3816. https://doi.org/10.1083/jcb.201704053.

Escrevente, C., Keller, S., Altevogt, P., & Costa, J. (2011). Interaction and uptake of exosomes by ovarian cancer cells. *BMC Cancer*, *11*, 108. https://doi.org/10.1186/1471-2407-11-108.

Fang, T., Lv, H., Lv, G., Li, T., Wang, C., Han, Q., et al. (2018). Tumor-derived exosomal miR-1247-3p induces cancer-associated fibroblast activation to foster lung metastasis of liver cancer. *Nature Communications*, *9*(1), 191. https://doi.org/10.1038/s41467-017-02583-0.

Franz, C. M., Jones, G. E., & Ridley, A. J. (2002). Cell migration in development and disease. *Developmental Cell*, *2*(2), 153–158. https://doi.org/10.1016/S1534-5807(02)00120-X.

French, K. C., Antonyak, M. A., & Cerione, R. A. (2017). Extracellular vesicle docking at the cellular port: Extracellular vesicle binding and uptake. *Seminars in Cell & Developmental Biology*, *67*, 48–55. https://doi.org/10.1016/j.semcdb.2017.01.002.

Gallagher, P. J., Herring, B. P., & Stull, J. T. (1997). Myosin light chain kinases. *Journal of Muscle Research and Cell Motility*, *18*(1), 1–16. https://doi.org/10.1023/a:1018616814417.

Gkretsi, V., & Stylianopoulos, T. (2018). Cell adhesion and matrix stiffness: Coordinating cancer cell invasion and metastasis. *Frontiers in Oncology*, *8*. https://doi.org/10.3389/fonc.2018.00145.

Goh, N. S. L., Desai, S. R., Anagnostopoulos, C., Hansell, D. M., Hoyles, R. K., Sato, H., et al. (2011). Increased epithelial permeability in pulmonary fibrosis in relation to disease progression. *The European Respiratory Journal*, *38*(1), 184–190. https://doi.org/10.1183/09031936.00010910.

Groschwitz, K. R., & Hogan, S. P. (2009). Intestinal barrier function: Molecular regulation and disease pathogenesis. *The Journal of Allergy and Clinical Immunology*, *124*(1), 3–22. https://doi.org/10.1016/j.jaci.2009.05.038.

György, B., Módos, K., Pállinger, E., Pálóczi, K., Pásztói, M., Misják, P., et al. (2011). Detection and isolation of cell-derived microparticles are compromised by protein complexes resulting from shared biophysical parameters. *Blood*, *117*(4), e39–e48. https://doi.org/10.1182/blood-2010-09-307595.

Harney, A. S., Arwert, E. N., Entenberg, D., Wang, Y., Guo, P., Qian, B.-Z., et al. (2015). Real-time imaging reveals local, transient vascular permeability, and tumor cell intravasation stimulated by TIE2hi macrophage–derived VEGFA. *Cancer Discovery*, *5*(9), 932–943. https://doi.org/10.1158/2159-8290.CD-15-0012.

Hart, M. J., Eva, A., Zangrilli, D., Aaronson, S. A., Evans, T., Cerione, R. A., et al. (1994). Cellular transformation and guanine nucleotide exchange activity are catalyzed by a common domain on the Dbl oncogene product. *The Journal of Biological Chemistry*, *269*(1), 62–65.

Hoon, J. L., Tan, M. H., & Koh, C.-G. (2016). The regulation of cellular responses to mechanical cues by rho GTPases. *Cells*, *5*(2). https://doi.org/10.3390/cells5020017.

Huaitong, X., Yuanyong, F., Yueqin, T., Peng, Z., Wei, S., & Kai, S. (2017). Microvesicles releasing by oral cancer cells enhance endothelial cell angiogenesis via Shh/RhoA signaling pathway. *Cancer Biology & Therapy*, *18*(10), 783–791. https://doi.org/10.1080/15384047.2017.1373213.

Huang, C., Hays, F. A., Tomasek, J. J., Benyajati, S., & Zhang, X. A. (2020). Tetraspanin CD82 interaction with cholesterol promotes extracellular vesicle–mediated release of ezrin to inhibit tumour cell movement. *Journal of Extracellular Vesicles*, *9*(1), 1692417. https://doi.org/10.1080/20013078.2019.1692417.

Huynh, J., Nishimura, N., Rana, K., Peloquin, J. M., Califano, J. P., Montague, C. R., et al. (2011). Age-related intimal stiffening enhances endothelial permeability and leukocyte transmigration. *Science Translational Medicine*, 3(112), 112ra122. https://doi.org/10.1126/scitranslmed.3002761.

Ivanov, A. I., Parkos, C. A., & Nusrat, A. (2010). Cytoskeletal regulation of epithelial barrier function during inflammation. *The American Journal of Pathology*, *177*(2), 512–524. https://doi.org/10.2353/ajpath.2010.100168.

Jiang, E., Xu, Z., Wang, M., Yan, T., Huang, C., Zhou, X., et al. (2019). Tumoral microvesicle–activated glycometabolic reprogramming in fibroblasts promotes the progression of oral squamous cell carcinoma. *The FASEB Journal*, *33*(4), 5690–5703. https://doi.org/10.1096/fj.201802226R.

Kawamoto, T., Ohga, N., Akiyama, K., Hirata, N., Kitahara, S., Maishi, N., et al. (2012). Tumor-derived microvesicles induce proangiogenic phenotype in endothelial cells via endocytosis. *PLoS One*, 7(3), e34045. https://doi.org/10.1371/journal.pone.0034045.

Kim, J.-H., & Asthagiri, A. R. (2011). Matrix stiffening sensitizes epithelial cells to EGF and enables the loss of contact inhibition of proliferation. *Journal of Cell Science*, *124*(8), 1280–1287. https://doi.org/10.1242/jcs.078394.

Kim, S.-Y., Kim, S., Bae, D.-J., Park, S.-Y., Lee, G.-Y., Park, G.-M., et al. (2017). Coordinated balance of Rac1 and RhoA plays key roles in determining phagocytic appetite. *PLoS One*, *12*(4). https://doi.org/10.1371/journal.pone.0174603.

Kraning-Rush, C. M., Califano, J. P., & Reinhart-King, C. A. (2012). Cellular traction stresses increase with increasing metastatic potential. *PLoS One*, 7(2), e32572. https://doi.org/10.1371/journal.pone.0032572.

Kreger, B. T., Dougherty, A. L., Greene, K. S., Cerione, R. A., & Antonyak, M. A. (2016). Microvesicle cargo and function changes upon induction of cellular transformation. *Journal of Biological Chemistry*, *291*(38), 19774–19785. https://doi.org/10.1074/jbc.M116.725705.

LaValley, D. J., Zanotelli, M. R., Bordeleau, F., Wang, W., Schwager, S. C., & Reinhart-King, C. A. (2017). Matrix stiffness enhances VEGFR-2 internalization, signaling, and proliferation in endothelial cells. *Convergent Science Physical Oncology*, *3*(4), 044001. https://doi.org/10.1088/2057-1739/aa9263.

Lee, H., Zhang, D., Zhu, Z., Dela Cruz, C. S., & Jin, Y. (2016). Epithelial cell-derived microvesicles activate macrophages and promote inflammation via microvesicle-containing microRNAs. *Scientific Reports*, *6*(1), 35250. https://doi.org/10.1038/srep35250.

Li, B., Antonyak, M. A., Zhang, J., & Cerione, R. A. (2012). RhoA triggers a specific signaling pathway that generates transforming microvesicles in cancer cells. *Oncogene*, *31*(45), 4740–4749. https://doi.org/10.1038/onc.2011.636.

Lorentzen, A., Bamber, J., Sadok, A., Elson-Schwab, I., & Marshall, C. J. (2011). An ezrin-rich, rigid uropod-like structure directs movement of amoeboid blebbing cells. *Journal of Cell Science*, *124*(Pt. 8), 1256–1267. https://doi.org/10.1242/jcs.074849.

Mandal, K., Aroush, D. R.-B., Graber, Z., Wu, B., Park, C. Y., Fredberg, J. J., et al. (2019). Soft hyaluronic gel promotes cell spreading, stress fibers, focal adhesion, membrane tension by phosphoinositide signaling, not traction force. *ACS Nano*, *13*(1), 203–214. https://doi.org/10.1021/acsnano.8b05286.

Menck, K., Bleckmann, A., Wachter, A., Hennies, B., Ries, L., Schulz, M., et al. (2017). Characterisation of tumour-derived microvesicles in cancer patients' blood and correlation with clinical outcome. *Journal of Extracellular Vesicles*, *6*(1). https://doi.org/10.1080/20013078.2017.1340745.

Menck, K., Scharf, C., Bleckmann, A., Dyck, L., Rost, U., Wenzel, D., et al. (2015). Tumor-derived microvesicles mediate human breast cancer invasion through differentially glycosylated EMMPRIN. *Journal of Molecular Cell Biology*, 7(2), 143–153. https://doi.org/10.1093/jmcb/mju047.

Mettlen, M., Pucadyil, T., Ramachandran, R., & Schmid, S. L. (2009). Dissecting dynamin's role in clathrin-mediated endocytosis. *Biochemical Society Transactions*, *37*(Pt. 5), 1022–1026. https://doi.org/10.1042/BST0371022.

Minciacchi, V. R., Freeman, M. R., & Di Vizio, D. (2015). Extracellular vesicles in cancer: Exosomes, microvesicles and the emerging role of large oncosomes. *Seminars in Cell & Developmental Biology*, *40*, 41–51. https://doi.org/10.1016/j.semcdb.2015.02.010.

Montecalvo, A., Larregina, A. T., Shufesky, W. J., Stolz, D. B., Sullivan, M. L. G., Karlsson, J. M., et al. (2012). Mechanism of transfer of functional microRNAs between mouse dendritic cells via exosomes. *Blood*, *119*(3), 756–766. https://doi.org/10.1182/blood-2011-02-338004.

Moorman, J. P., Luu, D., Wickham, J., Bobak, D. A., & Hahn, C. S. (1999). A balance of signaling by Rho family small GTPases RhoA, Rac1 and Cdc42 coordinates cytoskeletal morphology but not cell survival. *Oncogene*, *18*(1), 47–57. https://doi.org/10.1038/sj.onc.1202262.

Morelli, A. E., Larregina, A. T., Shufesky, W. J., Sullivan, M. L. G., Stolz, D. B., Papworth, G. D., et al. (2004). Endocytosis, intracellular sorting, and processing of exosomes by dendritic cells. *Blood*, *104*(10), 3257–3266. https://doi.org/10.1182/blood-2004-03-0824.

Mosier, J. A., Rahman-Zaman, A., Zanotelli, M. R., VanderBurgh, J. A., Bordeleau, F., Hoffman, B. D., et al. (2019). Extent of cell confinement in microtracks affects speed and results in differential matrix strains. *Biophysical Journal, 117*(9), 1692–1701. https://doi.org/10.1016/j.bpj.2019.09.024.

Mulcahy, L. A., Pink, R. C., & Carter, D. R. F. (2014). Routes and mechanisms of extracellular vesicle uptake. *Journal of Extracellular Vesicles, 3.* https://doi.org/10.3402/jev.v3.24641.

Muralidharan-Chari, V., Clancy, J., Plou, C., Romao, M., Chavrier, P., Raposo, G., et al. (2009). ARF6-regulated shedding of tumor cell-derived plasma membrane microvesicles. *Current Biology: CB, 19*(22), 1875–1885. https://doi.org/10.1016/j.cub.2009.09.059.

Nambiar, R., McConnell, R. E., & Tyska, M. J. (2009). Control of cell membrane tension by myosin-I. *Proceedings of the National Academy of Sciences of the United States of America, 106*(29), 11972–11977. https://doi.org/10.1073/pnas.0901641106.

Narumiya, S., Tanji, M., & Ishizaki, T. (2009). Rho signaling, ROCK and mDia1, in transformation, metastasis and invasion. *Cancer and Metastasis Reviews, 28*(1), 65–76. https://doi.org/10.1007/s10555-008-9170-7.

Nguyen, L. K., Kholodenko, B. N., & von Kriegsheim, A. (2016). Rac1 and RhoA: Networks, loops and bistability. *Small GTPases, 9*(4), 316–321. https://doi.org/10.1080/21541248.2016.1224399.

Nishida, K., Brune, K. A., Putcha, N., Mandke, P., O'Neal, W. K., Shade, D., et al. (2017). Cigarette smoke disrupts monolayer integrity by altering epithelial cell-cell adhesion and cortical tension. *American Journal of Physiology—Lung Cellular and Molecular Physiology, 313*(3), L581–L591. https://doi.org/10.1152/ajplung.00074.2017.

Ofir-Birin, Y., Abou karam, P., Rudik, A., Giladi, T., Porat, Z., & Regev-Rudzki, N. (2018). Monitoring extracellular vesicle cargo active uptake by imaging flow cytometry. *Frontiers in Immunology, 9.* https://doi.org/10.3389/fimmu.2018.01011.

Olivo, C., Vanni, C., Mancini, P., Silengo, L., Torrisi, M. R., Tarone, G., et al. (2000). Distinct involvement of Cdc42 and RhoA GTPases in actin organization and cell shape in untransformed and Dbl oncogene transformed NIH3T3 cells. *Oncogene, 19*(11), 1428–1436. https://doi.org/10.1038/sj.onc.1203440.

O'Neill, P. R., Castillo-Badillo, J. A., Meshik, X., Kalyanaraman, V., Melgarejo, K., & Gautam, N. (2018). Membrane flow drives an adhesion-independent amoeboid cell migration mode. *Developmental Cell, 46*(1). https://doi.org/10.1016/j.devcel.2018.05.029. 9-22.e4.

Paluch, E. K., & Raz, E. (2013). The role and regulation of blebs in cell migration. *Current Opinion in Cell Biology, 25*(5), 582–590. https://doi.org/10.1016/j.ceb.2013.05.005.

Paňková, K., Rösel, D., Novotný, M., & Brábek, J. (2010). The molecular mechanisms of transition between mesenchymal and amoeboid invasiveness in tumor cells. *Cellular and Molecular Life Sciences, 67*(1), 63–71. https://doi.org/10.1007/s00018-009-0132-1.

Parolini, I., Federici, C., Raggi, C., Lugini, L., Palleschi, S., De Milito, A., et al. (2009). Microenvironmental pH is a key factor for exosome traffic in tumor cells. *The Journal of Biological Chemistry, 284*(49), 34211–34222. https://doi.org/10.1074/jbc.M109.041152.

Paszek, M. J., Zahir, N., Johnson, K. R., Lakins, J. N., Rozenberg, G. I., Gefen, A., et al. (2005). Tensional homeostasis and the malignant phenotype. *Cancer Cell, 8*(3), 241–254. https://doi.org/10.1016/j.ccr.2005.08.010.

Peng, Y., Chen, Z., Chen, Y., Li, S., Jiang, Y., Yang, H., et al. (2019). ROCK isoforms differentially modulate cancer cell motility by mechanosensing the substrate stiffness. *Acta Biomaterialia, 88*, 86–101. https://doi.org/10.1016/j.actbio.2019.02.015.

Pertz, O., Hodgson, L., Klemke, R. L., & Hahn, K. M. (2006). Spatiotemporal dynamics of RhoA activity in migrating cells. *Nature*, *440*(7087), 1069–1072. https://doi.org/10.1038/nature04665.

Peyton, S. R., Kim, P. D., Ghajar, C. M., Seliktar, D., & Putnam, A. J. (2008). The effects of matrix stiffness and RhoA on the phenotypic plasticity of smooth muscle cells in a 3-D biosynthetic hydrogel system. *Biomaterials*, *29*(17), 2597–2607. https://doi.org/10.1016/j.biomaterials.2008.02.005.

Pollard, T. D., & Borisy, G. G. (2003). Cellular motility driven by assembly and disassembly of actin filaments. *Cell*, *112*(4), 453–465. https://doi.org/10.1016/s0092-8674(03)00120-x.

Polte, T. R., Eichler, G. S., Wang, N., & Ingber, D. E. (2004). Extracellular matrix controls myosin light chain phosphorylation and cell contractility through modulation of cell shape and cytoskeletal prestress. *American Journal of Physiology Cell Physiology*, *286*(3), C518–C528. https://doi.org/10.1152/ajpcell.00280.2003.

Raucher, D., & Sheetz, M. P. (1999). Membrane expansion increases endocytosis rate during mitosis. *The Journal of Cell Biology*, *144*(3), 497–506. https://doi.org/10.1083/jcb.144.3.497.

Reid, S. E., Kay, E. J., Neilson, L. J., Henze, A., Serneels, J., McGhee, E. J., et al. (2017). Tumor matrix stiffness promotes metastatic cancer cell interaction with the endothelium. *The EMBO Journal*, *36*(16), 2373–2389. https://doi.org/10.15252/embj.201694912.

Schillaci, O., Fontana, S., Monteleone, F., Taverna, S., Di Bella, M. A., Di Vizio, D., et al. (2017). Exosomes from metastatic cancer cells transfer amoeboid phenotype to non-metastatic cells and increase endothelial permeability: Their emerging role in tumor heterogeneity. *Scientific Reports*, 7(1), 1–15. https://doi.org/10.1038/s41598-017-05002-y.

Schneider, D. J., Speth, J. M., Penke, L. R., Wettlaufer, S. H., Swanson, J. A., & Peters-Golden, M. (2017). Mechanisms and modulation of microvesicle uptake in a model of alveolar cell communication. *Journal of Biological Chemistry*, *292*(51), 20897–20910. https://doi.org/10.1074/jbc.M117.792416.

Schwager, S. C., Bordeleau, F., Zhang, J., Antonyak, M. A., Cerione, R. A., & Reinhart-King, C. A. (2019). Matrix stiffness regulates microvesicle-induced fibroblast activation. *American Journal of Physiology-Cell Physiology*, *317*(1), C82–C92. https://doi.org/10.1152/ajpcell.00418.2018.

Sedgwick, A. E., Clancy, J. W., Balmert, M. O., & D'Souza-Schorey, C. (2015). Extracellular microvesicles and invadopodia mediate non-overlapping modes of tumor cell invasion. *Scientific Reports*, *5*, 14748. https://doi.org/10.1038/srep14748.

Soler, A. P., Miller, R. D., Laughlin, K. V., Carp, N. Z., Klurfeld, D. M., & Mullin, J. M. (1999). Increased tight junctional permeability is associated with the development of colon cancer. *Carcinogenesis*, *20*(8), 1425–1431. https://doi.org/10.1093/carcin/20.8.1425.

Sonbol, H. S. (2018). Extracellular matrix remodeling in human disease. *Journal of Microscopy and Ultrastructure*, *6*(3), 123–128. https://doi.org/10.4103/JMAU.JMAU_4_18.

Sugimoto, W., Itoh, K., Mitsui, Y., Ebata, T., Fujita, H., Hirata, H., et al. (2018). Substrate rigidity-dependent positive feedback regulation between YAP and ROCK2. *Cell Adhesion & Migration*, *12*(2), 101–108. https://doi.org/10.1080/19336918.2017.1338233.

Sundborger, A. C., & Hinshaw, J. E. (2014). Regulating dynamin dynamics during endocytosis. *F1000Prime Reports*, 6. https://doi.org/10.12703/P6-85.

Taddei, M. L., Giannoni, E., Morandi, A., Ippolito, L., Ramazzotti, M., Callari, M., et al. (2014). Mesenchymal to amoeboid transition is associated with stem-like features of melanoma cells. *Cell Communication and Signaling: CCS*, *12*, 24. https://doi.org/10.1186/1478-811X-12-24.

Thottacherry, J. J., Kosmalska, A. J., Kumar, A., Vishen, A. S., Elosegui-Artola, A., Pradhan, S., et al. (2018). Mechanochemical feedback control of dynamin independent endocytosis modulates membrane tension in adherent cells. *Nature Communications*, *9*(1), 1–14. https://doi.org/10.1038/s41467-018-06738-5.

Toribio, V., Morales, S., López-Martín, S., Cardeñes, B., Cabañas, C., & Yáñez-Mó, M. (2019). Development of a quantitative method to measure EV uptake. *Scientific Reports*, *9*(1), 1–14. https://doi.org/10.1038/s41598-019-47023-9.

Tricarico, C., Clancy, J., & D'Souza-Schorey, C. (2016). Biology and biogenesis of shed microvesicles. *Small GTPases*, *8*(4), 220–232. https://doi.org/10.1080/21541248.2016.1215283.

Tucher, C., Bode, K., Schiller, P., Claßen, L., Birr, C., Souto-Carneiro, M. M., et al. (2018). Extracellular vesicle subtypes released from activated or apoptotic T-lymphocytes carry a specific and stimulus-dependent protein cargo. *Frontiers in Immunology, 9*. https://doi.org/10.3389/fimmu.2018.00534.

Vader, P., Breakefield, X. O., & Wood, M. J. A. (2014). Extracellular vesicles: Emerging targets for cancer therapy. *Trends in Molecular Medicine*, *20*(7), 385–393. https://doi.org/10.1016/j.molmed.2014.03.002.

VanderBurgh, J. A., Hotchkiss, H., Potharazu, A., Taufalele, P. V., & Reinhart-King, C. A. (2018). Substrate stiffness heterogeneities disrupt endothelial barrier integrity in a micropillar model of heterogeneous vascular stiffening. *Integrative Biology: Quantitative Biosciences from Nano to Macro*, *10*(12), 734–746. https://doi.org/10.1039/c8ib00124c.

Wendler, F., Favicchio, R., Simon, T., Alifrangis, C., Stebbing, J., & Giamas, G. (2017). Extracellular vesicles swarm the cancer microenvironment: From tumor–stroma communication to drug intervention. *Oncogene*, *36*(7), 877. https://doi.org/10.1038/onc.2016.253.

Wolf, P. (1967). The nature and significance of platelet products in human plasma. *British Journal of Haematology*, *13*(3), 269–288. https://doi.org/10.1111/j.1365-2141.1967.tb08741.x.

Xuan, Y. H., Huang, B. B., Tian, H. S., Chi, L. S., Duan, Y. M., Wang, X., et al. (2014). High-glucose inhibits human fibroblast cell migration in wound healing via repression of bFGF-regulating JNK phosphorylation. *PLoS One*, *9*(9), e108182. https://doi.org/10.1371/journal.pone.0108182.

Zhao, X.-P., Wang, M., Song, Y., Song, K., Yan, T.-L., Wang, L., et al. (2015). Membrane microvesicles as mediators for melanoma-fibroblasts communication: Roles of the VCAM-1/VLA-4 axis and the ERK1/2 signal pathway. *Cancer Letters*, *360*(2), 125–133. https://doi.org/10.1016/j.canlet.2015.01.032.

CHAPTER EIGHT

The interplay of membrane cholesterol and substrate on vascular smooth muscle biomechanics

Hanna J. Sanyour†, Alex P. Rickel†, and Zhongkui Hong*

Department of Biomedical Engineering, University of South Dakota, Vermillion, SD, United States

*Corresponding author: e-mail address: zhongkui.hong@usd.edu

Contents

Abstract

Cardiovascular disease (CVD) remains the primary cause of death worldwide. Specifically, atherosclerosis is a CVD characterized as a slow progressing chronic inflammatory disease. During atherosclerosis, vascular walls accumulate cholesterol and cause fatty streak formation. The progressive changes in vascular wall stiffness exert alternating mechanical cues on vascular smooth muscle cells (VSMCs). The detachment of VSMCs in the media layer of the vessel and migration toward the intima is a critical step in atherosclerosis. VSMC phenotypic switching is a complicated process that modifies VSMC structure and biomechanical function. These changes affect the expression and function of cell adhesion molecules, thus impacting VSMC migration. Accumulating

† These two authors contributed equally.

Current Topics in Membranes, Volume 86
ISSN 1063-5823
https://doi.org/10.1016/bs.ctm.2020.08.003

evidence has shown cholesterol is capable of regulating cellular migration, proliferation, and spreading. However, the interaction and coordinated effects of both cellular cholesterol and the extracellular matrix (ECM) stiffness/composition on VSMC biomechanics remains to be elucidated.

1. Introduction

CVD in the United States had been rising steadily from the 1900s to the 1980s and declined into the 2010s. However, CVD is still responsible for more deaths each year than chronic pulmonary disease and cancer combined (Benjamin et al., 2019). Atherosclerosis is an inflammatory disease characterized the buildup of plaque and subsequent narrowing of the artery. During the progression of atherosclerosis, vascular smooth muscle cells (VSMCs) switch from their normal contractile phenotype to a synthetic one, where they proliferate and migrate toward the innermost layer of the artery (Rudijanto, 2007). This switch is governed by changes in extracellular matrix (ECM) stiffness and composition as well as the accumulation of cholesterol and lipids in the nascent plaque. Together, this causes alterations in VSMC biomechanics including adhesion to the ECM and neighboring cells through integrins and cadherins, respectively, VSMC stiffness, and cytoskeletal architecture. The purpose of this review is to further explore the alterations in VSMC biomechanics during atherosclerosis and possible cellular mechanisms governing these changes.

2. Alterations in VSMC biomechanics in the progression of atherosclerosis

In the United States, coronary heart disease is the leading cause of death (43.2%), trailed by stroke (16.9%), high blood pressure (9.8%), heart failure (9.3%), diseases of the arteries (3.0%), and several minor CVD causes combined (17.7%). The prevalence of CVD increases the older an individual gets. It is projected that there is a 20% risk of CVD in the early 20s, 50% chance at the age of 45, and over 90% of individuals will have some form of CVD over the age of 80. Moreover, the annual direct and indirect cost of CVD in the United States is estimated to be $351.2 billion. Between 2015 and 2035, total direct medical costs of CVD are predicted to surge from $318 billion to $749 billion (Benjamin et al., 2019; Mozaffarian et al., 2015). Thus, the need to understand and reduce CVD occurrence and progression is essential.

Blood vessels are made up of three layers. Moving from the outermost layer toward the inner lumen are the adventitia, the tunica media, and the tunica intima. The adventitia is primarily composed of collagen. The tunica media is composed of VSMCs which control internal vessel diameter and are separated by collagen and elastic fibers. The tunica intima, the innermost layer, contains a single layer of endothelial cells and is responsible for the smooth flow of blood. Depending on location, size, and type of the vessel, the amount of ECM protein composition is different within each layer (Xu & Shi, 2014).

Blood vessels are biological systems that detect and react to mechanical cues and exert forces that control their surroundings' mechanical properties (Humphrey, Dufresne, & Schwartz, 2014). Mechanobiology, also known as biomechanics, spans a range of biological organization ranging from molecules to cells to organisms that convert mechanical stimuli into biochemical cues. Cells maintain homeostasis during normal physiological function by continuously assessing and reading environmental cues especially their ECM mechanical properties. This process is called mechanosensing and is mediated by cell-ECM adhesion proteins known as integrins. Integrins detect and respond to ECM mechanical cues with cytoskeletal re-organization, force generation, and gene expression thus affecting the cell biomechanics (Humphrey et al., 2014). During the progression of atherosclerosis, vessel walls undergo extensive modification including intimal stiffening, calcification, and lipid streak formation altering the underlying ECM and activating mechanosensitive proteins in VSMCs. This will lead to disruptions in the cytoskeletal architecture and cellular adhesion in addition to changes in the mechanical and elastic properties of the cell, contributing to further disease progression. Ultimately, studying cellular biomechanics will further untangle the mechanisms governing atherosclerosis development.

2.1 Modifications in cell-ECM and cell-cell adhesions of VSMCs in the progression of atherosclerosis

Primarily, VSMCs retain a contractile phenotype as they reside within the media layer of the vascular wall. The "phenotypic switching" of VSMCs from a contractile to a synthetic/non-contractile phenotype (*i.e.*, migratory and proliferative) occurs in a number of CVDs. This phenotype transition is an important step in the progression of atherosclerosis, which entails the detachment, proliferation, and migration of VSMCs toward the intima of the blood vessel, thus participating in plaque formation (Rudijanto, 2007).

During phenotypic switching, VSMCs decrease their expression of contractile protein specific genes and/or enhanced expression of synthetic/proliferative VSMCS or other cell type markers in response to arterial injury or lipid infiltration within the arterial wall (Basatemur, Jorgensen, Clarke, Bennett, & Mallat, 2019). It is essential to recognize that phenotypic switching is not a binary process but rather a spectrum ranging from contractile to synthetic. There are several altered molecular mechanisms and events that lead to phenotypic switching. Transcription factors such as Myocardin (MYOCD) that drives VSMCs contraction-related genes are often altered in expression or activity (Parmacek, 2008). Another transcription factor, Krüppel-like factor 4 (KLF4), blocks the expression of contraction-related gene. VSMC-specific deletion of KLF4 in Apolipoprotein E knockout (Apoe$^{-/-}$) mice results in reduced numbers of VSMC-derived, macrophage-marker, and mesenchymal stem-cell-marker-positive cells in atherosclerotic plaques (Shankman et al., 2015). In addition to altered transcription factors, phenotypic switching is promoted by external stimuli. Transforming growth factor-β (TGFβ) and platelet-derived growth factor (PDGF) signaling were shown to promote synthetic VSMC phenotype formation (He et al., 2015). Also, growing evidence indicated that VSMC phenotype switching is prone to epigenetic regulation. Studies have suggested that VSCMs retain cellular memory of their contractility state after changing phenotype indicating histone-modifying enzymes and DNA methylation directly affects VSMC behavior (Gomez, Shankman, Nguyen, & Owens, 2013; Liu et al., 2013). Including an access state of inflammation and hypercholesterolemia, VSMCs will undergo a phenotypic shift and migrate toward the intima.

The migration of VSMCs is a complex process that entails the modification of cell adhesion protein molecules (Schwartz & Ginsberg, 2002). The detachment of cell-cell adhesions forms the initial step of VSMC migration toward the blood vessel intima (Johnson, 2014). N-cadherin is the major cell-cell adhesion molecule in VSMCs. The cleavage of N-cadherin-mediated cell-cell adhesion directly affects VSMC behavior including migration, proliferation as well as apoptosis leading to plaque rupture (Brown et al., 2010; Dwivedi, Slater, & George, 2009; Lyon, Johnson, White, Sala-Newby, & George, 2014). A number of studies have shown that cholesterol plays an important role in stabilizing cadherin-mediated cell-cell adhesion, facilitated by the formation of cholesterol-enriched plasma membrane microdomains such as lipid rafts (Baumgartner,

Weth, Gutberlet, Harms, & Groschner, 2014; Taulet et al., 2009). Furthermore, both *in vitro* and *ex vivo* studies have suggested that substrate stiffness may regulate the stability of cadherin-mediated cell-cell adheren junctions in endothelial cells (Huynh et al., 2011). *In vitro* studies demonstrated cell-cell adhesion in VSMCs is impacted in a substrate stiffness dependent manner. The alterations in cell density exerts a strong effect on cellular morphology. Thus, it is thought that cell-cell contacts are essential to maintaining vascular function and important in mediating VSMCs response to vascular injury (Sazonova et al., 2011). However, the impact of ECM stiffness on $VSMC_S$ cell-cell adhesion is not fully deciphered. During atherosclerosis, vascular walls experience gradual increase in stiffness. Mui et al., demonstrated that N-cadherin is induced in VSMCs in response to vascular injury using an *in vivo* model of tissue stiffening and proliferation (Mui et al., 2015). In addition, N-cadherin expression is regulated by stiffness-sensitive focal adhesion (FA) kinase (FAK)-p130Cas-Rac signaling pathway. The deletion of N-cadherin in VSMCs *in vivo*, strongly reduced vascular injury-induced cell cycling, while the inhibition of FAK in VSMCs reduced proliferation after injury and accompanied by reduced induction of N-cadherin. Thus, N-cadherin participates in cell-cycle control within the vasculature and mediates the degree of cell spreading (Mui et al., 2015).

The attachment of VSMCs to the ECM is mediated by a variety of integrins and integrin β_1 is predominantly expressed on VSMCs' surfaces (Hillis, Mlynski, Simpson, & MacLeod, 1998). $\alpha_5\beta_1$ is an important integrin that binds to fibronectin and promotes FA formation (Sun, Martinez-Lemus, Hill, & Meininger, 2008). In addition to integrin β_1, integrins β_3 and β_5 also conduct key roles in VSMC adhesion and migration (Kappert et al., 2001; Liaw et al., 1995). During phenotypic switching of VSMCs, integrins $\alpha_V\beta_3$ and $\alpha_V\beta_5$ are up-regulated, while $\alpha_5\beta_1$ is down-regulated (Kappert et al., 2001). Furthermore, the presence of $\alpha_V\beta_3$ has been demonstrated in atherosclerotic lesions (Hoshiga, Alpers, Smith, Giachelli, & Schwartz, 1995) and if blocked, reduces plaque growth (Choi et al., 2004) and VSMC migration toward the intima (Kappert et al., 2001). A previous study found that integrin adhesion was rapidly altered after stimulation with contractile or dilatory vasoactive agonists with an increased adhesion during contractile activation and decreased adhesion during relaxation. Parallel changes in cell stiffness was also observed as cell adhesion was altered (Hong et al., 2012). This supports the concept that rapid switching of integrin activity in VSMCs is associated with the cellular cytoskeletal structure. In addition to the effects on integrin-mediated cell-ECM adhesion, cadherin-mediated cell-cell adhesions also play a critical role in the progression of atherosclerosis.

2.2 VSMC cytoskeletal remodeling in the progression of atherosclerosis

VSMCs are dynamic cellular systems that constantly change their shape through coordinated rearrangement of their cellular components and position within the vascular wall to adapt to alterations in their chemical and mechanical environment (Martinez-Lemus, Hill, & Meininger, 2009). During the early stages of atherosclerosis, following the infiltration of monocytes and T lymphocytes into the vascular wall, leukocytes release several cytokines and reactive oxygen species that impact VSMC phenotype allowing their proliferation and migration. These changes subsequently promote the deposition of additional fibrous ECM proteins (*e.g.*, collagen) in the intima (Hillis & Flapan, 1998). During phenotypic switching, not only do VSMCs undergo quantitative changes in contractility and but also significantly remodel their cytoskeletal architecture (Worth, Rolfe, Song, & Campbell, 2001). In the contractile state, α-actin was primarily expressed while non-muscle β-actin became much more prevalent in the synthetic state along with redistribution of these and other contractile proteins (Worth et al., 2001). Earlier studies also showed that depleting cellular cholesterol reduced endothelial cell spreading compared to cholesterol-enriched cells (Hong et al., 2013; Sun et al., 2007). In our recent publication, we found that depleting cellular cholesterol and decreased substrate stiffness significantly reduced the cortical cytoskeletal alignment in VSMCs (Sanyour et al., 2019).

The mechanism governing cytoskeletal organization in response to substrate stiffness and cholesterol is not fully understood. Actin is linked to the ECM through FA complexes containing various proteins including talin, vinculin, and various integrins (Critchley, 2000). At sufficiently high forces, talin unfolds allowing vinculin to bind, stabilizing the FA and increasing force transmission to actin (del Rio et al., 2009; Yao et al., 2014). These mechanosensitive proteins likely have a role in cytoskeleton organization as talin has been shown to mediate the co-alignment of integrins in FAs and F-actin and the growth of FAs parallel to strain. Additionally, vinculin, through its link to actin, has been shown to coordinate cell polarization in response to stretching forces (Carisey et al., 2013). FAs are also known to cluster with lipid rafts, which are membrane domains rich in cholesterol and sphingolipids (Simons & Ikonen, 1997). Lipid rafts themselves may be mechanosensitive with the ganglioside GM1 having been shown to accumulate in lipid rafts in endothelial cells in response to deformation, prior to talin accumulation (Fuentes & Butler, 2012). Clustering of GM1 lipid rafts has also been associated with enhanced β1 integrin adhesion, cortical F-actin polymerization, and shear resistance in T cells (Mitchell, Brown, Woodside, Vanderslice, & McIntyre, 2009). Cholesterol depletion from lipid rafts has been shown to inhibit phosphorylation of Src and FAK and cause subsequent dislocation of FA components (Jeon et al., 2010). Actin disruption following cholesterol depletion has also been association with the reorganization of the phospholipid phosphatidylinositol 4,5-bisphosphate, a key F-actin regulatory molecule (Kwik et al., 2003). Furthermore the cholesterol metabolite 25-hydroxycholesterol has been implicated in activation of integrin-FAK signaling (Pokharel et al., 2019).

2.3 Progressive stiffening of vessel walls

The vessel wall experiences progressive stiffening with aging associated with changes to ECM composition (Dao, Essalihi, Bouvet, & Moreau, 2005; Gaballa et al., 1998). Increased vascular stiffness is associated with hypertension (Benetos, Laurent, Asmar, & Lacolley, 1997; Lemarie, Tharaux, & Lehoux, 2010) and atherosclerosis development (Glasser, 2000; Wang, Keith Jr., Struthers, & Feuerstein, 2008). The increase in vessel wall stiffness alters VSMCs' microenvironment. In fact, a number of recent studies have shown that VSMCs become intrinsically stiffer with aging and hypertension (Sehgel et al., 2013; Zhu et al., 2012). Both chemical and mechanical factors (Sazonova et al., 2011) lead to changes in VSMC characteristics (Brown et al., 2010; Isenberg, Dimilla, Walker, Kim, & Wong, 2009; McDaniel et al., 2007; Peyton, Kim, Ghajar, Seliktar, & Putnam, 2008; Peyton & Putnam, 2005; Stegemann, Hong, & Nerem, 2005). A possible mechanism governing VSMC stiffness increase is through a mechanotransduction involving the FA complex. Through FAs, cells sense and respond to mechanical changes in their surrounding environment. Our recent study showed that VSMCs cultured on collagen 1 (COL-1) coated gel substrates with variable stiffness are able to sense the substrate stiffness and respond with changes in cell stiffness and adhesion (Fig. 1) (Sanyour et al., 2019).

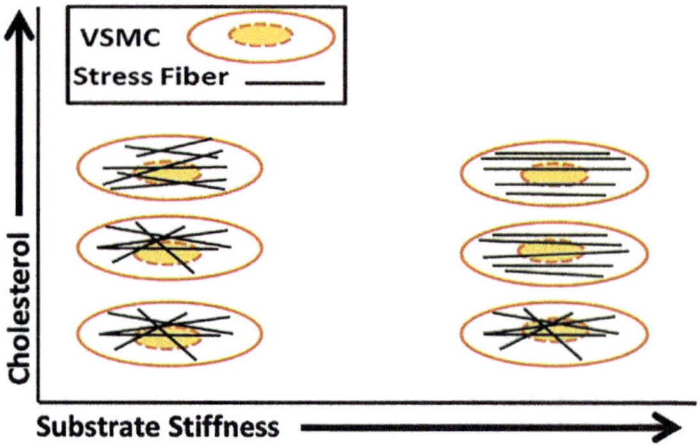

Fig. 1 Effect of substrate stiffness and cholesterol on VSMC cytoskeleton. On low stiffness substrates, both normal and cholesterol-depleted VSMCs present a disorganized cytoskeleton orientation, while cholesterol-enriched cells present a more organized cytoskeleton. On stiff substrates, the cytoskeleton becomes more orientated for normal and cholesterol-enriched VSMCs.

2.4 VSMC migration in the progression of atherosclerosis

As atherosclerosis progresses, VSMCs proliferate, migrate, and secrete new ECM proteins that contribute to fibrous plaque formation (Glass & Witztum, 2001; Willis, Pierre-Paul, Sumpio, & Gahtan, 2004). The process of cell migration can be summarized as the protrusion of leading edge lamellipodia, attachment of the lamellipodia to the substrate through integrin interaction with ECM, formation of FAs, stress fiber alignment, and contraction. This is accompanied by the disassembly of FAs and disengagement of integrin-substrate binding at the trailing edge of cells (Horwitz & Parsons, 1999; Stossel, 1993) as cells move forward. We have also shown that VSMC migration behavior is dependent upon both the underlying substrate stiffness and the ECM protein composition. VSMCs were observed to migrate further on stiffer fibronectin-coated polyacrylamide (PA) gels, while the opposite behavior was observed on COL-1 coated PA gels (Rickel, Sanyour, Leyda, & Hong, 2020). This is vital to understanding atherosclerosis progression as both stiffness and protein compositions of atherosclerotic lesions vary with different regions and over time (Doddapattar et al., 2015; Lakatta, 2003; Tracqui et al., 2011; Wang et al., 2006).

3. The role of lipids and cholesterol in driving cardiovascular disease

Generally, cholesterol synthesis is divided into two major components: pre-squalene cholesterol synthesis and post-squalene cholesterol synthesis. Pre-squalene cholesterol synthesis contributes to both sterol and isoprenoid formation, whereas post-squalene cholesterol synthesis commits to vitamin D and sterol synthesis. Changes in cellular cholesterol homeostasis and cholesterol accumulation possess a significant clinical relevance. Not only is cholesterol a major factor involved in fatty deposition in the atherosclerotic lesion, but also plays an important role in regulating cellular mechanics (Byfield, Tikku, Rothblat, Gooch, & Levitan, 2006; Khatibzadeh, Spector, Brownell, & Anvari, 2013; Kowalsky, Byfield, & Levitan, 2008; Kwik et al., 2003; Levitan & Shentu, 2011; Sun et al., 2007), spreading (Hong et al., 2013), and migration (Nagao, Qin, Grosheva, Maxfield, & Pierini, 2007; Ramprasad et al., 2007; Sekine et al., 2010; Yan, Xia, Duan, Li, & Mei, 2011). However, reports about the effect of cholesterol on cell mechanics and migration are varied and controversial and even contradictory in different cell types underscoring the need for further study. In addition, a further complication is the vascular wall mechanical environment that undergoes progressive stiffening with aging and hypertension, therefore, may accelerate the development of atherosclerosis (Dao et al., 2005; Glasser, 2000; Wang et al., 2008). As the major cellular component of blood vessel, VSMCs carry out vital roles at each stage of atherosclerosis development (Willis et al., 2004).

A recent study reported that the amount of excess cholesterol loaded into human atherosclerotic VSMCs was much larger than previously known, with about 40% of the total foam cells in the atherosclerotic plaque being VSMC-derived (Allahverdian, Chehroudi, McManus, Abraham, & Francis, 2014). Similarly, arterial VSMCs isolated from cholesterol-fed atherosclerotic rabbits showed a progressive elevation in the membrane unesterified cholesterol-phospholipid mole ratio (Chen, Preston Mason, & Tulenko, 1995; Tulenko, Chen, Mason, & Mason, 1998). These findings suggest that dietary cholesterol induces changes in the structure and organization of VSMC plasma membrane *in vivo*. We and others propose that these alterations may cause functional and phenotypic switching of aortic VSMCs during atherogenesis. Another important finding is that excess membrane cholesterol decreases membrane fluidity (Gleason, Medow, & Tulenko, 1991) and alters Ca^{2+} channels in arterial VSMCs (Bialecki & Tulenko, 1989), changing Ca^{2+} movements and increasing cytosolic Ca^{2+} levels. Furthermore, changes in cortical stiffness and adhesion following contractile activation have been shown to occur in sync with changes in cytosolic Ca^{2+} (Huang, Sun, Hill, & Meininger, 2018). These changes may have important effects on vascular wall properties. In relation to atherosclerosis, Ca^{2+} is known to participate in signal transduction that promotes VSMC proliferation and migration

(Mason & Jacob, 2003). In contrast to VSMCs, most prior studies in relation to the effect of cholesterol on the vascular cell biomechanics have focused on the endothelial cells (Byfield, Aranda-Espinoza, Romanenko, Rothblat, & Levitan, 2004; Hong et al., 2013; Kowalsky et al., 2008; Levitan & Shentu, 2011; Sun et al., 2007) and neglected possible direct effects of cholesterol on VSMCs.

4. Atomic Force Microscope, a powerful tool to study live VSMC biomechanics

Atomic Force Microscope (AFM) is a unique instrument utilized for a plethora of applications. These include contact imaging, tapping imaging, viscoelastic measurements, single force molecule measurement, lithography, and several other modes (Binnig, Quate, & Gerber, 1986; Krieg et al., 2019). AFM is an ideal tool for understanding the scale and size of conducting biomechanical testing of cells. AFM can be used to indent the VSMC surface and conduct nanoscale resolution imaging (Hong et al., 2012). Contact mode imaging is a powerful method that generates high-resolution three-dimensional (3D) topographical images of samples or cells (Hong et al., 2016). This is achieved by applying a constant force across the surface to indent approximately 180–300 nm into the VSMCs. The obtained images will show cortical actin and stress fibers that can be analyzed for their orientation and density (Sanyour et al., 2019). Nanoindentation allows for the determination of viscoelastic (stiffness) properties of VSMCs. In addition, AFM can be used to conduct single molecule spectroscopy (Neumann, 2008). Essentially, an AFM stylus tip is coated with a protein that is complementary to cellular surface adhesion receptors, such as integrins or cadherins. This allows for the characterization and quantification of cell-cell and cell-ECM adhesion forces. Recently, we were able to develop an automated approach to filter background noise and extract dynamic biomedical signals/oscillations from large sets of real-time AFM data (Sanyour, Childs, Meininger, & Hong, 2018). Earlier work describes conducting AFM studies to measure the elastic modulus (*E*-modulus) (Zhu et al., 2012), stress fiber orientation (Gupta et al., 2015), and adhesion of fibronectin to integrins (Helenius, Heisenberg, Gaub, & Muller, 2008) cultured on different substrate stiffnesses. Nevertheless, the combined effect of substrate stiffness and cellular lipid/cholesterol regulations is yet to be assessed. Deciphering the impact of both cholesterol management and substrate stiffness on VSMC biomechanics will better define VSMCs' experience in the progression of atherosclerosis. Ultimately, this might provide a better understating of the mechanical mechanisms underlying atherosclerosis.

5. Impact of vascular ECM stiffness and cholesterol on VSMC biomechanics

To study the impact of cholesterol on cellular mechanics, cyclodextrins (CD) are widely used to manipulate cell membrane cholesterol by directly removing cholesterol from the cell membrane (Zidovetzki & Levitan, 2007). CDs are made up of α-(1-4)-linked D-glycopyranose units in a cyclic arrangement. Because of their water soluble nature and hydrophobic cavity, they have been used as hydrophobic molecules and drugs carriers (Zidovetzki & Levitan, 2007). Methyl-β-cyclodextrin (MβCD) in particular is widely used to either deplete or enrich cellular membrane cholesterol. However, CD's are also reported to have toxic effects on cells and induce cell death (Zidovetzki & Levitan, 2007).

VSMCs experience a wide variety of environments of varying stiffness and cholesterol concentration within an atherosclerotic plaque (Fig. 2). In a recent study, we reported the combinatorial effects of substrate stiffness and MβCD mediated cholesterol depletion and enrichment on VSMC cytoskeletal remodeling and biomechanics. Rat VSMCs were cultured on COL-1 coated polyacrylamide (PA) gels with a tunable stiffness. AFM was employed to measure cell and PA gel stiffness, α5β1-fibronectin adhesion, and contact mode was used to image cortical cytoskeleton orientation. To further confirm AFM results, confocal microscopy was used to image the entire cell cytoskeleton (Fig. 3E). Upon cholesterol depletion, VSMC stiffness and α5β1-fibronectin adhesion were reduced with the

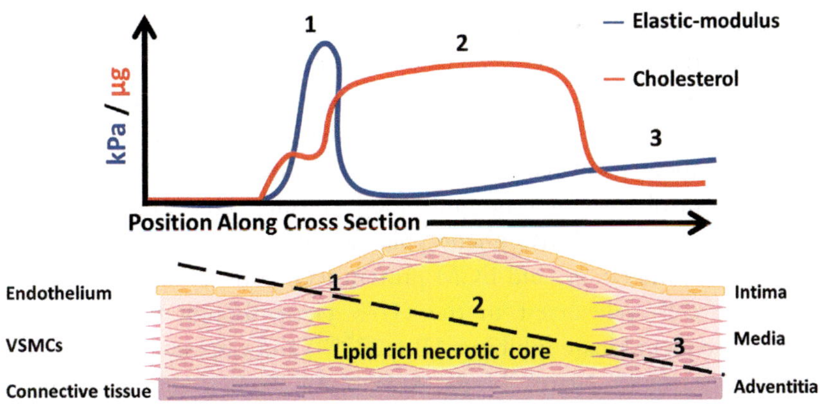

Fig. 2 The different microenvironments of an atherosclerotic plaque. The graph in the upper panel follows the black dotted line in the lower panel showing: (1) the fibrous cap is stiff and enriched with cholesterol. (2) The plaque core is rich in cholesterol and very soft. (3) Normal arterial wall structure and cholesterol composition.

cells displaying a more disorganized cytoskeleton (Fig. 3A–C, green). Cholesterol enrichment with a MβCD-cholesterol saturate solution increased VSMC stiffness, α5β1 adhesion, and cytoskeletal orientation compared to cholesterol-depleted VSMCs (Fig. 3A–C, red). Furthermore, decreasing substrate stiffness resulted in a reduced VSMC stiffness (Fig. 3A), α5β1 adhesion (Fig. 3B), and whole cytoskeletal organization on the softest substrate (Fig. 3D and E) (Sanyour et al., 2019).

3-Hydroxy-3-methyl-glutaryl coenzyme A (HMG-CoA) reductase inhibitors, commonly known as statins, block cholesterol synthesis by inhibiting HMG-CoA reductase, the rate-limiting enzyme in the mevalonate pathway. While statin therapy is known to decrease the incidence of clinical cardiovascular disease through the lowering of serum cholesterol levels (Scandinavian Simvastatin Survival Study Group, 1994; Shepherd et al., 1995), evidence suggests that statins exert additional benefits including decreased VSMC proliferation (Bellosta et al., 2004; Corsini, Pazzucconi, Pfister, Paoletti, & Sirtori, 1996). Apart from lowering cholesterol biosynthesis, statins also halt the synthesis of various metabolic intermediates associated with the mevalonate pathway, such as isoprenoid synthesis (Ward, Watts, & Eckel, 2019). Cholesterol-independent effects of statins potentially mediate pleiotropic effects on cardiovascular physiology by modifying plaque stability, signaling pathways, platelet aggregation, and VSMC function (Oesterle, Laufs, & Liao, 2017). In vascular disease, inflammation plays a key role spanning initial disease events such as endothelial damage and macrophage recruitment to further distal events such as plaque rupture and stability (Libby, Ridker, & Maseri, 2002). Hence, statins were explored for their properties beyond their cholesterol lowering properties as anti-inflammatories. Statins have been extensively reviewed for over two decades and demonstrated to have clinical lipid-independent anti-inflammatory effects. Several different clinical anti-inflammatory mechanisms have been described (Jain & Ridker, 2005). Under normal conditions, VSMCs control vascular tone and blood pressure. Due to inflammatory vascular injury and VSMC phenotypic switching, VSMCs increase their proliferation, migration, and ECM protein secretions. Statins were shown to induce cell-cycle arrest in the G1/S phase though factors such as cyclin dependent kinase inhibitor 1B ($p27^{Kip1}$) and caspases to inhibit their proliferation (Laufs, Marra, Node, & Liao, 1999). In addition, Sakamoto et al. demonstrated that fluvastatin prevented VSMC hyperplasia, proliferation, and phenotype modulation by ERK1/2 and p38MAPK inactivation (Sakamoto, Murata, Chuma, Hori, & Ozaki, 2005). We previously

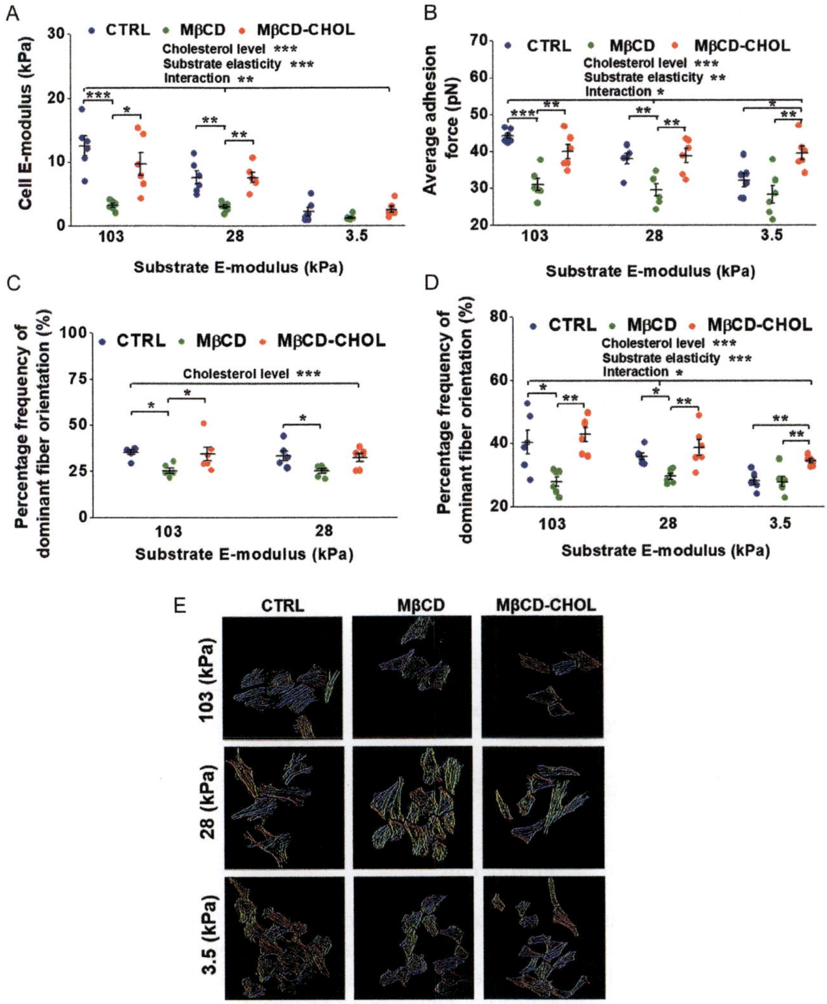

Fig. 3 Effects of MβCD mediated cholesterol depletion and enrichment on VSMCs cultured on substrates of varying stiffness. (A) Cholesterol depletion and decreasing substrate stiffness significantly lowered VSMC *E*-modulus while cholesterol-enriched VSMCs had higher stiffness than cholesterol depleted ones. (B) α5B1-fibronectin adhesion force was significantly reduced upon cholesterol depletion and decreasing substrate stiffness. (C) AFM imaging of submembraneous actin showed cholesterol depletion decreased cytoskeletal organization while enriched VSMCs showed greater organization. (D) Quantification of confocal imaging of the whole cell cytoskeleton revealed VSMCs on the softest substrate had significantly reduced cytoskeletal organization. (E) Representative color map images of actin fiber orientation computed from confocal images. Different colors represent different actin orientation angles. VSMCs on the stiffer substrates showed more uniform color, indicative of a more orientated actin cytoskeleton. Cholesterol-depleted VSMCs showed more variable colors and a more disperse actin alignment, while cholesterol-enriched VSMCs had more uniform color and actin orientation. *Figure is adopted from Sanyour H. J., Li N., Rickel A. P., Childs J. D., Kinser C. N. & Hong Z. (2019). Membrane cholesterol and substrate stiffness co-ordinate to induce the remodelling of the cytoskeleton and the alteration in the biomechanics of vascular smooth muscle cells.* Cardiovascular Research, 115, *1369–1380.*

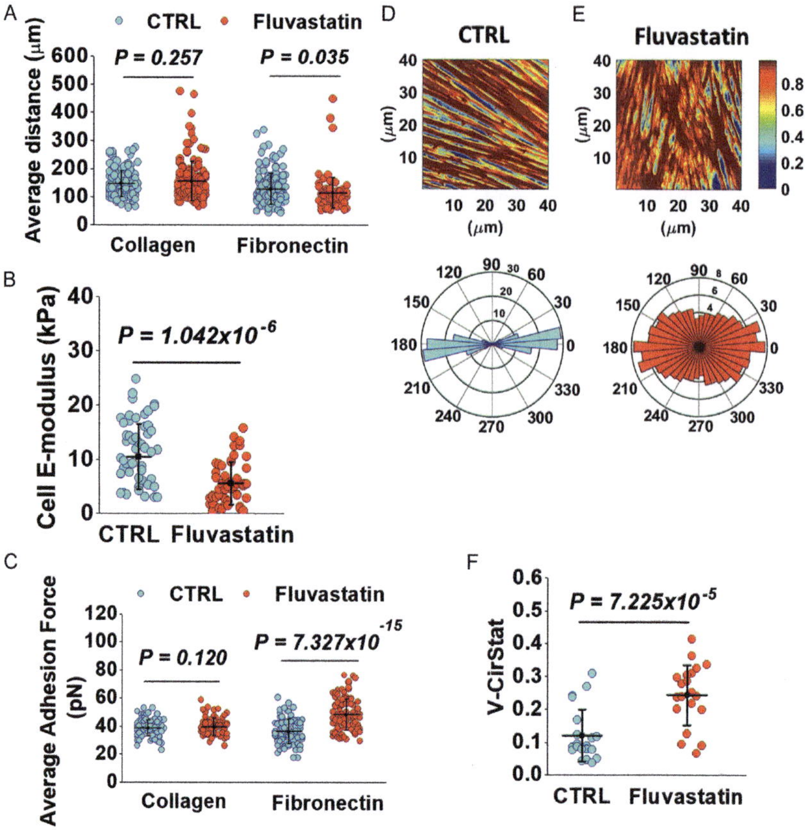

Fig. 4 Effects of fluvastatin-mediated cholesterol depletion. (A) Fluvastatin-treated VSMCs migrating on fibronectin had significantly reduced migration distance while those on COL-1 had no difference in migration distance. (B) Fluvastatin significantly reduced VSMC *E*-modulus. (C) Adhesion to fibronectin significantly increased with fluvastatin treated while adhesion to COL-1 did not change. (D and E) (Top) Representative AFM height images of submembraneous actin for control and fluvastatin treatments, respectively. (Bottom) Corresponding circular distribution plots of actin orientation angles normalized to the dominant orientation angle. (F) Fluvastatin treatment significantly reduced submembraneous actin organization. *Figure is adopted from Sanyour H. J., Li N., Rickel A. P., Torres H. M., Anderson R. H., Miles M. R., Childs J. D., Francis K. R., Tao J. & Hong Z. (2020). Statin mediated cholesterol depletion exerts coordinated effects on the alterations in rat vascular smooth muscle cell biomechanics and migration.* The Journal of Physiology, *598, 1505–1522.*

demonstrated that total VSMCs and aortic tissue cholesterol was reduced and their biomechanical properties altered following fluvastatin treatment (Sanyour et al., 2020). AFM analysis showed fluvastatin lowered VSMC *E*-modulus and caused them to display a more disorganized submembraneous actin structure (Fig. 4B and D–F). Our results also revealed fluvastatin-mediated cholesterol depletion had differential effects on VSMC responses to COL-1 and fibronectin. VSMC migration distance decreased upon cholesterol depletion on a fibronectin-coated surface while adhesion to fibronectin increased (Fig. 4A and C). No changes were observed for migration distance on COL-1 or VSMC adhesion to COL-1 (Fig. 4A and C) (Sanyour et al., 2020). Yet, our preliminary ELISA (enzyme-linked immunosorbent assay) experiments (not shown) suggested a none-significant effect on proinflammatory tumor necrosis factor (TNF) and interleukin 1 (IL-1) secretion in VSMCs and aortic tissue-culture media after 3 days of statin treatment. Therefore, deciphering the exact cholesterol dependent/independent effects of statins in modifying VSMC biomechanics, is essential.

6. Conclusions and future directions

VSMCs undergo a myriad of changes as CVD and especially atherosclerosis progress. Recent discoveries have expanded the role of cholesterol in the progression of atherosclerosis and indicates the need for further exploration. Since VSMCs experience a wide variety of microenvironments *in vivo*, novel *in vitro* models need to be developed in order to study the mechanical contribution of VSMCs in CVD development. Using different inhibitors of cholesterol biosynthesis precursors to modulate cholesterol homeostasis might elucidate specific sterols that regulate VSMC functions. Interrogating the changes in VSMCs from atherosclerosis disease models is a logical and essential future direction. Our recent findings that both cholesterol and ECM stiffness coordinate and contribute to the changes in VSMC biomechanics sheds light on the complicated dynamics involved in atherosclerosis development (Fig. 5). Understanding and deciphering

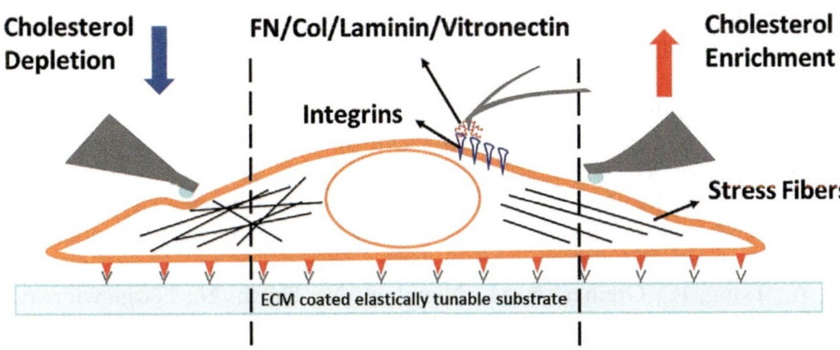

Fig. 5 VSMC biomechanics obtained from AFM. (Left) VSMC stiffness and cytoskeletal orientation decreases following cholesterol depletion. (Center) AFM stylus tip is coated with protein to measure adhesion forces. (Right) VSMC stiffness and cytoskeletal orientation increases following cholesterol enrichment.

the molecular crosstalk mechanisms between the ECM and cholesterol will enhance the discovery of new targets that potentially hinder atherosclerosis development.

Acknowledgment

This work was supported, in part, by American Heart Association 15SDG25420001 (to Z. H.), South Dakota Board of Regents UP1600205 (to Z. H.), and NIH R15HL147214 (Z. H.).

References

Allahverdian, S., Chehroudi, A. C., McManus, B. M., Abraham, T., & Francis, G. A. (2014). Contribution of intimal smooth muscle cells to cholesterol accumulation and macrophage-like cells in human atherosclerosis. *Circulation, 129*, 1551–1559.

Basatemur, G. L., Jorgensen, H. F., Clarke, M. C. H., Bennett, M. R., & Mallat, Z. (2019). Vascular smooth muscle cells in atherosclerosis. *Nature Reviews Cardiology, 16*, 727–744.

Baumgartner, W., Weth, A., Gutberlet, J., Harms, G., & Groschner, K. (2014). Localization of VE-cadherin in plasmalemmal cholesterol rich microdomains and the effects of cholesterol depletion on VE-cadherin mediated cell–cell adhesion. *Biochimica et Biophysica Acta (BBA)—Molecular and Cell Biology of Lipids, 1841*, 1725–1732.

Bellosta, S., Arnaboldi, L., Gerosa, L., Canavesi, M., Parente, R., Baetta, R., et al. (2004). Statins effect on smooth muscle cell proliferation. *Seminars in Vascular Medicine, 4*, 347–356.

Benetos, A., Laurent, S., Asmar, R. G., & Lacolley, P. (1997). Large artery stiffness in hypertension. *Journal of Hypertension Supplement, 15*, S89–S97.

Benjamin, E. J., Muntner, P., Alonso, A., Bittencourt, M. S., Callaway, C. W., Carson, A. P., et al. (2019). Heart disease and stroke statistics-2019 update: A report from the American Heart Association. *Circulation, 139*, e56–e528.

Bialecki, R. A., & Tulenko, T. N. (1989). Excess membrane cholesterol alters calcium channels in arterial smooth muscle. *American Journal of Physiology-Cell Physiology, 257*, C306–C314.

Binnig, G., Quate, C. F., & Gerber, C. (1986). Atomic force microscope. *Physical Review Letters, 56*, 930.

Brown, X. Q., Bartolak-Suki, E., Williams, C., Walker, M. L., Weaver, V. M., & Wong, J. Y. (2010). Effect of substrate stiffness and PDGF on the behavior of vascular smooth muscle cells: Implications for atherosclerosis. *Journal of Cellular Physiology, 225*, 115–122.

Byfield, F. J., Aranda-Espinoza, H., Romanenko, V. G., Rothblat, G. H., & Levitan, I. (2004). Cholesterol depletion increases membrane stiffness of aortic endothelial cells. *Biophysical Journal, 87*, 3336–3343.

Byfield, F. J., Tikku, S., Rothblat, G. H., Gooch, K. J., & Levitan, I. (2006). OxLDL increases endothelial stiffness, force generation, and network formation. *Journal of Lipid Research, 47*, 715–723.

Carisey, A., Tsang, R., Greiner, A. M., Nijenhuis, N., Heath, N., Nazgiewicz, A., et al. (2013). Vinculin regulates the recruitment and release of core focal adhesion proteins in a force-dependent manner. *Current Biology, 23*, 271–281.

Chen, M., Preston Mason, R., & Tulenko, T. N. (1995). Atherosclerosis alters the composition, structure and function of arterial smooth muscle cell plasma membranes. *Biochimica et Biophysica Acta, Molecular Basis of Disease, 1272*, 101–112.

Choi, E. T., Khan, M. F., Leidenfrost, J. E., Collins, E. T., Boc, K. P., Villa, B. R., et al. (2004). β_3-Integrin mediates smooth muscle cell accumulation in neointima after carotid ligation in mice. *Circulation, 109*, 1564.

Corsini, A., Pazzucconi, F., Pfister, P., Paoletti, R., & Sirtori, C. R. (1996). Inhibitor of proliferation of arterial smooth-muscle cells by fluvastatin. *The Lancet, 348*, 1584.

Critchley, D. R. (2000). Focal adhesions—The cytoskeletal connection. *Current Opinion in Cell Biology, 12*, 133–139.

Dao, H. H., Essalihi, R., Bouvet, C., & Moreau, P. (2005). Evolution and modulation of age-related medial elastocalcinosis: Impact on large artery stiffness and isolated systolic hypertension. *Cardiovascular Research, 66*, 307–317.

del Rio, A., Perez-Jimenez, R., Liu, R., Roca-Cusachs, P., Fernandez, J. M., & Sheetz, M. P. (2009). Stretching single talin rod molecules activates vinculin binding. *Science, 323*, 638–641.

Doddapattar, P., Gandhi, C., Prakash, P., Dhanesha, N., Grumbach, I. M., Dailey, M. E., et al. (2015). Fibronectin splicing variants containing extra domain a promote atherosclerosis in mice through toll-like receptor 4. *Arteriosclerosis, Thrombosis, and Vascular Biology, 35*, 2391–2400.

Dwivedi, A., Slater, S. C., & George, S. J. (2009). MMP-9 and -12 cause N-cadherin shedding and thereby beta-catenin signalling and vascular smooth muscle cell proliferation. *Cardiovascular Research, 81*, 178–186.

Fuentes, D. E., & Butler, P. J. (2012). Coordinated mechanosensitivity of membrane rafts and focal adhesions. *Cellular and Molecular Bioengineering, 5*, 143–154.

Gaballa, M. A., Jacob, C. T., Raya, T. E., Liu, J., Simon, B., & Goldman, S. (1998). Large artery remodeling during aging: Biaxial passive and active stiffness. *Hypertension, 32*, 437–443.

Glass, C. K., & Witztum, J. L. (2001). Atherosclerosis. The road ahead. *Cell, 104*, 503–516.

Glasser, S. P. (2000). On arterial physiology, pathophysiology of vascular compliance, and cardiovascular disease. *Heart Disease, 2*, 375–379.

Gleason, M. M., Medow, M. S., & Tulenko, T. N. (1991). Excess membrane cholesterol alters calcium movements, cytosolic calcium levels, and membrane fluidity in arterial smooth muscle cells. *Circulation Research, 69*, 216–227.

Gomez, D., Shankman, L. S., Nguyen, A. T., & Owens, G. K. (2013). Detection of histone modifications at specific gene loci in single cells in histological sections. *Nature Methods, 10*, 171–177.

Gupta, M., Sarangi, B. R., Deschamps, J., Nematbakhsh, Y., Callan-Jones, A., Margadant, F., et al. (2015). Adaptive rheology and ordering of cell cytoskeleton govern matrix rigidity sensing. *Nature Communications, 6*, 7525.

He, C., Medley, S. C., Hu, T., Hinsdale, M. E., Lupu, F., Virmani, R., et al. (2015). PDGFRβ signalling regulates local inflammation and synergizes with hypercholesterolaemia to promote atherosclerosis. *Nature Communications, 6*, 7770.

Helenius, J., Heisenberg, C.-P., Gaub, H. E., & Muller, D. J. (2008). Single-cell force spectroscopy. *Journal of Cell Science, 121*, 1785–1791.

Hillis, G., & Flapan, A. (1998). Cell adhesion molecules in cardiovascular disease: A clinical perspective. *Heart, 79*, 429–431.

Hillis, G. S., Mlynski, R. A., Simpson, J. G., & MacLeod, A. M. (1998). The expression of $\beta1$ integrins in human coronary artery. *Basic Research in Cardiology, 93*, 295–302.

Hong, Z., Ersoy, I., Sun, M., Bunyak, F., Hampel, P., Hong, Z., et al. (2013). Influence of membrane cholesterol and substrate elasticity on endothelial cell spreading behavior. *Journal of Biomedical Materials Research. Part A, 101*, 1994–2004.

Hong, Z., Sun, Z., Li, Z., Mesquitta, W.-T., Trzeciakowski, J. P., & Meininger, G. A. (2012). Coordination of fibronectin adhesion with contraction and relaxation in microvascular smooth muscle. *Cardiovascular Research, 96*, 73–80.

Hong, K., Zhao, G., Hong, Z., Sun, Z., Yang, Y., Clifford, P. S., et al. (2016). Mechanical activation of angiotensin II type 1 receptors causes actin remodelling and myogenic responsiveness in skeletal muscle arterioles. *The Journal of Physiology*, *594*, 7027–7047.

Horwitz, A. R., & Parsons, J. T. (1999). Cell migration—Movin' on. *Science*, *286*, 1102–1103.

Hoshiga, M., Alpers, C. E., Smith, L. L., Giachelli, C. M., & Schwartz, S. M. (1995). Integrin expression in normal and atherosclerotic artery. *Circulation Research*, 77, 1129.

Huang, H., Sun, Z., Hill, M. A., & Meininger, G. A. (2018). A calcium mediated mechanism coordinating vascular smooth muscle cell adhesion during KCl activation. *Frontiers in Physiology*, *9*, 1810.

Humphrey, J. D., Dufresne, E. R., & Schwartz, M. A. (2014). Mechanotransduction and extracellular matrix homeostasis. *Nature Reviews. Molecular Cell Biology*, *15*, 802–812.

Huynh, J., Nishimura, N., Rana, K., Peloquin, J. M., Califano, J. P., Montague, C. R., et al. (2011). Age-related intimal stiffening enhances endothelial permeability and leukocyte transmigration. *Science Translational Medicine*, *3*, 112ra122.

Isenberg, B. C., Dimilla, P. A., Walker, M., Kim, S., & Wong, J. Y. (2009). Vascular smooth muscle cell durotaxis depends on substrate stiffness gradient strength. *Biophysical Journal*, *97*, 1313–1322.

Jain, M. K., & Ridker, P. M. (2005). Anti-inflammatory effects of statins: Clinical evidence and basic mechanisms. *Nature Reviews. Drug Discovery*, *4*, 977–987.

Jeon, J. H., Kim, S. K., Kim, H. J., Chang, J., Ahn, C. M., & Chang, Y. S. (2010). Lipid raft modulation inhibits NSCLC cell migration through delocalization of the focal adhesion complex. *Lung Cancer*, *69*, 165–171.

Johnson, J. L. (2014). Emerging regulators of vascular smooth muscle cell function in the development and progression of atherosclerosis. *Cardiovascular Research*, *103*, 452–460.

Kappert, K., Blaschke, F., Meehan, W. P., Kawano, H., Grill, M., Fleck, E., et al. (2001). Integrins αvβ3 and αvβ5 mediate VSMC migration and are elevated during neointima formation in the rat aorta. *Basic Research in Cardiology*, *96*, 42–49.

Khatibzadeh, N., Spector, A. A., Brownell, W. E., & Anvari, B. (2013). Effects of plasma membrane cholesterol level and cytoskeleton F-actin on cell protrusion mechanics. *PLoS One*, *8*, e57147.

Kowalsky, G. B., Byfield, F. J., & Levitan, I. (2008). oxLDL facilitates flow-induced realignment of aortic endothelial cells. *American Journal of Physiology. Cell Physiology*, *295*, C332–C340.

Krieg, M., Fläschner, G., Alsteens, D., Gaub, B. M., Roos, W. H., Wuite, G. J. L., et al. (2019). Atomic force microscopy-based mechanobiology. *Nature Reviews. Physics*, *1*, 41–57.

Kwik, J., Boyle, S., Fooksman, D., Margolis, L., Sheetz, M. P., & Edidin, M. (2003). Membrane cholesterol, lateral mobility, and the phosphatidylinositol 4,5-bisphosphate-dependent organization of cell actin. *Proceedings of the National Academy of Sciences of the United States of America*, *100*, 13964–13969.

Lakatta, E. G. (2003). Arterial and cardiac aging: Major shareholders in cardiovascular disease enterprises: Part III: Cellular and molecular clues to heart and arterial aging. *Circulation*, *107*, 490–497.

Laufs, U., Marra, D., Node, K., & Liao, J. K. (1999). 3-Hydroxy-3-methylglutaryl-CoA reductase inhibitors attenuate vascular smooth muscle proliferation by preventing rho GTPase-induced down-regulation of p27(Kip1). *The Journal of Biological Chemistry*, *274*, 21926–21931.

Lemarie, C. A., Tharaux, P. L., & Lehoux, S. (2010). Extracellular matrix alterations in hypertensive vascular remodeling. *Journal of Molecular and Cellular Cardiology*, *48*, 433–439.

Levitan, I., & Shentu, T.-P. (2011). Impact of oxLDL on cholesterol-rich membrane rafts. *Journal of Lipids*, *2011*, 730209.

Liaw, L., Skinner, M. P., Raines, E. W., Ross, R., Cheresh, D. A., Schwartz, S. M., et al. (1995). The adhesive and migratory effects of osteopontin are mediated via distinct cell surface integrins. Role of alpha v beta 3 in smooth muscle cell migration to osteopontin in vitro. *The Journal of Clinical Investigation, 95*, 713–724.

Libby, P., Ridker, P. M., & Maseri, A. (2002). Inflammation and atherosclerosis. *Circulation, 105*, 1135–1143.

Liu, R., Jin, Y., Tang, W. H., Qin, L., Zhang, X., Tellides, G., et al. (2013). Ten-eleven translocation-2 (TET2) is a master regulator of smooth muscle cell plasticity. *Circulation, 128*, 2047–2057.

Lyon, C. A., Johnson, J. L., White, S., Sala-Newby, G. B., & George, S. J. (2014). EC4, a truncation of soluble N-cadherin, reduces vascular smooth muscle cell apoptosis and markers of atherosclerotic plaque instability. *Molecular Therapy—Methods & Clinical Development, 1*, 14004.

Martinez-Lemus, L. A., Hill, M. A., & Meininger, G. A. (2009). The plastic nature of the vascular wall: A continuum of remodeling events contributing to control of arteriolar diameter and structure. *Physiology (Bethesda), 24*, 45–57.

Mason, R. P., & Jacob, R. F. (2003). Membrane microdomains and vascular biology: Emerging role in atherogenesis. *Circulation, 107*, 2270–2273.

McDaniel, D. P., Shaw, G. A., Elliott, J. T., Bhadriraju, K., Meuse, C., Chung, K. H., et al. (2007). The stiffness of collagen fibrils influences vascular smooth muscle cell phenotype. *Biophysical Journal, 92*, 1759–1769.

Mitchell, J. S., Brown, W. S., Woodside, D. G., Vanderslice, P., & McIntyre, B. W. (2009). Clustering T-cell GM1 lipid rafts increases cellular resistance to shear on fibronectin through changes in integrin affinity and cytoskeletal dynamics. *Immunology and Cell Biology, 87*, 324–336.

Mozaffarian, D., Benjamin, E. J., Go, A. S., Arnett, D. K., Blaha, M. J., Cushman, M., et al. (2015). Heart disease and stroke statistics—2015 Update: A report from the American Heart Association. *Circulation, 131*, e29–322.

Mui, K. L., Bae, Y. H., Gao, L., Liu, S.-L., Xu, T., Radice, G. L., et al. (2015). N-cadherin induction by ECM stiffness and FAK overrides the spreading requirement for proliferation of vascular smooth muscle cells. *Cell Reports, 10*, 1477–1486.

Nagao, T., Qin, C., Grosheva, I., Maxfield, F. R., & Pierini, L. M. (2007). Elevated cholesterol levels in the plasma membranes of macrophages inhibit migration by disrupting RhoA regulation. *Arteriosclerosis, Thrombosis, and Vascular Biology, 27*, 1596–1602.

Neumann, T. (2008). Determining the elastic modulus of biological samples using atomic force microscopy. In *JPK instruments application report* (pp. 1–9).

Oesterle, A., Laufs, U., & Liao, J. K. (2017). Pleiotropic effects of statins on the cardiovascular system. *Circulation Research, 120*, 229–243.

Parmacek, M. S. (2008). Myocardin: Dominant driver of the smooth muscle cell contractile phenotype. *Arteriosclerosis, Thrombosis, and Vascular Biology, 28*, 1416–1417.

Peyton, S. R., Kim, P. D., Ghajar, C. M., Seliktar, D., & Putnam, A. J. (2008). The effects of matrix stiffness and RhoA on the phenotypic plasticity of smooth muscle cells in a 3-D biosynthetic hydrogel system. *Biomaterials, 29*, 2597–2607.

Peyton, S. R., & Putnam, A. J. (2005). Extracellular matrix rigidity governs smooth muscle cell motility in a biphasic fashion. *Journal of Cellular Physiology, 204*, 198–209.

Pokharel, S. M., Shil, N. K., Gc, J. B., Colburn, Z. T., Tsai, S. Y., Segovia, J. A., et al. (2019). Integrin activation by the lipid molecule 25-hydroxycholesterol induces a proinflammatory response. *Nature Communications, 10*, 1482.

Ramprasad, O. G., Srinivas, G., Rao, K. S., Joshi, P., Thiery, J. P., Dufour, S., et al. (2007). Changes in cholesterol levels in the plasma membrane modulate cell signaling and regulate cell adhesion and migration on fibronectin. *Cell Motility and the Cytoskeleton, 64*, 199–216.

Rickel, A. P., Sanyour, H. J., Leyda, N. A., & Hong, Z. (2020). Extracellular matrix proteins and substrate stiffness synergistically regulate vascular smooth muscle cell migration and cortical cytoskeleton organization. *ACS Applied Bio Materials*, *3*, 2360–2369.

Rudijanto, A. (2007). The role of vascular smooth muscle cells on the pathogenesis of atherosclerosis. *Acta Medica Indonesiana*, *39*, 86–93.

Sakamoto, K., Murata, T., Chuma, H., Hori, M., & Ozaki, H. (2005). Fluvastatin prevents vascular hyperplasia by inhibiting phenotype modulation and proliferation through extracellular signal-regulated kinase 1 and 2 and p38 mitogen-activated protein kinase inactivation in organ-cultured artery. *Arteriosclerosis, Thrombosis, and Vascular Biology*, *25*, 327–333.

Sanyour, H., Childs, J., Meininger, G. A., & Hong, Z. (2018). Spontaneous oscillation in cell adhesion and stiffness measured using atomic force microscopy. *Scientific Reports*, *8*, 2899.

Sanyour, H. J., Li, N., Rickel, A. P., Childs, J. D., Kinser, C. N., & Hong, Z. (2019). Membrane cholesterol and substrate stiffness co-ordinate to induce the remodelling of the cytoskeleton and the alteration in the biomechanics of vascular smooth muscle cells. *Cardiovascular Research*, *115*, 1369–1380.

Sanyour, H. J., Li, N., Rickel, A. P., Torres, H. M., Anderson, R. H., Miles, M. R., et al. (2020). Statin mediated cholesterol depletion exerts coordinated effects on the alterations in rat vascular smooth muscle cell biomechanics and migration. *The Journal of Physiology*, *598*, 1505–1522.

Sazonova, O. V., Lee, K. L., Isenberg, B. C., Rich, C. B., Nugent, M. A., & Wong, J. Y. (2011). Cell-cell interactions mediate the response of vascular smooth muscle cells to substrate stiffness. *Biophysical Journal*, *101*, 622–630.

Scandinavian Simvastatin Survival Study Group. (1994). Randomised trial of cholesterol lowering in 4444 patients with coronary heart disease: The Scandinavian Simvastatin Survival Study (4S). *The Lancet*, *344*, 1383–1389.

Schwartz, M. A., & Ginsberg, M. H. (2002). Networks and crosstalk: Integrin signalling spreads. *Nature Cell Biology*, *4*, E65–E68.

Sehgel, N. L., Zhu, Y., Sun, Z., Trzeciakowski, J. P., Hong, Z., Hunter, W. C., et al. (2013). Increased vascular smooth muscle cell stiffness: A novel mechanism for aortic stiffness in hypertension. *American Journal of Physiology. Heart and Circulatory Physiology*, *305*, H1281–H1287.

Sekine, Y., Demosky, S. J., Stonik, J. A., Furuya, Y., Koike, H., Suzuki, K., et al. (2010). High-density lipoprotein induces proliferation and migration of human prostate androgen-independent cancer cells by an ABCA1-dependent mechanism. *Molecular Cancer Research*, *8*, 1284–1294.

Shankman, L. S., Gomez, D., Cherepanova, O. A., Salmon, M., Alencar, G. F., Haskins, R. M., et al. (2015). KLF4-dependent phenotypic modulation of smooth muscle cells has a key role in atherosclerotic plaque pathogenesis. *Nature Medicine*, *21*, 628–637.

Shepherd, J., Cobbe, S. M., Ford, I., Isles, C. G., Lorimer, A. R., MacFarlane, P. W., et al. (1995). Prevention of coronary heart disease with pravastatin in men with hypercholesterolemia. West of Scotland Coronary Prevention Study Group. *The New England Journal of Medicine*, *333*, 1301–1307.

Simons, K., & Ikonen, E. (1997). Functional rafts in cell membranes. *Nature*, *387*, 569–572.

Stegemann, J. P., Hong, H., & Nerem, R. M. (2005). Mechanical, biochemical, and extracellular matrix effects on vascular smooth muscle cell phenotype. *Journal of Applied Physiology*, *98*, 2321–2327.

Stossel, T. P. (1993). On the crawling of animal cells. *Science, 260*, 1086–1094.

Sun, Z., Martinez-Lemus, L. A., Hill, M. A., & Meininger, G. A. (2008). Extracellular matrix-specific focal adhesions in vascular smooth muscle produce mechanically active adhesion sites. *American Journal of Physiology. Cell Physiology, 295*, C268–C278.

Sun, M., Northup, N., Marga, F., Huber, T., Byfield, F. J., Levitan, I., et al. (2007). The effect of cellular cholesterol on membrane-cytoskeleton adhesion. *Journal of Cell Science, 120*, 2223–2231.

Taulet, N., Comunale, F., Favard, C., Charrasse, S., Bodin, S., & Gauthier-Rouviere, C. (2009). N-cadherin/p120 catenin association at cell-cell contacts occurs in cholesterol-rich membrane domains and is required for RhoA activation and myogenesis. *The Journal of Biological Chemistry, 284*, 23137–23145.

Tracqui, P., Broisat, A., Toczek, J., Mesnier, N., Ohayon, J., & Riou, L. (2011). Mapping elasticity moduli of atherosclerotic plaque in situ via atomic force microscopy. *Journal of Structural Biology, 174*, 115–123.

Tulenko, T. N., Chen, M., Mason, P. E., & Mason, R. P. (1998). Physical effects of cholesterol on arterial smooth muscle membranes: Evidence of immiscible cholesterol domains and alterations in bilayer width during atherogenesis. *Journal of Lipid Research, 39*, 947–956.

Wang, X., Keith, J. C., Jr., Struthers, A. D., & Feuerstein, G. Z. (2008). Assessment of arterial stiffness, a translational medicine biomarker system for evaluation of vascular risk. *Cardiovascular Therapeutics, 26*, 214–223.

Wang, M., Zhao, D., Spinetti, G., Zhang, J., Jiang, L. Q., Pintus, G., et al. (2006). Matrix metalloproteinase 2 activation of transforming growth factor-beta1 (TGF-beta1) and TGF-beta1-type II receptor signaling within the aged arterial wall. *Arteriosclerosis, Thrombosis, and Vascular Biology, 26*, 1503–1509.

Ward, N. C., Watts, G. F., & Eckel, R. H. (2019). Statin toxicity. *Circulation Research, 124*, 328–350.

Willis, A. I., Pierre-Paul, D., Sumpio, B. E., & Gahtan, V. (2004). Vascular smooth muscle cell migration: Current research and clinical implications. *Vascular and Endovascular Surgery, 38*, 11–23.

Worth, N. F., Rolfe, B. E., Song, J., & Campbell, G. R. (2001). Vascular smooth muscle cell phenotypic modulation in culture is associated with reorganisation of contractile and cytoskeletal proteins. *Cell Motility and the Cytoskeleton, 49*, 130–145.

Xu, J., & Shi, G.-P. (2014). Vascular wall extracellular matrix proteins and vascular diseases. *Biochimica et Biophysica Acta (BBA) - Molecular Basis of Disease, 1842*, 2106–2119.

Yan, P., Xia, C., Duan, C., Li, S., & Mei, Z. (2011). Biological characteristics of foam cell formation in smooth muscle cells derived from bone marrow stem cells. *International Journal of Biological Sciences*, 7, 937–946.

Yao, M., Goult, B. T., Chen, H., Cong, P., Sheetz, M. P., & Yan, J. (2014). Mechanical activation of vinculin binding to talin locks talin in an unfolded conformation. *Scientific Reports, 4*, 4610.

Zhu, Y., Qiu, H., Trzeciakowski, J. P., Sun, Z., Li, Z., Hong, Z., et al. (2012). Temporal analysis of vascular smooth muscle cell elasticity and adhesion reveals oscillation waveforms that differ with aging. *Aging Cell, 11*, 741–750.

Zidovetzki, R., & Levitan, I. (2007). Use of cyclodextrins to manipulate plasma membrane cholesterol content: Evidence, misconceptions and control strategies. *Biochimica et Biophysica Acta (BBA) - Biomembranes, 1768*, 1311–1324.

CHAPTER NINE

Roles of microglial membranes in Alzheimer's disease

Jae-Won Shin[a,b] and James C. Lee[a,*]
[a]Department of Bioengineering, University of Illinois at Chicago, College of Medicine, Chicago, IL, United States
[b]Department of Pharmacology, University of Illinois at Chicago, College of Medicine, Chicago, IL, United States
*Corresponding author: e-mail address: leejam@uic.edu

Contents

Abstract

The majority of Alzheimer's disease (AD) risk genes are highly and selectively expressed by microglia in the brain. Several of these genes are related to lipid and cholesterol metabolism, lipid synthesis, lipid transport, endocytosis, exocytosis and phagocytosis. Therefore, studying the roles of cellular membrane biophysics in microglial function should improve our understanding of the AD pathology. In this chapter, we discuss how lipid rafts and membrane-cytoskeleton adhesion impact microglial-mediated oxidative stress and clearance of amyloid-β peptide (Aβ). We also discuss potential roles of lipid membrane-bound extracellular vesicles as carriers of pathological factors to promote inflammation and cytotoxicity.

1. Introduction

Microglia are the major neuroimmune cells. They provide three essential functions including constant sensing of their environment changes, promoting neuronal health and operation, providing defense function and neuroprotection (Hickman, Izzy, Sen, Morsett, & El Khoury, 2018).

Current Topics in Membranes, Volume 86
ISSN 1063-5823
https://doi.org/10.1016/bs.ctm.2020.08.007

Microglial functions play pivotal roles in driving the pathology of Alzheimer's disease (AD). For example, microglia mediate early synapse loss in AD (Hong et al., 2016; Rajendran & Paolicelli, 2018). The presence of triggering receptor expressed on myeloid cells 2 (TREM2) in microglia is essential for microglia to surround and alter amyloid-β peptide (Aβ) plaque structure, thereby limiting neuritic damage to protect from AD (Wang et al., 2016). Microglia mediate synaptic remodeling, inflammation, and oxidative stress (Bisht, Sharma, & Tremblay, 2018), spreading of tau (Asai et al., 2015), and clearance of Aβ (Paolicelli et al., 2017).

In microglia-specific transcriptomic studies, genome wide association study (GWAS) shows that variants of highly expressed microglial transcripts are associated with an increased risk of AD, suggesting microglia play an important role in both the development and onset of the disease (Gosselin et al., 2017; Heneka, Kummer, & Latz, 2014; Karch & Goate, 2015). Importantly, more than 50% of genes associated with noncoding GWAS risk alleles for AD are preferentially expressed in microglia (Gosselin et al., 2017). In addition, numerous AD risk genes expressed by microglia are related to lipid and cholesterol metabolism, lipid synthesis, lipid transport, endocytosis, exocytosis and phagocytosis (McQuade & Blurton-Jones, 2019). Therefore, investigating the roles of cellular membrane biophysics in microglial function should further our understanding of the AD pathology. In this chapter, we discuss how physical properties of cellular membranes impact microglial-mediated oxidative stress and clearance of Aβ. We will also discuss potential roles of extracellular vesicles in promoting pathological spreading from microglia.

2. The roles of lipid rafts in microglial-mediated oxidative stress

Specialized regions of plasma membranes, such as lipid rafts and caveolae, govern the states of reactive oxygen species (ROS) producing systems (Nordzieke & Medrano-Fernandez, 2018). Many redox enzymatic systems exert their activity at the membrane. An integral membrane protein, NADPH oxidases (NOX), produce ROS (DeYulia Jr., Carcamo, Borquez-Ojeda, Shelton, & Golde, 2005). While xanthine oxidases (XOR) generate ROS at the outer part of the membrane (Radi, Rubbo, Bush, & Freeman, 1997), they generate nitric oxide synthases (NOS) at the inner part (Sessa, Barber, & Lynch, 1993). Since increased activation of NOX have been observed in AD brains, and microglial-mediated ROS are the major source of oxidative stress (Ansari & Scheff, 2011; Bruce-Keller et al., 2010; Choi, Aid, Kim, Jackson, & Bosetti, 2012; Shimohama et al., 2000), here we focus on the effects of lipid rafts on NOX activity in microglia.

Microglia produce ROS mainly through activation of NOX (Wilkinson & Landreth, 2006), which is an enzyme complex comprised of the membrane bound cytochrome *b*558 (p22^{phox} and the enzymatic subunit, gp91^{phox}), cytosolic subunits (p47^{phox}, p67^{phox}, and p40^{phox}), and the Rac G-protein (Babior, 2000, 2004). Phosphorylation of cytosolic subunits and activation of Rac result in their translocation to the membrane forming the active NOX complex with cytochrome *b*558, and subsequently producing superoxide. NOX activation to produce ROS can be stimulated by amyloid-β peptide (Aβ) (Bianca, Dusi, Bianchini, Dal Pra, & Rossi, 1999; Qin et al., 2002). Consistently, a significant translocation of p47^{phox} and p67^{phox} from the cytosol to the membrane was found in AD brains (Shimohama et al., 2000). Immunohistochemical evaluations of NOX expression indicate that gp91^{phox} in microglia was upregulated specifically in a vulnerable brain region of mild cognitive impairment (MCI) patients (Bruce-Keller et al., 2010). Increased activation and upregulation of NOX in frontal and temporal cortex were found in the MCI and AD groups (Ansari & Scheff, 2011). Another study also showed a twofold increase in p47phox, but not gp91phox in the brain of AD patients (Choi et al., 2012). Increased expression of NOX was also found in the 5xFAD early onset mouse model of AD (Landel et al., 2014). These findings suggest that NOX participate in the early pathogenesis of AD.

Lipid rafts are dynamic structures. They are assemblies of mainly saturated lipids and cholesterol. Modifications of lipid raft structures and composition with nonpathological aging (Díaz, Fabelo, Ferrer, & Marín, 2018) and at different diseased stages including AD and Parkinson's diseases (PD) (Mesa-Herrera, Taoro-González, Valdés-Baizabal, Diaz, & Marín, 2019) have been observed. In normal aging, major changes in the levels of plasmalogens, polyunsaturated fatty acids (especially docosahexaenoic acid (DHA) and arachidonic acid (ARA)), total polar lipids (mainly phosphatidylinositol, sphingomyelin, sulfatides, and cerebrosides), and total neutral lipids (particularly cholesterol and sterol esters) were detected in brain cortex (Díaz et al., 2018). These membrane and lipid raft modifications may impact activities of membrane molecules including NOX and caveolae (Fig. 1).

Numerous studies have reported the roles of caveolin-containing lipid rafts in NOX activities. For example, caveolae can serve as the membrane

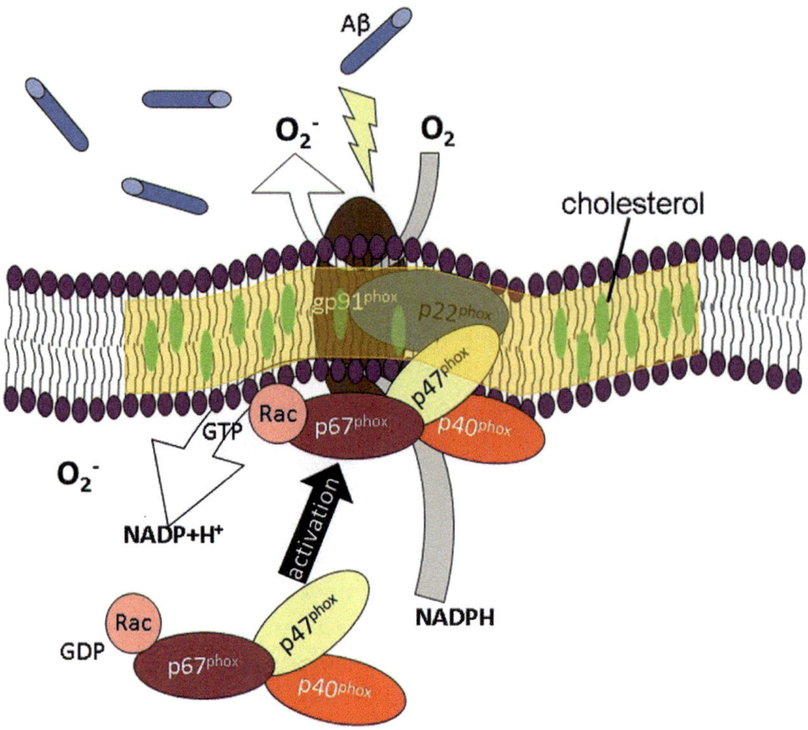

Fig. 1 A schematic depicts Aβ induces activation of NADPH oxidase (NOX), in which the cytosolic subunits of NOX translocate to the membrane subunits located at the lipid raft. Therefore, it is reasonable to hypothesize that alterations in nano-structure and composition of lipid rafts affect NOX activation for producing superoxides.

target for $p47^{phox}$, a cytosolic subunit of NOX in macrophages (Wang et al., 2018), and act as mediators of strain-induced ROS generation through NOX in glomerular mesangial cells (Zhang, Peng, Gao, Ingram, & Krepinsky, 2010). In turn, caveolae function is also dependent of NOX activity. NOX deficiency alters caveolin phosphorylation in vascular smooth muscle cells (Basset et al., 2009). High glucose downregulates caveolae in monocytes through oxidative stress from NOX (Hayashi, Juliet, Miyazaki, Ignarro, & Iguchi, 2007). However, the roles of caveolae in NOX activities in microglia, and their relations to the AD pathology have yet to be fully elucidated.

Besides caveolin, lipid raft modifications due to changes in the membrane cholesterol content have been reported to modulate NOX activity in microglia. Disruption of lipid rafts with methyl-β-cyclodextrin, fumonisin B1 or nystatin inhibit NOX-mediated ROS production and its downstream caspase-1 pathway in microglia stimulated with the pro-inflammatory lysophosphatidylcholine (LPC) (Schilling & Eder, 2010). Consistently, prolonged application of oxysterol significantly enhances lysophosphatidylserine (LPS)-induced association of cytosolic NOX subunit, $p47^{phox}$, with detergent-resistant microdomains (DRMs) in BV-2 cells, thereby increasing ROS production (Rackova, 2013). (−)-Epigallocatechin-3-gallate (EGCG) inhibits function of voltage-gated proton channels probably through its ability to disturb lipid rafts by sequestering cholesterol and subsequently inhibits NOX-dependent ROS generation (Jin, Park, & Song, 2013).

In addition to cholesterol, polyunsaturated fatty acids (PUFAs), especially docosahexaenoic acid (DHA), have recently been studied for their ability to incorporate into cellular membranes, alter physical properties of membranes, and subsequently alter cellular pathways. In fact, DHA supplement has been found to be beneficial to AD patients and animal models. DHA-enriched diet reduces amyloid burden in an aged AD mouse model (Lim et al., 2005). It has been shown to slow the progression of AD in E4FAD mouse (Ma et al., 2018). Therefore, the mechanisms underlying the beneficial effects of DHA are of research interests.

In fact, DHA can incorporate into and fluidize plasma membranes in neuronal cells, resulting in an upregulation of a neurotrophic α-secretase-dependent amyloid precursor protein processing (Eckert et al., 2011; Yang, Sheng, Sun, & Lee, 2011). Although there is no direct evidence showing the effects of DHA on lipid rafts in microglia resulting in subsequent inhibition of NADPH oxidase activity, numerous studies have demonstrated the ability of DHA to inhibit microglial-mediated oxidative stress. For example, omega-3 fish oils, including DHA, inhibit ROS production induced by lipopolysaccharide (LPS) (Hadad & Levy, 2017; Yang et al., 2018, 2020), interferon-γ (IFNγ) (Hadad & Levy, 2017) and oligomeric amyloid-β peptide (oAβ) in microglial cells (Geng et al., 2020). However, these studies lack examinations of how DHA alters lipid rafts in microglia, and the roles of lipid rafts in the inhibiting effects of DHA on microglial-mediated oxidative stress have yet to be fully elucidated.

3. The roles of membrane-cytoskeleton adhesion in microglial-mediated clearance of Aβ

As the resident macrophage cells, microglia are key cells responsible for scavenging cell debris in the brain, plaques, damaged neurons and synapses, and infectious agents (Gehrmann, Matsumoto, & Kreutzberg, 1995).

The clearance of Aβ by microglial cells has become an important AD research direction. Aβ phagocytosis has been found to be mediated by a range of receptors in microglia (Zuroff, Daley, Black, & Koronyo-Hamaoui, 2017), including CD14-Toll-like receptors (TLRs) to decrease phagocytosis of fibrillar $A\beta_{42}$ (Reed-Geaghan, Savage, Hise, & Landreth, 2009; Udan, Ajit, Crouse, & Nichols, 2008), macrophage scavenger receptor 1 (SCARA1) to facilitate both soluble and fibrillar $A\beta_{42}$ uptake (Chung, Brazil, Irizarry, Hyman, & Maxfield, 2001; El Khoury et al., 1996; Frenkel et al., 2013; Yang, Shiao, et al., 2011), CD36 to mediate phagocytosis of fibrillar $A\beta_{42}$ through interactions with two distinct receptor complexes acting as a functional unit (Koenigsknecht & Landreth, 2004; Wilkinson & El Khoury, 2012; Yu & Ye, 2015), and a functional triggering receptor expressed on myeloid cells 2 protein (TREM2) required for microglia to appropriately phagocytose Aβ (Guerreiro et al., 2013; Jonsson et al., 2013). TREM2-mediated uptake of lipoprotein-Aβ complexes and modeling the ApoE-Aβ interactions have been observed in vivo (Yeh, Wang, Tom, Gonzalez, & Sheng, 2016). Increased expression of CD33, a member of the sialic acid-binding immunoglobulin-like lectins (SIGLECS) family was shown to significantly reduce the ability to phagocytose Aβ (Hollingworth et al., 2011; Naj et al., 2011).

In addition to receptor-mediated Aβ phagocytosis, endocytosis as well as exocytosis are not only biochemical but also mechanical processes governed by the physical properties of plasma membranes and the membrane-cytoskeleton adhesion (Fig. 2) (Sheetz, 2001). Yet cellular and mechanical pathways dictating Aβ internalization in microglia are still not fully understood. From the viewpoint of energetics, the formation of an endocytic vesicle is similar to membrane tether formation using modern cell mechanics techniques, such as atomic force microscopy (AFM) or laser tweezers (Sheetz, 2001). There are three mechanical factors governing the formation of an endocytic vesicle. They are in-plane bilayer membrane tension, the membrane bending stiffness, and the membrane-cytoskeleton connectivity (Sheetz, 2001). In turn, the major mechanical energy required to overcome and then produce an endocytic vesicle is the membrane-cytoskeleton adhesion (Fig. 2). Under normal conditions, the main changes in membrane tether force are the consequence of changes in membrane-cytoskeleton adhesion (Dai & Sheetz, 1999; Raucher et al., 2000). A marked rise in the membrane tether force has been found to parallel the drop in endocytosis rate (Raucher & Sheetz, 1999). Decrease in tether force paralleled the rise in endocytosis rate (Dai, Sheetz, Wan, & Morris, 1998;

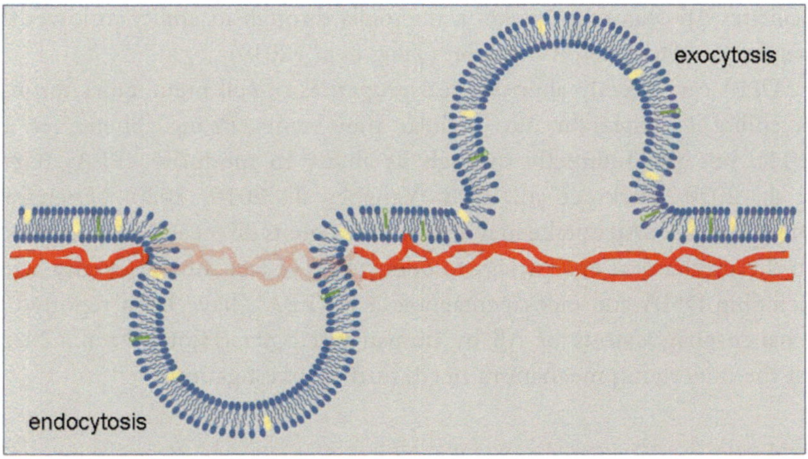

Fig. 2 A schematic describes the role of membrane-cytoskeleton adhesion in endocytosis and exocytosis. Mechanical energy is required to overcome the membrane-cytoskeleton adhesion for detaching the membrane from cytoskeleton during the formation of invagination to produce intracellular vesicles for endocytosis and outward membrane budding for exocytosis.

Raucher & Sheetz, 1999). Therefore, measuring membrane tether force is a powerful method to study the mechanical pathway of endocytosis, especially macropinocytosis.

In fact, microglia have been reported to mediate the clearance of soluble Aβ through fluid phase macropinocytosis (Mandrekar et al., 2009). Therefore, it is reasonable to hypothesize that cell signaling molecules involving in lipid metabolism, such as cytosolic phospholipase A_2 (cPLA_2), play a role in microglial-mediated clearance of Aβ. Indeed aberrant cPLA_2 activity has been observed in AD brains (Stephenson et al., 1999). As activated cPLA_2 is known for its ability to hydrolyze membrane phospholipids, cPLA_2 has been found to be involved in Aβ-induced alterations in the physical properties of cellular membranes, such as membrane molecular order, in astrocytes (Hicks et al., 2008). In addition, cPLA_2 has also been found to mediate actin rearrangements (Liu, Zhao, Fang, & Chen, 2012; Moes, Boonstra, & Regan-Klapisz, 2010). Both dynamic rearrangements of cytoskeleton and alterations of cellular membranes due to cPLA_2 activity may subsequent alter the mechanical energetics of endocytosis. In fact, cPLA_2 has been reported to drive recycling through the clathrin-independent endocytic route (Capestrano et al., 2014). Subsequently, we recently reported that cPLA_2

facilitates Aβ oligomer uptake in microglia through its ability to lower the membrane-cytoskeleton adhesion (Teng et al., 2019).

DHA can directly alter physical properties of cell membranes through its ability to integrate into cellular membranes (Yang, Sheng, et al., 2011), but also indirectly through its ability to modulate $cPLA_2$ (Geng et al., 2020; Hicks et al., 2008; Yang et al., 2019, 2020). Therefore, DHA can affect Aβ uptake in microglia through its direct and indirect effects on the mechanical energetics of endocytosis. In fact, omega-3 fatty acids, including DHA and eicosapentaenoic acid (EPA), have been reported to enhance phagocytosis of Aβ by human microglia (Hjorth et al., 2013), yet the underlying mechanism needs further investigations.

4. Roles of microglial-derived extracellular vesicles in AD

Extracellular vesicles (EVs) are lipid membrane-bound, cell-secreted nanoparticles that play important roles in intercellular communication (van Niel, D'Angelo, & Raposo, 2018). EVs are generally classified into exosomes (50–150 nm in size), which are formed by fusion of multivesicular bodies, and larger microvesicles (150–500 nm in size), which are formed by outward budding of the plasma membrane (Meldolesi, 2018). Interestingly, plasma membrane-derived vesicle budding is known to be driven by phosphatidylserine and phosphatidylethanolamine asymmetry (Hugel, Martínez, Kunzelmann, & Freyssinet, 2005) suggesting potential roles of lipids in EV release.

In neurodegenerative diseases, activation of microglia could result in EV release to promote inflammation and neurotoxicity. For example, microglial EVs were found to be neurotoxic and myelinotoxic in the presence of Aβ (Agosta et al., 2014). Accumulation of pro-inflammatory EVs from microglia was also observed in rat brains of post focal cerebral ischemia, which was inhibited by GW4869 that suppresses inward budding of multivesicular bodies (Gao et al., 2020). Glutaminase C regulates microglial activation and pro-inflammatory exosome release (Gao et al., 2019). Interestingly, the miR-124-3p level in microglial exosomes from injured brain was significantly altered after repetitive mild traumatic brain injury (TBI)—this miRNA was then transferred into hippocampal neurons and able to alleviate post-traumatic neurodegeneration by targeting the Rela/ApoE signaling pathway (Ge et al., 2020). While the roles of microglial-derived EVs have been recognized in brain injury and AD, how cell and membrane mechanics affect EV production in microglia has yet to be fully elucidated.

A recent study shows that EVs have the natural ability to undergo transport in confined tissue environments in part by tuning EV membrane deformability via water flux (Lenzini, Bargi, Chung, & Shin, 2020). Thus, microglial-derived EVs may also facilitate the transport of larger proteins within the brain, such as neurotoxic AD protein aggregates. Aβ Protofibrils are trafficked through microglial-derived EVs (Gouwens et al., 2018). BIN1 favors the spreading of tau via EVs (Crotti et al., 2019). Depletion of microglia and inhibition of their EV synthesis halt Tau propagation (Asai et al., 2015). In turn, microglial-EVs are also involved in clearance of these neurotoxic AD protein aggregates. Statin has been reported to trigger clearance of extracellular Aβ by microglia through the secretion of insulin-degrading enzyme in a nonconventional pathway in association with EVs (Glebov & Walter, 2012; Tamboli et al., 2010).

5. Conclusion

Microglia are reservoirs of lipids and their regulators, which impact biophysical properties of cell membrane. Understanding how lipid rafts and membrane-cytoskeleton adhesion of microglia regulate oxidative stress and clearance of Aβ will help deepen a fundamental insight into AD pathology from the perspective of cellular biophysics. Since EVs are lipid membrane-bound particles, understanding how EVs navigate in brain tissues and transport secreted factors will provide a novel angle to understand AD pathology, which can be helpful to devise effective treatment for the disease.

Acknowledgment

This work was supported, in part, by NIH R01AG044404 (J.C.L.), and NIH R01HL141255 (J.W.S.).

References

Agosta, F., Dalla Libera, D., Spinelli, E. G., Finardi, A., Canu, E., Bergami, A., et al. (2014). Myeloid microvesicles in cerebrospinal fluid are associated with myelin damage and neuronal loss in mild cognitive impairment and Alzheimer disease. *Annals of Neurology*, *76*, 813–825.

Ansari, M. A., & Scheff, S. W. (2011). NADPH-oxidase activation and cognition in Alzheimer disease progression. *Free Radical Biology & Medicine*, *51*, 171–178.

Asai, H., Ikezu, S., Tsunoda, S., Medalla, M., Luebke, J., Haydar, T., et al. (2015). Depletion of microglia and inhibition of exosome synthesis halt tau propagation. *Nature Neuroscience*, *18*, 1584–1593.

Babior, B. M. (2000). Phagocytes and oxidative stress. *The American Journal of Medicine, 109,* 33–44.

Babior, B. M. (2004). NADPH oxidase. *Current Opinion in Immunology, 16,* 42–47.

Basset, O., Deffert, C., Foti, M., Bedard, K., Jaquet, V., Ogier-Denis, E., et al. (2009). NADPH oxidase 1 deficiency alters caveolin phosphorylation and angiotensin II-receptor localization in vascular smooth muscle. *Antioxidants & Redox Signaling, 11,* 2371–2384.

Bianca, V. D., Dusi, S., Bianchini, E., Dal Pra, I., & Rossi, F. (1999). Beta-amyloid activates the O-2 forming NADPH oxidase in microglia, monocytes, and neutrophils. A possible inflammatory mechanism of neuronal damage in Alzheimer's disease. *The Journal of Biological Chemistry, 274,* 15493–15499.

Bisht, K., Sharma, K., & Tremblay, M. (2018). Chronic stress as a risk factor for Alzheimer's disease: Roles of microglia-mediated synaptic remodeling, inflammation, and oxidative stress. *Neurobiology of Stress, 9,* 9–21.

Bruce-Keller, A. J., Gupta, S., Parrino, T. E., Knight, A. G., Ebenezer, P. J., Weidner, A. M., et al. (2010). NOX activity is increased in mild cognitive impairment. *Antioxidants & Redox Signaling, 12,* 1371–1382.

Capestrano, M., Mariggio, S., Perinetti, G., Egorova, A. V., Iacobacci, S., Santoro, M., et al. (2014). Cytosolic phospholipase A(2)epsilon drives recycling through the clathrin-independent endocytic route. *Journal of Cell Science, 127,* 977–993.

Choi, S. H., Aid, S., Kim, H. W., Jackson, S. H., & Bosetti, F. (2012). Inhibition of NADPH oxidase promotes alternative and anti-inflammatory microglial activation during neuroinflammation. *Journal of Neurochemistry, 120,* 292–301.

Chung, H., Brazil, M. I., Irizarry, M. C., Hyman, B. T., & Maxfield, F. R. (2001). Uptake of fibrillar beta-amyloid by microglia isolated from MSR-A (type I and type II) knockout mice. *Neuroreport, 12,* 1151–1154.

Crotti, A., Sait, H. R., McAvoy, K. M., Estrada, K., Ergun, A., Szak, S., et al. (2019). BIN1 favors the spreading of Tau via extracellular vesicles. *Scientific Reports, 9,* 9477.

Dai, J., & Sheetz, M. P. (1999). Membrane tether formation from blebbing cells. *Biophysical Journal, 77,* 3363–3370.

Dai, J., Sheetz, M. P., Wan, X., & Morris, C. E. (1998). Membrane tension in swelling and shrinking molluscan neurons. *The Journal of neuroscience : the official journal of the Society for Neuroscience, 18,* 6681–6692.

DeYulia, G. J., Jr., Carcamo, J. M., Borquez-Ojeda, O., Shelton, C. C., & Golde, D. W. (2005). Hydrogen peroxide generated extracellularly by receptor-ligand interaction facilitates cell signaling. *Proceedings of the National Academy of Sciences of the United States of America, 102,* 5044–5049.

Díaz, M., Fabelo, N., Ferrer, I., & Marín, R. (2018). "Lipid raft aging" in the human frontal cortex during nonpathological aging: Gender influences and potential implications in Alzheimer's disease. *Neurobiology of Aging, 67,* 42–52.

Eckert, G. P., Chang, S., Eckmann, J., Copanaki, E., Hagl, S., Hener, U., et al. (2011). Liposome-incorporated DHA increases neuronal survival by enhancing non-amyloidogenic APP processing. *Biochimica et Biophysica Acta, 1808,* 236–243.

El Khoury, J., Hickman, S. E., Thomas, C. A., Cao, L., Silverstein, S. C., & Loike, J. D. (1996). Scavenger receptor-mediated adhesion of microglia to beta-amyloid fibrils. *Nature, 382,* 716–719.

Frenkel, D., Wilkinson, K., Zhao, L., Hickman, S. E., Means, T. K., Puckett, L., et al. (2013). Scara1 deficiency impairs clearance of soluble amyloid-beta by mononuclear phagocytes and accelerates Alzheimer's-like disease progression. *Nature Communications, 4,* 2030.

Gao, G., Li, C., Zhu, J., Wang, Y., Huang, Y., Zhao, S., et al. (2020). Glutaminase 1 regulates neuroinflammation after cerebral ischemia through enhancing microglial activation and pro-inflammatory exosome release. *Frontiers in Immunology, 11,* 161.

Gao, G., Zhao, S., Xia, X., Li, C., Li, C., Ji, C., et al. (2019). Glutaminase C regulates microglial activation and pro-inflammatory exosome release: Relevance to the pathogenesis of Alzheimer's disease. *Frontiers in Cellular Neuroscience*, *13*, 264.

Ge, X., Guo, M., Hu, T., Li, W., Huang, S., Yin, Z., et al. (2020). Increased microglial exosomal miR-124-3p alleviates neurodegeneration and improves cognitive outcome after rmTBI. *Molecular Therapy: The Journal of the American Society of Gene Therapy*, *28*, 503–522.

Gehrmann, J., Matsumoto, Y., & Kreutzberg, G. W. (1995). Microglia: Intrinsic immuneffector cell of the brain. *Brain Research. Brain Research Reviews*, *20*, 269–287.

Geng, X., Yang, B., Li, R., Teng, T., Ladu, M. J., Sun, G. Y., et al. (2020). Effects of docosahexaenoic acid and its peroxidation product on amyloid-beta peptide-stimulated microglia. *Molecular Neurobiology*, *57*, 1085–1098.

Glebov, K., & Walter, J. (2012). Statins in unconventional secretion of insulin-degrading enzyme and degradation of the amyloid-β peptide. *Neuro-Degenerative Diseases*, *10*, 309–312.

Gosselin, D., Skola, D., Coufal, N. G., Holtman, I. R., Schlachetzki, J. C. M., Sajti, E., et al. (2017). An environment-dependent transcriptional network specifies human microglia identity. *Science*, *356*(6344), eaal3222.

Gouwens, L. K., Ismail, M. S., Rogers, V. A., Zeller, N. T., Garrad, E. C., Amtashar, F. S., et al. (2018). Aβ42 protofibrils interact with and are trafficked through microglial-derived microvesicles. *ACS Chemical Neuroscience*, *9*, 1416–1425.

Guerreiro, R., Wojtas, A., Bras, J., Carrasquillo, M., Rogaeva, E., Majounie, E., et al. (2013). TREM2 variants in Alzheimer's disease. *The New England Journal of Medicine*, *368*, 117–127.

Hadad, N., & Levy, R. (2017). Combination of EPA with carotenoids and polyphenol synergistically attenuated the transformation of microglia to M1 phenotype via inhibition of NF-κB. *Neuromolecular Medicine*, *19*, 436–451.

Hayashi, T., Juliet, P. A., Miyazaki, A., Ignarro, L. J., & Iguchi, A. (2007). High glucose downregulates the number of caveolae in monocytes through oxidative stress from NADPH oxidase: Implications for atherosclerosis. *Biochimica et Biophysica Acta*, *1772*, 364–372.

Heneka, M. T., Kummer, M. P., & Latz, E. (2014). Innate immune activation in neurodegenerative disease. *Nature Reviews. Immunology*, *14*, 463–477.

Hickman, S., Izzy, S., Sen, P., Morsett, L., & El Khoury, J. (2018). Microglia in neurodegeneration. *Nature Neuroscience*, *21*, 1359–1369.

Hicks, J. B., Lai, Y., Sheng, W., Yang, X., Zhu, D., Sun, G. Y., et al. (2008). Amyloid-beta peptide induces temporal membrane biphasic changes in astrocytes through cytosolic phospholipase A2. *Biochimica et Biophysica Acta*, *1778*, 2512–2519.

Hjorth, E., Zhu, M., Toro, V. C., Vedin, I., Palmblad, J., Cederholm, T., et al. (2013). Omega-3 fatty acids enhance phagocytosis of Alzheimer's disease-related amyloid-β42 by human microglia and decrease inflammatory markers. *Journal of Alzheimer's Disease: JAD*, *35*, 697–713.

Hollingworth, P., Harold, D., Sims, R., Gerrish, A., Lambert, J. C., Carrasquillo, M. M., et al. (2011). Common variants at ABCA7, MS4A6A/MS4A4E, EPHA1, CD33 and CD2AP are associated with Alzheimer's disease. *Nature Genetics*, *43*, 429–435.

Hong, S., Beja-Glasser, V. F., Nfonoyim, B. M., Frouin, A., Li, S., Ramakrishnan, S., et al. (2016). Complement and microglia mediate early synapse loss in Alzheimer mouse models. *Science*, *352*, 712–716.

Hugel, B., Martínez, M. C., Kunzelmann, C., & Freyssinet, J. M. (2005). Membrane microparticles: Two sides of the coin. *Physiology (Bethesda, Md.)*, *20*, 22–27.

Jin, S., Park, M., & Song, J. H. (2013). (−)-Epigallocatechin-3-gallate inhibits voltage-gated proton currents in BV2 microglial cells. *European Journal of Pharmacology*, *698*, 154–160.

Jonsson, T., Stefansson, H., Steinberg, S., Jonsdottir, I., Jonsson, P. V., Snaedal, J., et al. (2013). Variant of TREM2 associated with the risk of Alzheimer's disease. *The New England Journal of Medicine*, *368*, 107–116.

Karch, C. M., & Goate, A. M. (2015). Alzheimer's disease risk genes and mechanisms of disease pathogenesis. *Biological Psychiatry*, *77*, 43–51.

Koenigsknecht, J., & Landreth, G. (2004). Microglial phagocytosis of fibrillar beta-amyloid through a beta1 integrin-dependent mechanism. *The Journal of Neuroscience: The Official Journal of the Society for Neuroscience*, *24*, 9838–9846.

Landel, V., Baranger, K., Virard, I., Loriod, B., Khrestchatisky, M., Rivera, S., et al. (2014). Temporal gene profiling of the 5XFAD transgenic mouse model highlights the importance of microglial activation in Alzheimer's disease. *Molecular Neurodegeneration*, *9*, 33.

Lenzini, S., Bargi, R., Chung, G., & Shin, J.-W. (2020). Matrix mechanics and water permeation regulate extracellular vesicle transport. *Nature Nanotechnology*, *15*, 217–223.

Lim, G. P., Calon, F., Morihara, T., Yang, F., Teter, B., Ubeda, O., et al. (2005). A diet enriched with the omega-3 fatty acid docosahexaenoic acid reduces amyloid burden in an aged Alzheimer mouse model. *The Journal of Neuroscience*, *25*, 3032–3040.

Liu, D. X., Zhao, W. D., Fang, W. G., & Chen, Y. H. (2012). cPLA2alpha-mediated actin rearrangements downstream of the Akt signaling is required for Cronobacter sakazakii invasion into brain endothelial cells. *Biochemical and Biophysical Research Communications*, *417*, 925–930.

Ma, Q. L., Teng, E., Zuo, X., Jones, M., Teter, B., Zhao, E. Y., et al. (2018). Neuronal pentraxin 1: A synaptic-derived plasma biomarker in Alzheimer's disease. *Neurobiology of Disease*, *114*, 120–128.

Mandrekar, S., Jiang, Q., Lee, C. Y., Koenigsknecht-Talboo, J., Holtzman, D. M., & Landreth, G. E. (2009). Microglia mediate the clearance of soluble Abeta through fluid phase macropinocytosis. *The Journal of Neuroscience: The Official Journal of the Society for Neuroscience*, *29*, 4252–4262.

McQuade, A., & Blurton-Jones, M. (2019). Microglia in Alzheimer's disease: Exploring how genetics and phenotype influence risk. *Journal of Molecular Biology*, *431*, 1805–1817.

Meldolesi, J. (2018). Exosomes and ectosomes in intercellular communication. *Current Biology*, *28*, R435–R444.

Mesa-Herrera, F., Taoro-González, L., Valdés-Baizabal, C., Diaz, M., & Marín, R. (2019). Lipid and lipid raft alteration in aging and neurodegenerative diseases: A window for the development of new biomarkers. *International Journal of Molecular Sciences*, *20*(15), 3810.

Moes, M., Boonstra, J., & Regan-Klapisz, E. (2010). Novel role of cPLA(2)alpha in membrane and actin dynamics. *Cellular and Molecular Life Sciences: CMLS*, *67*, 1547–1557.

Naj, A. C., Jun, G., Beecham, G. W., Wang, L. S., Vardarajan, B. N., Buros, J., et al. (2011). Common variants at MS4A4/MS4A6E, CD2AP, CD33 and EPHA1 are associated with late-onset Alzheimer's disease. *Nature Genetics*, *43*, 436–441.

Nordzieke, D. E., & Medrano-Fernandez, I. (2018). The plasma membrane: A platform for intra- and intercellular redox signaling. *Antioxidants (Basel, Switzerland)*, 7(11), 168.

Paolicelli, R. C., Jawaid, A., Henstridge, C. M., Valeri, A., Merlini, M., Robinson, J. L., et al. (2017). TDP-43 depletion in microglia promotes amyloid clearance but also induces synapse loss. *Neuron*, *95*, 297–308.e296.

Qin, L., Liu, Y., Cooper, C., Liu, B., Wilson, B., & Hong, J. S. (2002). Microglia enhance beta-amyloid peptide-induced toxicity in cortical and mesencephalic neurons by producing reactive oxygen species. *Journal of Neurochemistry*, *83*, 973–983.

Rackova, L. (2013). Cholesterol load of microglia: Contribution of membrane architecture changes to neurotoxic power? *Archives of Biochemistry and Biophysics*, *537*, 91–103.

Radi, R., Rubbo, H., Bush, K., & Freeman, B. A. (1997). Xanthine oxidase binding to glycosaminoglycans: Kinetics and superoxide dismutase interactions of immobilized xanthine oxidase-heparin complexes. *Archives of Biochemistry and Biophysics, 339*, 125–135.

Rajendran, L., & Paolicelli, R. C. (2018). Microglia-mediated synapse loss in Alzheimer's disease. *The Journal of Neuroscience: The Official Journal of the Society for Neuroscience, 38*, 2911–2919.

Raucher, D., & Sheetz, M. P. (1999). Membrane expansion increases endocytosis rate during mitosis. *The Journal of Cell Biology, 144*, 497–506.

Raucher, D., Stauffer, T., Chen, W., Shen, K., Guo, S., York, J. D., et al. (2000). Phosphatidylinositol 4,5-bisphosphate functions as a second messenger that regulates cytoskeleton-plasma membrane adhesion. *Cell, 100*, 221–228.

Reed-Geaghan, E. G., Savage, J. C., Hise, A. G., & Landreth, G. E. (2009). CD14 and toll-like receptors 2 and 4 are required for fibrillar A{beta}-stimulated microglial activation. *The Journal of Neuroscience: The Official Journal of the Society for Neuroscience, 29*, 11982–11992.

Schilling, T., & Eder, C. (2010). Importance of lipid rafts for lysophosphatidylcholine-induced caspase-1 activation and reactive oxygen species generation. *Cellular Immunology, 265*, 87–90.

Sessa, W. C., Barber, C. M., & Lynch, K. R. (1993). Mutation of N-myristoylation site converts endothelial cell nitric oxide synthase from a membrane to a cytosolic protein. *Circulation Research, 72*, 921–924.

Sheetz, M. P. (2001). Cell control by membrane-cytoskeleton adhesion. *Nature Reviews. Molecular Cell Biology, 2*, 392–396.

Shimohama, S., Tanino, H., Kawakami, N., Okamura, N., Kodama, H., Yamaguchi, T., et al. (2000). Activation of NADPH oxidase in Alzheimer's disease brains. *Biochemical and Biophysical Research Communications, 273*, 5–9.

Stephenson, D., Rash, K., Smalstig, B., Roberts, E., Johnstone, E., Sharp, J., et al. (1999). Cytosolic phospholipase A2 is induced in reactive glia following different forms of neurodegeneration. *GLIA, 27*, 110–128.

Tamboli, I. Y., Barth, E., Christian, L., Siepmann, M., Kumar, S., Singh, S., et al. (2010). Statins promote the degradation of extracellular amyloid {beta}-peptide by microglia via stimulation of exosome-associated insulin-degrading enzyme (IDE) secretion. *The Journal of Biological Chemistry, 285*, 37405–37414.

Teng, T., Dong, L., Ridgley, D. M., Ghura, S., Tobin, M. K., Sun, G. Y., et al. (2019). Cytosolic phospholipase A2 facilitates oligomeric amyloid-beta peptide association with microglia via regulation of membrane-cytoskeleton connectivity. *Molecular Neurobiology, 56*, 3222–3234.

Udan, M. L., Ajit, D., Crouse, N. R., & Nichols, M. R. (2008). Toll-like receptors 2 and 4 mediate Abeta(1-42) activation of the innate immune response in a human monocytic cell line. *Journal of Neurochemistry, 104*, 524–533.

van Niel, G., D'Angelo, G., & Raposo, G. (2018). Shedding light on the cell biology of extracellular vesicles. *Nature Reviews Molecular Cell Biology, 19*, 213–228.

Wang, J., Bai, Y., Zhao, X., Ru, J., Kang, N., Tian, T., et al. (2018). oxLDL-mediated cellular senescence is associated with increased NADPH oxidase p47phox recruitment to caveolae. *Bioscience Reports, 38*(3). BSR20180283.

Wang, Y., Ulland, T. K., Ulrich, J. D., Song, W., Tzaferis, J. A., Hole, J. T., et al. (2016). TREM2-mediated early microglial response limits diffusion and toxicity of amyloid plaques. *The Journal of Experimental Medicine, 213*, 667–675.

Wilkinson, K., & El Khoury, J. (2012). Microglial scavenger receptors and their roles in the pathogenesis of Alzheimer's disease. *International Journal of Alzheimer's Disease, 2012*, 489456.

Wilkinson, B. L., & Landreth, G. E. (2006). The microglial NADPH oxidase complex as a source of oxidative stress in Alzheimer's disease. *Journal of Neuroinflammation*, *3*, 30.

Yang, B., Fritsche, K. L., Beversdorf, D. Q., Gu, Z., Lee, J. C., Folk, W. R., et al. (2019). Yin-Yang mechanisms regulating lipid peroxidation of docosahexaenoic acid and arachidonic acid in the central nervous system. *Frontiers in Neurology*, *10*, 642.

Yang, B., Li, R., Liu, P. N., Geng, X., Mooney, B. P., Chen, C., et al. (2020). Quantitative proteomics reveals docosahexaenoic acid-mediated neuroprotective effects in lipopolysaccharide-stimulated microglial cells. *Journal of Proteome Research*, *19*(6), 2236–2246.

Yang, B., Li, R., Michael Greenlief, C., Fritsche, K. L., Gu, Z., Cui, J., et al. (2018). Unveiling anti-oxidative and anti-inflammatory effects of docosahexaenoic acid and its lipid peroxidation product on lipopolysaccharide-stimulated BV-2 microglial cells. *Journal of Neuroinflammation*, *15*, 202.

Yang, X., Sheng, W., Sun, G. Y., & Lee, J. C. (2011). Effects of fatty acid unsaturation numbers on membrane fluidity and α-secretase-dependent amyloid precursor protein processing. *Neurochemistry International*, *58*, 321–329.

Yang, C. N., Shiao, Y. J., Shie, F. S., Guo, B. S., Chen, P. H., Cho, C. Y., et al. (2011). Mechanism mediating oligomeric Abeta clearance by naive primary microglia. *Neurobiology of Disease*, *42*, 221–230.

Yeh, F. L., Wang, Y., Tom, I., Gonzalez, L. C., & Sheng, M. (2016). TREM2 binds to apolipoproteins, including APOE and CLU/APOJ, and thereby facilitates uptake of amyloid-beta by microglia. *Neuron*, *91*, 328–340.

Yu, Y., & Ye, R. D. (2015). Microglial Abeta receptors in Alzheimer's disease. *Cellular and Molecular Neurobiology*, *35*, 71–83.

Zhang, Y., Peng, F., Gao, B., Ingram, A. J., & Krepinsky, J. C. (2010). Mechanical strain-induced RhoA activation requires NADPH oxidase-mediated ROS generation in caveolae. *Antioxidants & Redox Signaling*, *13*, 959–973.

Zuroff, L., Daley, D., Black, K. L., & Koronyo-Hamaoui, M. (2017). Clearance of cerebral Abeta in Alzheimer's disease: Reassessing the role of microglia and monocytes. *Cellular and Molecular Life Sciences*, *74*(12), 2167–2201.

9780128210215